Student Solutions Manual

Laurel Technical Services

Prealgebra

THIRD EDITION

K. Elayn Martin-Gay

Upper Saddle River, NJ 07458

Executive Editor: Karin E. Wagner
Special Projects Manager: Barbara A. Murray
Production Editor: Wendy A. Perez
Supplement Cover Manager: Paul Gourhan
Supplement Cover Designer: PM Workshop Inc.
Manufacturing Manager: Trudy Pisciotti

Printed in the United States of America

10 9 8 7 6 5

ISBN 0-13-026459-8

Prentice-Hall International (UK) Limited, London
Prentice-Hall of Australia Pty. Limited, Sydney
Prentice-Hall Canada, Inc., Toronto
Prentice-Hall Hispanoamericana, S.A., Mexico
Prentice-Hall of India Private Limited, New Delhi
Pearson Education Asia Pte. Ltd., Singapore
Prentice-Hall of Japan, Inc., Tokyo
Editora Prentice-Hall do Brazil, Ltda., Rio de Janeiro

Table of Contents

Chapter 1

Pretest

1. The place value of the 7 in 5732 is hundreds.

2. 23,490 is written as twenty-three thousand, four hundred ninety.

3. $$\begin{array}{r} \scriptstyle 1 \\ 58 \\ +\ 29 \\ \hline 87 \end{array}$$

4. $$\begin{array}{r} 413 \\ \times\ 9 \\ \hline 3717 \end{array}$$

5. $$\begin{array}{r} 857 \\ -\ 231 \\ \hline 626 \end{array}$$

   Check:
   $$\begin{array}{r} 626 \\ +\ 231 \\ \hline 857 \end{array}$$

6. $$\begin{array}{r} 51 \\ -\ 19 \\ \hline 32 \end{array}$$

   Check:
   $$\begin{array}{r} \scriptstyle 1 \\ 32 \\ +\ 19 \\ \hline 51 \end{array}$$

7. To find how many more pages Karen must read, we subtract the number of pages she has read from the number of pages in the book.
   $$\begin{array}{r} 329 \\ -\ 193 \\ \hline 136 \end{array}$$
   Karen has 136 pages left to read.

8. To round 9045 to the nearest 10, observe that the digit in the ones place is 5. Since this digit is at least 5, we need to add 1 to the digit in the tens place. The number 9045, rounded to the nearest ten is 9050.

9. $$\begin{array}{rlr} 382 & \text{rounds to} & 400 \\ 436 & \text{rounds to} & 400 \\ 2084 & \text{rounds to} & 2100 \\ +\ 176 & \text{rounds to} & +\ 200 \\ \hline & & 3100 \end{array}$$
   The estimated sum is 3100.

10. $9(3 + 11) = 9 \cdot 3 + 9 \cdot 11$

11. The perimeter is the sum of the lengths of the sides.
    $$\begin{array}{r} \scriptstyle 1 \\ 8 \\ 11 \\ +\ 6 \\ \hline 25 \end{array}$$
    The perimeter is 25 inches.

12. The area of a rectangle is the product of its length and its width.
    (23 yards)(8 yards) = 184 square yards

13. The total number of seats is the product of the number of rows and the number of seats in each row.
    (32 rows)(18 seats per row) = 576 seats

14. $$\begin{array}{r} 243 \\ 9\overline{)\ 2187} \\ \underline{-18} \\ 38 \\ \underline{-36} \\ 27 \\ \underline{-27} \\ 0 \end{array}$$
    Check: $243 \cdot 9 = 2187$

15. $\frac{5361}{12}$

$$\begin{array}{r} 446 \text{ R9} \\ 12\overline{)\,5361} \\ \underline{-48} \\ 56 \\ \underline{-48} \\ 81 \\ \underline{-72} \\ 9 \end{array}$$

Check: $446 \cdot 12 + 9 = 5361$

16. There are 7 numbers.

$$\begin{array}{r} 29 \\ 36 \\ 84 \\ 41 \\ 6 \\ 12 \\ \underline{+\ 65} \\ 273 \end{array} \quad \text{average} = \frac{273}{7} = 39$$

The average is 39.

17. There are 7 factors of 9, so $9 \cdot 9 \cdot 9 \cdot 9 \cdot 9 \cdot 9 \cdot 9 = 9^7$.

18. $7^4 = 7 \cdot 7 \cdot 7 \cdot 7 = 2401$

19. $36 + 18 \div 6 = 36 + 3 = 39$

20. $$\frac{3a+2b}{5} = \frac{3(4)+2(9)}{5} = \frac{12+18}{5} = \frac{30}{5} = 6$$

21. In words: 10 decreased by a number
Translate: $10 - x$

Section 1.1

Practice Problems

1. The place value of the 7 in 72,589,620 is ten-millions.

2. The place value of the 7 in 67,890 is thousands.

3. The place value of the 7 in 50,722 is hundreds.

4. 67 is written as sixty-seven.

5. 395 is written as three hundred ninety-five.

6. 321,670,200 is written as three hundred twenty-one million, six hundred seventy thousand, two hundred.

7. Twenty-nine in standard form is 29.

8. Seven hundred ten in standard form is 710.

9. Twenty-six thousand seventy-one in standard form is 26,071.

10. Six thousand, five hundred seven in standard form is 6507.

11. 1,047,608
$= 1{,}000{,}000 + 40{,}000 + 7000 + 600 + 8$

12. a. $0 < 19$

b. $18 < 32$

c. $107 > 103$

13. a. Read from left to right across the line marked Austria until the "Bronze" column is reached. We find that Austria has won 53 bronze medals.

b. USSR/Russia has won 107 gold medals and Norway has won 83. Every other country has won fewer than 70. So, only USSR/Russia and Norway have won more than 70 gold medals.

Exercise Set 1.1

1. The place value of the 5 in 352 is tens.

3. The place value of the 5 in 5890 is thousands.

5. The place value of the 5 in 62,500,000 is hundred-thousands.

7. The place value of the 5 in 5,070,099 is millions.

9. 5420 is written as five thousand, four hundred twenty.

11. 26,990 is written as twenty-six thousand, nine hundred ninety.

13. 1,620,000 is written as one million, six hundred twenty thousand.

15. 53,520,170 is written as fifty-three million, five hundred twenty thousand, one hundred seventy.

17. 4,992,838 is written as four million, nine hundred ninety-two thousand, eight hundred thirty-eight.

19. 620,000 is written as six hundred twenty thousand.

21. 3893 is written as three thousand, eight hundred ninety-three.

23. Six thousand, five hundred eight in standard form is 6508.

25. Twenty-nine thousand, nine hundred in standard form is 29,900.

27. Six million, five hundred four thousand, nineteen in standard form is 6,504,019.

29. Three million, fourteen in standard form is 3,000,014.

31. One thousand, eight hundred twenty-one in standard form is 1821.

33. Sixty-three million, one hundred thousand in standard form is 63,100,000.

35. One hundred million in standard form is 100,000,000.

37. $406 = 400 + 6$

39. $5290 = 5000 + 200 + 90$

41. $62{,}407 = 60{,}000 + 2000 + 400 + 7$

43. $30{,}680 = 30{,}000 + 600 + 80$

45. $39{,}680{,}000 = 30{,}000{,}000 + 9{,}000{,}000 + 600{,}000 + 80{,}000$

47. $1006 = 1000 + 6$

49. $3 < 8$

51. $9 > 0$

53. $6 > 2$

55. $22 > 0$

57. 4000 is written as four thousand.

59. 4145 = 4000 + 100 + 40 + 5

61. The Nile is the longest river in the world.

63. The Pomeranian has the least AKC registrations, with thirty-eight thousand, five hundred forty.

65. There are fewer Rottweilers than German Shepherds registered.

67. The largest number is achieved when the largest number available is used for each place value when reading from left to right. Thus, the largest number possible is 55,543.

69. Answers may vary.

Section 1.2

Practice Problems

1. $\begin{array}{r} 7235 \\ +\ 542 \\ \hline 7777 \end{array}$

2. $\begin{array}{r} {\scriptstyle 11\ 11} \\ 27,364 \\ +\ 92,977 \\ \hline 120,341 \end{array}$

3. $\begin{aligned} 11+7+8+9+13 &= 11+9+7+13+8 \\ &= (11+9)+(7+13)+8 \\ &= 20+20+8 \\ &= 40+8 \\ &= 48 \end{aligned}$

4. $\begin{array}{r} {\scriptstyle 112} \\ 19 \\ 5042 \\ 638 \\ +\ 526 \\ \hline 6225 \end{array}$

5. To find the perimeter, add the lengths of the sides.
5 centimeters + 8 centimeters + 10 centimeters + 4 centimeters = 27 centimeters

6. To find the perimeter of the mall, add the lengths of the sides, all three of which have length 532 feet.

$\begin{array}{r} 532 \\ 532 \\ +\ 532 \\ \hline 1596 \end{array}$

The perimeter of the building is 1596 feet.

7. The phrase "increased by" suggests we add. The addition statement
old salary + increase = new salary translates to:

$\begin{array}{r} 73,176 \\ +\ 2\ 196 \\ \hline 75,372 \end{array}$

The new yearly salary is $75,372.

8. The number of thimbles that Elham has is the sum of the numbers of different types of thimbles that she has.

	13
glass thimbles	42
steel thimbles	17
porcelain thimbles	37
silver thimbles	9
plastic thimbles	+ 15
total thimbles	120

Elham has 120 thimbles.

Exercise Set 1.2

1. $\begin{array}{r} 14 \\ +\ 22 \\ \hline 36 \end{array}$

3. $\begin{array}{r} 62 \\ +\ 30 \\ \hline 92 \end{array}$

5. $\begin{array}{r} 12 \\ 13 \\ +\ 24 \\ \hline 49 \end{array}$

7.
$$\begin{array}{r} 5267 \\ +\ 132 \\ \hline 5399 \end{array}$$

9.
$$\begin{array}{r} {}^{1} \\ 53 \\ +\ 64 \\ \hline 117 \end{array}$$

11.
$$\begin{array}{r} {}^{1} \\ 22 \\ +\ 49 \\ \hline 71 \end{array}$$

13.
$$\begin{array}{r} {}^{11} \\ 38 \\ +\ 79 \\ \hline 117 \end{array}$$

15.
$$\begin{array}{r} {}^{2} \\ 8 \\ 9 \\ 2 \\ 5 \\ +\ 1 \\ \hline 25 \end{array}$$

17.
$$\begin{array}{r} {}^{2} \\ 6 \\ 21 \\ 14 \\ 9 \\ +\ 12 \\ \hline 62 \end{array}$$

19.
$$\begin{array}{r} {}^{22} \\ 81 \\ 17 \\ 23 \\ 79 \\ +\ 12 \\ \hline 212 \end{array}$$

21.
$$\begin{array}{r} {}^{1} \\ 62 \\ 18 \\ +\ 14 \\ \hline 94 \end{array}$$

23.
$$\begin{array}{r} {}^{1} \\ 40 \\ 800 \\ +\ 70 \\ \hline 910 \end{array}$$

25.
$$\begin{array}{r} {}^{111} \\ 7542 \\ 49 \\ +\ 682 \\ \hline 8273 \end{array}$$

27.
$$\begin{array}{r} {}^{1\ 12} \\ 24 \\ 9006 \\ 489 \\ +\ 2407 \\ \hline 11,926 \end{array}$$

29.
$$\begin{array}{r} {}^{1\ 2} \\ 627 \\ 628 \\ +\ 629 \\ \hline 1884 \end{array}$$

31.
$$\begin{array}{r} {}^{111} \\ 6820 \\ 4271 \\ +\ 5626 \\ \hline 16,717 \end{array}$$

33.
$$\begin{array}{r} {}^{111} \\ 507 \\ 593 \\ +\ 10 \\ \hline 1110 \end{array}$$

35.
$$\begin{array}{r} 4200 \\ 2107 \\ +\ 2692 \\ \hline 8999 \end{array}$$

37.
$$\begin{array}{r} {}^{11\ 22} \\ 49 \\ 628 \\ 5\ 762 \\ +\ 29,462 \\ \hline 35,901 \end{array}$$

39. $\begin{array}{r} {}^{1\,2\,2\,2\,1}\\ 121{,}742 \\ 57{,}279 \\ 6\ 586 \\ +\ 426{,}782 \\ \hline 612{,}389 \end{array}$

41. $8+3+5+7+5+1 = 8+1+3+7+5+5$
$= 9+10+10$
$= 29$
The perimeter is 29 inches.

43. $8 + 10 + 7 = 8 + 7 + 10 = 15 + 10 = 25$
The perimeter is 25 feet.

45. Opposite sides of a rectangle have the same length.
$4 + 8 + 4 + 8 = 12 + 12 = 24$
The perimeter is 24 inches.

47. All the sides of a square have the same length.
$2 + 2 + 2 + 2 = 4 + 4 = 8$
The perimeter is 8 yards.

49. $\begin{array}{r} {}^{1\,1}\\ 285 \\ +\ 98 \\ \hline 383 \end{array}$

It is 383 miles from Kansas City to Colby.

51. $\begin{array}{r} {}^{2\,1}\\ 70 \\ 78 \\ 90 \\ +\ 102 \\ \hline 340 \end{array}$

He needs 340 feet of wire.

53. $\begin{array}{r} {}^{1\,1}\\ 105{,}600 \\ +\ 17{,}500 \\ \hline 123{,}100 \end{array}$

There were 123,100 people employed in 1997.

55. $\begin{array}{r} 5031 \\ +\ 1918 \\ \hline 6949 \end{array}$

Cats had staged 6949 performances.

57. $\begin{array}{r} {}^{1\,1}\\ 12{,}166 \\ +\ \ 973 \\ \hline 13{,}139 \end{array}$

There were 13,139 transplants involving a kidney performed in 1998.

59. Key word: total

upper	1430	
middle	675	
+ lower	+ 320	
total height	2425	feet

61. $\begin{array}{r} 1795 \\ +\ 11{,}460 \\ \hline 13{,}255 \end{array}$

The total highway mileage in Alaska is 13,255 miles.

63. Texas has the most Wal-Mart stores.

65. $\begin{array}{r} {}^{1}\\ 166 \\ 130 \\ +\ 293 \\ \hline 589 \end{array}$

Texas, Florida, and Illinois have the most Wal-Mart stores, with a total of 589 stores.

67. $\begin{array}{r} {}^{17\ 5}\\ 85 \\ 81 \\ 124 \\ 166 \\ 103 \\ 130 \\ 89 \\ 84 \\ 121 \\ 100 \\ 104 \\ 84 \\ 99 \\ +\ 293 \\ \hline 1663 \end{array}$

There are 1663 stores in the states given in the table.

69. Answers may vary.

71.
$$\begin{array}{r} \scriptstyle 1111\ 121\ 1 \\ 56,468,980 \\ 1,236,785 \\ +\ 986,768,000 \\ \hline 1,044,473,765 \end{array}$$

Section 1.3

Practice Problems

1. a. $14 - 9 = 5$ because $5 + 9 = 14$.

b. $9 - 9 = 0$ because $0 + 9 = 9$.

c. $4 - 0 = 4$ because $4 + 0 = 4$.

2. a.
$$\begin{array}{r} 4689 \\ -\ 253 \\ \hline 4436 \end{array}$$

Check:
$$\begin{array}{r} 4436 \\ +\ 253 \\ \hline 4689 \end{array}$$

b.
$$\begin{array}{r} 981 \\ -\ 630 \\ \hline 351 \end{array}$$

Check:
$$\begin{array}{r} 351 \\ +\ 630 \\ \hline 981 \end{array}$$

3. a.
$$\begin{array}{r} 227 \\ -\ 175 \\ \hline 52 \end{array}$$

Check
$$\begin{array}{r} \scriptstyle 1\ \ \\ 52 \\ +\ 175 \\ \hline 227 \end{array}$$

b.
$$\begin{array}{r} 1136 \\ -\ 914 \\ \hline 222 \end{array}$$

Check:
$$\begin{array}{r} \scriptstyle 1\ \ \\ 222 \\ +\ 914 \\ \hline 1136 \end{array}$$

c.
$$\begin{array}{r} 8627 \\ -\ 4119 \\ \hline 4508 \end{array}$$

Check:
$$\begin{array}{r} \scriptstyle 1\ \\ 4508 \\ +\ 4119 \\ \hline 8627 \end{array}$$

4. a.
$$\begin{array}{r} 400 \\ -\ 164 \\ \hline 236 \end{array}$$

Check:
$$\begin{array}{r} \scriptstyle 11\ \\ 236 \\ +\ 164 \\ \hline 400 \end{array}$$

b.
$$\begin{array}{r} 200 \\ -\ 45 \\ \hline 155 \end{array}$$

Check:
$$\begin{array}{r} \scriptstyle 11\ \\ 155 \\ +\ 45 \\ \hline 200 \end{array}$$

c.
$$\begin{array}{r} 1000 \\ -\ 762 \\ \hline 238 \end{array}$$

Check:
$$\begin{array}{r} \scriptstyle 11\ \\ 238 \\ +\ 762 \\ \hline 1000 \end{array}$$

5.

radius of Earth	6378
less 2981	− 2981
radius of Mars	3397

The radius of Mars is 3397 kilometers.

6.

original price	92
sale price	− 47
amount taken off	45

$45 was taken off the original price.

7. a.

	1 3
walking	37
swimming	27
bicycle riding	23
bowling	18
fishing	+ 17
total	122

122 people total responded to the survey.

b.

walking	37
bowling	− 18
difference	19

19 more people responded that walking rather than bowling is their favorite activity.

Mental Math

1. $9 - 2 = 7$

2. $6 - 6 = 0$

3. $5 - 0 = 5$

4. $44 - 22 = 22$

5. $93 - 93 = 0$

6. $700 - 400 = 300$

7. $700 - 300 = 400$

8. $700 - 700 = 0$

9. $600 - 100 = 500$

10. $600 - 0 = 600$

Exercise Set 1.3

1. $\begin{array}{r} 67 \\ -\ 23 \\ \hline 44 \end{array}$

Check: $\begin{array}{r} 44 \\ +\ 23 \\ \hline 67 \end{array}$

3. $\begin{array}{r} 82 \\ -\ 22 \\ \hline 60 \end{array}$

Check: $\begin{array}{r} 60 \\ +\ 22 \\ \hline 82 \end{array}$

5. $\begin{array}{r} 389 \\ -\ 124 \\ \hline 265 \end{array}$

Check: $\begin{array}{r} 265 \\ +\ 124 \\ \hline 389 \end{array}$

7. $\begin{array}{r} 677 \\ -\ 423 \\ \hline 254 \end{array}$

Check: $\begin{array}{r} 254 \\ +\ 423 \\ \hline 677 \end{array}$

9. $\begin{array}{r} 998 \\ -\ 453 \\ \hline 545 \end{array}$

Check: $\begin{array}{r} 545 \\ +\ 453 \\ \hline 998 \end{array}$

11. $$\begin{array}{r} 749 \\ -\ 149 \\ \hline 600 \end{array}$$

Check: $$\begin{array}{r} 600 \\ +\ 149 \\ \hline 749 \end{array}$$

13. $$\begin{array}{r} 62 \\ -\ 37 \\ \hline 25 \end{array}$$

Check: $$\begin{array}{r} {\scriptstyle 1} \\ 25 \\ +\ 37 \\ \hline 62 \end{array}$$

15. $$\begin{array}{r} 70 \\ -\ 25 \\ \hline 45 \end{array}$$

Check: $$\begin{array}{r} {\scriptstyle 1} \\ 45 \\ +\ 25 \\ \hline 70 \end{array}$$

17. $$\begin{array}{r} 938 \\ -\ 792 \\ \hline 146 \end{array}$$

Check: $$\begin{array}{r} {\scriptstyle 1} \\ 146 \\ +\ 792 \\ \hline 938 \end{array}$$

19. $$\begin{array}{r} 922 \\ -\ 634 \\ \hline 288 \end{array}$$

Check: $$\begin{array}{r} {\scriptstyle 11} \\ 288 \\ +\ 634 \\ \hline 922 \end{array}$$

21. $$\begin{array}{r} 600 \\ -\ 432 \\ \hline 168 \end{array}$$

Check: $$\begin{array}{r} {\scriptstyle 11} \\ 168 \\ +\ 432 \\ \hline 600 \end{array}$$

23. $$\begin{array}{r} 42 \\ -\ 36 \\ \hline 6 \end{array}$$

Check: $$\begin{array}{r} {\scriptstyle 1} \\ 6 \\ +\ 36 \\ \hline 42 \end{array}$$

25. $$\begin{array}{r} 923 \\ -\ 476 \\ \hline 447 \end{array}$$

Check: $$\begin{array}{r} {\scriptstyle 11} \\ 447 \\ +\ 476 \\ \hline 923 \end{array}$$

27. $$\begin{array}{r} 6283 \\ -\ 560 \\ \hline 5723 \end{array}$$

Check: $$\begin{array}{r} {\scriptstyle 1} \\ 5723 \\ +\ 560 \\ \hline 6283 \end{array}$$

29. $$\begin{array}{r} 533 \\ -\ 29 \\ \hline 504 \end{array}$$

Check: $$\begin{array}{r} {\scriptstyle 1} \\ 504 \\ +\ 29 \\ \hline 533 \end{array}$$

31. $$\begin{array}{r} 200 \\ -\ 111 \\ \hline 89 \end{array}$$

Check: $$\begin{array}{r} {\scriptstyle 11} \\ 89 \\ +\ 111 \\ \hline 200 \end{array}$$

33. $$\begin{array}{r} 1983 \\ -\ 1904 \\ \hline 79 \end{array}$$

Check: $$\begin{array}{r} {\scriptstyle 1} \\ 79 \\ +\ 1904 \\ \hline 1983 \end{array}$$

35. $$\begin{array}{r} 56,422 \\ -\ 16,508 \\ \hline 39,914 \end{array}$$

Check: $$\begin{array}{r} {\scriptstyle 11\ \ 1} \\ 39,914 \\ +\ 16,508 \\ \hline 56,422 \end{array}$$

37. $$\begin{array}{r} 50,000 \\ -\ 17,289 \\ \hline 32,711 \end{array}$$

Check: $$\begin{array}{r} {\scriptstyle 11\ 11} \\ 32,711 \\ +\ 17,289 \\ \hline 50,000 \end{array}$$

39. $$\begin{array}{r} 7020 \\ -\ 1979 \\ \hline 5041 \end{array}$$

Check: $$\begin{array}{r} {\scriptstyle 111} \\ 5041 \\ +\ 1979 \\ \hline 7020 \end{array}$$

41. $$\begin{array}{r} 51,111 \\ -\ 19,898 \\ \hline 31,213 \end{array}$$

Check: $$\begin{array}{r} {\scriptstyle 11\ 11} \\ 31,213 \\ +\ 19,898 \\ \hline 51,111 \end{array}$$

43. $$\begin{array}{r} 9 \\ -\ 5 \\ \hline 4 \end{array}$$

Check: $$\begin{array}{r} 4 \\ +\ 5 \\ \hline 9 \end{array}$$

45. $$\begin{array}{r} 41 \\ -\ 21 \\ \hline 20 \end{array}$$

Check: $$\begin{array}{r} 20 \\ +\ 21 \\ \hline 41 \end{array}$$

47. $$\begin{array}{r} 63 \\ -\ 56 \\ \hline 7 \end{array}$$

Check: $$\begin{array}{r} {\scriptstyle 1} \\ 7 \\ +\ 56 \\ \hline 63 \end{array}$$

49. $$\begin{array}{r} 63 \\ -\ 7 \\ \hline 56 \end{array}$$

Ronnie ate 56 crawfish.

51. $$\begin{array}{r} 503 \\ -\ 239 \\ \hline 264 \end{array}$$

Dyllis must read 264 more pages.

53.
$$\begin{array}{r} 20{,}320 \\ -\ 14{,}255 \\ \hline 6\,065 \end{array}$$
Mt. McKinley is 6065 feet higher than Long's Peak.

55. We need to find the total amount of the checks.
$$\begin{array}{r} \scriptstyle 1 \\ 27 \\ 101 \\ +\ 236 \\ \hline 364 \end{array}$$
The checks total $364.
$$\begin{array}{r} 539 \\ -\ 364 \\ \hline 175 \end{array}$$
Buhler will have $175 left in his account.

57.
$$\begin{array}{r} 645 \\ -\ 287 \\ \hline 358 \end{array}$$
The distance between Hays and Denver is 358 miles.

59.
$$\begin{array}{r} 914 \\ -\ 525 \\ \hline 389 \end{array}$$
Prunella will have $389 left in her savings account.

61.
$$\begin{array}{r} 547 \\ -\ 99 \\ \hline 448 \end{array}$$
The sale price is $448.

63. We need to find the total number of women enrolled in mathematics classes.

	11
Basic Mathematics	78
College Algebra	185
Calculus	+ 23
Total	286

$$\begin{array}{r} 459 \\ -\ 286 \\ \hline 173 \end{array}$$
There were 173 men enrolled in mathematics classes at FHSU.

65. The total number of votes cast for Jo was:
$$\begin{array}{r} \scriptstyle 1\,2\,1 \\ 276 \\ 362 \\ 201 \\ +\ 179 \\ \hline 1018 \end{array}$$
The total number of votes cast for Trudy was:
$$\begin{array}{r} \scriptstyle 2\,1 \\ 295 \\ 122 \\ 312 \\ +\ 182 \\ \hline 911 \end{array}$$
Since more votes were cast for Jo than for Trudy, Jo won the election.
$$\begin{array}{r} 1018 \\ -\ 911 \\ \hline 107 \end{array}$$
Jo won by 107 votes.

67.
$$\begin{array}{r} 100{,}000 \\ -\ 94{,}080 \\ \hline 5\,920 \end{array}$$
The Dole Plantation maze is 5920 square feet larger.

69.
$$\begin{array}{r} 1{,}130{,}000 \\ -\ 633{,}000 \\ \hline 497{,}000 \end{array}$$
North Korea's fighting force is 497,000 soldiers larger than South Korea's fighting force.

71.
$$\begin{array}{r} 13{,}152 \\ -\ 6\,473 \\ \hline 6\,679 \end{array}$$
The patient's WBC declined by 6679.

73. Atlanta, Hartsfield International is busiest.

75.
$$\begin{array}{r} 61{,}216 \\ -\ 60{,}483 \\ \hline 733 \end{array}$$
Los Angeles International has 733 thousand or 733,000 more passengers per year than Dallas/Ft. Worth International.

77. General Motors and Procter & Gamble spent more than $1500 million on ads.

79.
$$\begin{array}{r} 1,264,353,200 \\ -\ 809,878,700 \\ \hline 454,474,500 \end{array}$$
Philip Morris spent $454,474,500 more on ads than Walt Disney.

81.
$$\begin{array}{r} 2,121,040,900 \\ 1,724,259,700 \\ 1,410,748,700 \\ 1,264,353,200 \\ +\ 1,147,589,200 \\ \hline 7,667,991,700 \end{array}$$
The top five companies spent $7,667,991,700 on ads.

83.
$$\begin{array}{r} 5269 \\ -\ 2385 \\ \hline 2884 \end{array}$$

85. Answers may vary.

Section 1.4

Practice Problems

1. a. To round 46 to the nearest ten, observe that the digit in the ones place is 6. Since this digit is at least 5, we need to add 1 to the digit in the tens place. The number 46 rounded to the nearest ten is 50.

b. To round 731 to the nearest ten, observe that the digit in the ones place is 1. Since this digit is less than 5, we do not add 1 to the digit in the tens place. The number 731 rounded to the nearest ten is 730.

c. To round 125 to the nearest ten, observe that the digit in the ones place is 5. Since this digit is at least 5, we need to add 1 to the digit in the tens place. The number 125 rounded to the nearest ten is 130.

2. a. To round 56,702 to the nearest thousand, observe that the digit in the hundreds place is 7. Since this digit is at least 5, we need to add 1 to the digit in the thousands place. The number 56,702 rounded to the nearest thousand is 57,000.

b. To round 7444 to the nearest thousand, observe that the digit in the hundreds place is 4. Since this digit is less than 5, we do not add 1 to the digit in the thousands place. The number 7444 rounded to the nearest thousand is 7000.

c. To round 291,500 to the nearest thousand, observe that the digit in the hundreds place is 5. Since this digit at least 5, we need to add 1 to the digit in the thousands place. The number 291,500 rounded to the nearest thousand is 292,000.

3. a. To round 2777 to the nearest hundred, observe that the digit in the tens place is 7. Since this digit is at least 5, we need to add 1 to the digit in the hundreds place. The number 2777 rounded to the nearest hundred is 2800.

b. To round 38,152 to the nearest hundred, observe that the digit in the tens place is 5. Since this digit is at least 5, we need to add 1 to the digit in the hundreds place. The number 38,152 rounded to the nearest hundred is 38,200.

c. To round 762,955 to the nearest hundred, observe that the digit in the tens place is 5. Since this digit is at least 5, we need to add 1 to the digit in the hundreds place. The number 762,955 rounded to the nearest hundred is 763,000.

4.
$$\begin{array}{rlr} 79 & \text{rounds to} & 80 \\ 35 & \text{rounds to} & 40 \\ 42 & \text{rounds to} & 40 \\ 21 & \text{rounds to} & 20 \\ +\ 98 & \text{rounds to} & +\ 100 \\ \hline & & 280 \end{array}$$

The estimated sum is 280. (The exact sum is 275.)

5.
$$\begin{array}{rlr} 4725 & \text{rounds to} & 5000 \\ -\ 2879 & \text{rounds to} & -\ 3000 \\ \hline & & 2000 \end{array}$$

The estimated difference is 2000. (The exact difference is 1846.)

6.
$$\begin{array}{rlr} 11 & \text{rounds to} & 10 \\ 16 & \text{rounds to} & 20 \\ 19 & \text{rounds to} & 20 \\ +\ 31 & \text{rounds to} & +\ 30 \\ \hline & & 80 \end{array}$$

It is approximately 80 miles from Grove to Hays. (The exact distance is 77 miles.)

7.
$$\begin{array}{rlr} 120,624 & \text{rounds to} & 120,000 \\ 22,866 & \text{rounds to} & 20,000 \\ +\ 45,970 & \text{rounds to} & +\ 50,000 \\ \hline & & 190,000 \end{array}$$

The approximate number of cases reported was 190,000.

Exercise Set 1.4

1. To round 632 to the nearest ten, observe that the digit in the ones place is 2. Since this digit is less than 5, we do not add 1 to the digit in the tens place. The number 632 rounded to the nearest ten is 630.

3. To round 635 to the nearest ten, observe that the digit in the ones place is 5. Since this digit is at least 5, we need to add 1 to the digit in the tens place. The number 635 rounded to the nearest ten is 640.

5. To round 792 to the nearest ten, observe that the digit in the ones place is 2. Since this digit is less than 5, we do not add 1 to the digit in the tens place. The number 792 rounded to the nearest ten is 790.

7. To round 395 to the nearest ten, observe that the digit in the ones place is 5. Since this digit is at least 5, we need to add 1 to the digit in the tens place. The number 395 rounded to the nearest ten is 400.

9. To round 1096 to the nearest ten, observe that the digit in the ones place is 6. Since this digit is at least 5, we need to add 1 to the digit in the tens place. The number 1096 rounded to the nearest ten is 1100.

11. To round 42,682 to the nearest thousand, observe that the digit in the hundreds place is 6. Since this digit is at least 5, we need to add 1 to the digit in the thousands place. The number 42,682 rounded to the nearest thousand is 43,000.

13. To round 248,695 to the nearest hundred, observe that the digit in the tens place is 9. Since this digit is at least 5, we need to add 1 to the digit in the hundreds place. The number 248,695 rounded to the nearest hundred is 248,700.

15. To round 36,499 to the nearest thousand, observe that the digit in the hundreds place is 4. Since this digit is less than 5, we do not add 1 to the digit in the thousands place. The number 36,499 rounded to the nearest thousand is 36,000.

17. To round 99,995 to the nearest ten, observe that the digit in the ones place is 5. Since this digit is at least 5, we need to add 1 to the digit in the tens place. The number 99,995 rounded to the nearest ten is 100,000.

19. To round 59,725,642 to the nearest ten-million, observe that the digit in the millions place is 9. Since this digit is at least 5, we need to add 1 to the digit in the ten-millions place. The number 59,725,642 rounded to the nearest ten-million is 60,000,000.

		Ten	Hundred	Thousand
21.	5281	5280	5300	5000
23.	9444	9440	9400	9000
25.	14,876	14,880	14,900	15,000

27. To round 11,187 to the nearest thousand, observe that the digit in the hundreds place is 1. Since this digit is less than 5, we do not add 1 to the digit in the thousands place. The number 11,187 rounded to the nearest thousand is 11,000.

29. To round 72,625 to the nearest thousand, observe that the digit in the hundreds place is 6. Since this digit is at least 5, we need to add 1 to the digit in the thousands place. The number 72,625 rounded to the nearest thousand is 73,000.

31. To round 167,122,000 to the nearest ten-million, observe that the digit in the millions place is 7. Since this digit is at least 5, we need to add 1 to the digit in the ten-millions place. The number 167,122,000 rounded to the nearest ten-million is 170,000,000.

33. To round $371,971,500 to the nearest hundred-thousand, observe that the digit in the ten-thousands place is 7. Since this digit is at least 5, we need to add 1 to the digit in the hundred-thousands place. The number $371,971,500 rounded to the nearest hundred-thousand is $372,000,000.

35.

29	rounds to	30
35	rounds to	40
42	rounds to	40
+ 16	rounds to	+ 20
		130

The estimated sum is 130.

37.

649	rounds to	650
− 272	rounds to	− 270
		380

The estimated difference is 380.

39.

1812	rounds to	1800
1776	rounds to	1800
+ 1945	rounds to	+ 1900
		5500

The estimated sum is 5500.

41.

1774	rounds to	1800
− 1492	rounds to	− 1500
		300

The estimated difference is 300.

43.

2995	rounds to	3000
1649	rounds to	1600
+ 3940	rounds to	+ 3900
		8500

The estimated sum is 8500.

45. 362 + 419 is approximately
360 + 420 = 780.
The answer of 781 is correct.

47. 432 + 679 + 198 is approximately
400 + 700 + 200 = 1300.
The answer of 1139 is incorrect.

49. 7806 + 5150 is approximately
7800 + 5200 = 13,000.
The answer of 12,956 is correct.

51. 31,439 + 18,781 is approximately
31,000 + 19,000 = 50,000.
The answer is 50,220 is correct.

53.

799	rounds to	800
1299	rounds to	1300
+ 999	rounds to	+ 1000
		3100

The total cost is approximately $3100.

55.
$$\begin{array}{rlr} 19 & \text{rounds to} & 20 \\ 27 & \text{rounds to} & 30 \\ +\ 34 & \text{rounds to} & +\ 30 \\ \hline & & 80 \end{array}$$

The distance from Stockton to LaCrosse is approximately 80 miles.

57.
$$\begin{array}{rlr} 29{,}028 & \text{rounds to} & 29{,}000 \\ -\ 4039 & \text{rounds to} & -\ 4\ 000 \\ \hline & & 25{,}000 \end{array}$$

The difference in elevation is approximately 25,000 feet.

59.
$$\begin{array}{rlr} 7{,}420{,}166 & \text{rounds to} & 7{,}400{,}000 \\ -\ 1{,}786{,}691 & \text{rounds to} & -\ 1{,}800{,}000 \\ \hline & & 5{,}600{,}000 \end{array}$$

The population of New York City is approximately 5,600,000 greater than that of Houston.

61.
$$\begin{array}{rlr} 41{,}126{,}233 & \text{rounds to} & 41{,}000{,}000 \\ -\ 27{,}174{,}898 & \text{rounds to} & -\ 27{,}000{,}000 \\ \hline & & 14{,}000{,}000 \end{array}$$

Johnson won the election by approximately 14,000,000 votes.

63.
$$\begin{array}{rlr} 794{,}000 & \text{rounds to} & 794{,}000 \\ -\ 540{,}930 & \text{rounds to} & -\ 541{,}000 \\ \hline & & 253{,}000 \end{array}$$

The Head Start enrollment increased by approximately 253,000 children.

65. 1,264,353,200 rounds to 1,300,000,000. Philip Morris spent approximately $1,300,000,000.

67. 1,410,748,700 rounds to 1,411,000,000. Daimler Chrysler spent approximately $1,411,000,000.

69. The smallest possible number that rounds to 8600 is 8550.

71. The largest possible number that rounds to 1,500,000 is 1,549,999.

Section 1.5

Practice Problems

1. a. $3 \times 0 = 0$

b. $4(1) = 4$

c. $(0)(34) = 0$

d. $1 \cdot 76 = 76$

2. a. $5(2+3) = 5 \cdot 2 + 5 \cdot 3$

b. $9(8+7) = 9 \cdot 8 + 9 \cdot 7$

c. $3(6+1) = 3 \cdot 6 + 3 \cdot 1$

3. a.
$$\begin{array}{r} {}^{2} \\ 36 \\ \times\ 4 \\ \hline 144 \end{array}$$

b.
$$\begin{array}{r} {}^{1} \\ 92 \\ \times\ 9 \\ \hline 828 \end{array}$$

4. a.
$$\begin{array}{r} 594 \\ \times\ 72 \\ \hline 1\ 188 \\ 41{,}580 \\ \hline 42{,}768 \end{array}$$

b.
$$\begin{array}{r} 306 \\ \times\ 81 \\ \hline 306 \\ 24{,}480 \\ \hline 24{,}786 \end{array}$$

5. a.
$$\begin{array}{r} 726 \\ \times\ 142 \\ \hline 1\ 452 \\ 29{,}040 \\ 72{,}600 \\ \hline 103{,}092 \end{array}$$

b.
$$\begin{array}{r} 4 \\ \times\ 288 \\ \hline 32 \\ 320 \\ 800 \\ \hline 1152 \end{array}$$

6.
$$\begin{array}{r} 360 \\ \times\ 280 \\ \hline 0 \\ 28,800 \\ 72,000 \\ \hline 100,800 \end{array}$$

The area is 100,800 square miles.

7.
$$\begin{array}{r} 240 \\ \times\ 15 \\ \hline 1200 \\ 2400 \\ \hline 3600 \end{array}$$

The printer can print 3600 characters in 15 seconds.

8. The cost of 4 plain shirts is $4 \cdot 6 = \$24$.
The cost of 5 striped shirts is $5 \cdot 7 = \$35$.
The total cost is $24 + 35 = \$59$.

9.
$$\begin{array}{rcr} 259 & \text{rounds to} & 300 \\ \times\ 195 & \text{rounds to} & \times\ 200 \\ \hline & & 60,000 \end{array}$$

There are approximately 60,000 words on 195 pages.

Mental Math

1. $1 \cdot 24 = 24$
2. $55 \cdot 1 = 55$
3. $0 \cdot 19 = 0$
4. $27 \cdot 0 = 0$
5. $8 \cdot 0 \cdot 9 = 0$
6. $7 \cdot 6 \cdot 0 = 0$
7. $87 \cdot 1 = 87$
8. $1 \cdot 41 = 41$

Exercise Set 1.5

1. $4(3 + 9) = 4 \cdot 3 + 4 \cdot 9$

3. $2(4 + 6) = 2 \cdot 4 + 2 \cdot 6$

5. $10(11 + 7) = 10 \cdot 11 + 10 \cdot 7$

7.
$$\begin{array}{r} 42 \\ \times\ 6 \\ \hline 252 \end{array}$$

9.
$$\begin{array}{r} 624 \\ \times\ 3 \\ \hline 1872 \end{array}$$

11.
$$\begin{array}{r} 277 \\ \times\ 6 \\ \hline 1662 \end{array}$$

13.
$$\begin{array}{r} 1062 \\ \times\ 5 \\ \hline 5310 \end{array}$$

15.
$$\begin{array}{r} 298 \\ \times\ 14 \\ \hline 1192 \\ 2980 \\ \hline 4172 \end{array}$$

17.
$$\begin{array}{r} 231 \\ \times\ 47 \\ \hline 1\ 617 \\ 9\ 240 \\ \hline 10,857 \end{array}$$

19.
$$\begin{array}{r} 809 \\ \times\ 14 \\ \hline 3236 \\ 8090 \\ \hline 11,326 \end{array}$$

21. $$\begin{array}{r} 620 \\ \times\ 40 \\ \hline 0 \\ 24{,}800 \\ \hline 24{,}800 \end{array}$$

23. $(998)(12)(0) = 0$

25. $(590)(1)(10) = 5900$

27. $$\begin{array}{r} 1\ 234 \\ \times\ 48 \\ \hline 9\ 872 \\ 49{,}360 \\ \hline 59{,}232 \end{array}$$

29. $$\begin{array}{r} 609 \\ \times\ 234 \\ \hline 2\ 436 \\ 18{,}270 \\ 121{,}800 \\ \hline 142{,}506 \end{array}$$

31. $$\begin{array}{r} 5621 \\ \times\ 324 \\ \hline 22{,}484 \\ 112{,}420 \\ 1{,}686{,}300 \\ \hline 1{,}821{,}204 \end{array}$$

33. $$\begin{array}{r} 1941 \\ \times\ 235 \\ \hline 9\ 705 \\ 58{,}230 \\ 388{,}200 \\ \hline 456{,}135 \end{array}$$

35. $$\begin{array}{r} 589 \\ \times\ 110 \\ \hline 0 \\ 5\ 890 \\ 58{,}900 \\ \hline 64{,}790 \end{array}$$

37. $$\begin{array}{rlr} 576 & \text{rounds to} & 600 \\ \times\ 354 & \text{rounds to} & \times\ 400 \\ \hline & & 240{,}000 \end{array}$$

576×354 is approximately 240,000.

39. $$\begin{array}{rlr} 604 & \text{rounds to} & 600 \\ \times\ 451 & \text{rounds to} & \times\ 500 \\ \hline & & 300{,}000 \end{array}$$

604×451 is approximately 300,000.

41. $$\begin{aligned} \text{Area} &= \text{length} \cdot \text{width} \\ &= (9 \text{ meters})(7 \text{ meters}) \\ &= 63 \text{ square meters} \end{aligned}$$

The area is 63 square meters.

43. $$\begin{aligned} \text{Area} &= \text{length} \cdot \text{width} \\ &= (30 \text{ feet})(13 \text{ feet}) \\ &= 390 \text{ square feet} \end{aligned}$$

The area is 390 square feet.

45. $$\begin{array}{r} 125 \\ \times\ 3 \\ \hline 375 \end{array}$$

There are 375 calories in 3 tablespoons of olive oil.

47. $$\begin{array}{r} 37 \\ \times\ 35 \\ \hline 185 \\ 1110 \\ \hline 1295 \end{array}$$

The books cost a total of $1295.

49. $$\begin{array}{r} 12 \\ \times\ 8 \\ \hline 96 \end{array}$$

$2 \times 96 = 192$
There are 192 cans in a case.

51.
$$\begin{array}{r} 90 \\ \times\ 110 \\ \hline 0 \\ 900 \\ 9000 \\ \hline 9900 \end{array}$$
The area of the plot is 9900 square feet.

53.
$$\begin{array}{r} 776 \\ \times\ 639 \\ \hline 6\ 984 \\ 23,280 \\ 465,600 \\ \hline 495,864 \end{array}$$
The floor area is 495,864 square meters.

55.
$$\begin{array}{r} 62 \\ \times\ 94 \\ \hline 248 \\ 5580 \\ \hline 5828 \end{array}$$
The screen contains a total of 5828 pixels.

57.
$$\begin{array}{r} 60 \\ \times\ 25 \\ \hline 300 \\ 1200 \\ \hline 1500 \end{array}$$
There are 1500 characters in 25 lines.

59.
$$\begin{array}{r} 160 \\ \times\ 8 \\ \hline 1280 \end{array}$$
There are 1280 calories in 8 ounces of peanuts.

61.
$$\begin{array}{r} 7927 \\ \times\ 9 \\ \hline 71,343 \end{array}$$
Saturn has a diameter of 71,343 miles.

63.
$$\begin{array}{r} 576 \\ \times\ 39 \\ \hline 5\ 184 \\ 17,280 \\ \hline 22,464 \end{array}$$
The herd needs 22,464 acres.

65.
$$\begin{array}{r} 700,000 \\ \times\ 31 \\ \hline 700,000 \\ 21,000,000 \\ \hline 21,700,000 \end{array}$$
21,700,000 quarts of milk would be used in March.

67.
$$\begin{array}{rcr} 752,077 & \text{rounds to} & 752,000 \\ \times\ 4571 & \text{rounds to} & \times\ 5000 \\ \cline{3-3} & & 3,760,000,000 \end{array}$$
The total cost is about \$3.76 billion.

69. $5 \times 10 = 50$
50 students chose grapes as their favorite fruit.

71. The apple (chosen by $9 \times 10 = 90$ students) and the orange (chosen by $7 \times 10 = 70$ students) were the most popular.

73. The result of multiplying 3 by the digit in the first blank is a number ending in 6. Only 2 works. The result of multiplying the digit in the second blank by 42 is 378, so the digit in the second blank is 9.
$$\begin{array}{r} 4\underline{2} \\ \times\ \underline{9}3 \\ \hline 126 \\ 3780 \\ \hline 3906 \end{array}$$

75. Answers may vary.

77. $154 \times 2 = 308$ and $58 \times 3 = 174$

$$\begin{array}{r} 204 \\ 308 \\ +\ 174 \\ \hline 686 \end{array}$$

Cynthia Cooper scored 686 points in the 1999 season.

79. Answers will vary.

Section 1.6

Practice Problems

1. a. $8\overline{)48}$ with quotient 6, because $6 \cdot 8 = 48$.

b. $35 \div 5 = 7$, because $7 \cdot 5 = 35$.

c. $\frac{49}{7} = 7$, because $7 \cdot 7 = 49$.

2. a. $\frac{8}{8} = 1$, because $1 \cdot 8 = 8$.

b. $3 \div 1 = 3$, because $3 \cdot 1 = 3$.

c. $1\overline{)12}$ with quotient 12, because $12 \cdot 1 = 12$.

d. $2 \div 1 = 2$, because $2 \cdot 1 = 2$.

e. $\frac{5}{1} = 5$, because $5 \cdot 1 = 5$.

3. a. $\frac{0}{7} = 0$, because $0 \cdot 7 = 0$.

b. $5\overline{)0}$ with quotient 0, because $0 \cdot 5 = 0$.

c. $0 \div 6 = 0$, because $0 \cdot 6 = 0$.

d. $9 \div 0$ is undefined, because any number multiplied by 0 is 0 and not 9.

4. a.

$$\begin{array}{r} 897 \\ 6\overline{)\,5382} \\ \underline{-48} \\ 58 \\ \underline{-54} \\ 42 \\ \underline{-42} \\ 0 \end{array}$$

Check: $897 \cdot 6 = 5382$

b.

$$\begin{array}{r} 553 \\ 4\overline{)\,2212} \\ \underline{-20} \\ 21 \\ \underline{-20} \\ 12 \\ \underline{-12} \\ 0 \end{array}$$

Check: $553 \cdot 4 = 2212$

5. a.

$$\begin{array}{r} 799 \\ 3\overline{)\,2397} \\ \underline{-21} \\ 29 \\ \underline{-27} \\ 27 \\ \underline{-27} \\ 0 \end{array}$$

Check: $799 \cdot 3 = 2397$

b.

$$\begin{array}{r} 360 \\ 7\overline{)\,2520} \\ \underline{-21} \\ 42 \\ \underline{-42} \\ 00 \\ \underline{-0} \\ 0 \end{array}$$

Check: $360 \cdot 7 = 2520$

6. a.

$$\begin{array}{r} 189 \text{ R } 4 \\ 5\overline{)\,949}\phantom{\text{ R } 4} \\ \underline{-5}\phantom{\text{ R } 4} \\ 44\phantom{\text{ R } 4} \\ \underline{-40}\phantom{\text{ R } 4} \\ 49\phantom{\text{ R } 4} \\ \underline{-45}\phantom{\text{ R } 4} \\ 4\phantom{\text{ R } 4} \end{array}$$

Check: $189 \cdot 5 + 4 = 949$

b.

$$\begin{array}{r} 733 \text{ R } 1 \\ 6\overline{)\,4399} \\ \underline{-42} \\ 19 \\ \underline{-18} \\ 19 \\ \underline{-18} \\ 1 \end{array}$$

Check: $733 \cdot 6 + 1 = 4399$

7. a.

$$\begin{array}{r} 8168 \text{ R } 1 \\ 5\overline{)\,40841} \\ \underline{-40} \\ 08 \\ \underline{-5} \\ 34 \\ \underline{-30} \\ 41 \\ \underline{-40} \\ 1 \end{array}$$

Check: $8168 \cdot 5 + 1 = 40{,}841$

b.

$$\begin{array}{r} 3204 \text{ R } 2 \\ 7\overline{)\,22430} \\ \underline{-21} \\ 14 \\ \underline{-14} \\ 03 \\ \underline{-0} \\ 30 \\ \underline{-28} \\ 2 \end{array}$$

Check: $3204 \cdot 7 + 2 = 22{,}430$

8.

$$\begin{array}{r} 302 \text{ R } 2 \\ 19\overline{)\,5740} \\ \underline{-57} \\ 04 \\ \underline{-0} \\ 40 \\ \underline{-38} \\ 2 \end{array}$$

Check: $302 \cdot 19 + 2 = 5740$

9.

$$\begin{array}{r} 67 \text{ R } 40 \\ 247\overline{)\,16589} \\ \underline{-1482} \\ 1769 \\ \underline{-1729} \\ 40 \end{array}$$

Check: $67 \cdot 247 + 40 = 16{,}589$

10.

$$\begin{array}{r} 40 \\ 3\overline{)\,120} \\ \underline{-12} \\ 00 \\ \underline{-0} \\ 0 \end{array}$$

Each person got 40 diskettes.

11.

$$\begin{array}{r} 32 \text{ R } 3 \\ 6\overline{)\,195} \\ \underline{-18} \\ 15 \\ \underline{-12} \\ 3 \end{array}$$

195 sandwiches yield 32 full packages, with 3 sandwiches left over.

12.

$$\begin{array}{r} 20 \text{ R } 17 \\ 24\overline{)\,497} \\ \underline{-48} \\ 17 \\ \underline{-0} \\ 17 \end{array}$$

20 full boxes will be shipped, with 17 calculators left over.

13. Add the scores, then divide by 7, the number of scores.

$$\begin{array}{r} 5 \\ 7 \\ 20 \\ 6 \\ 9 \\ 3 \\ \underline{+\ 48} \\ 98 \end{array} \qquad \begin{array}{r} 14 \\ 7\overline{)\,98} \\ \underline{-7} \\ 28 \\ \underline{-28} \\ 0 \end{array}$$

The average elapsed time is 14 minutes.

Mental Math

1. $40 \div 8 = 5$
2. $72 \div 9 = 8$
3. $45 \div 5 = 9$
4. $24 \div 3 = 8$
5. $0 \div 5 = 0$
6. $0 \div 8 = 0$

7. $9 \div 1 = 9$

8. $12 \div 1 = 12$

9. $\frac{16}{16} = 1$

10. $\frac{49}{49} = 1$

11. $\frac{25}{5} = 5$

12. $\frac{45}{9} = 5$

13. $6 \div 0$ is undefined

14. $\frac{12}{0}$ is undefined

15. $7 \div 1 = 7$

16. $6 \div 6 = 1$

17. $0 \div 4 = 0$

18. $7 \div 0$ is undefined

19. $16 \div 2 = 8$

20. $18 \div 3 = 6$

Exercise Set 1.6

1.
$$\begin{array}{r} 12 \\ 9\overline{)108} \\ \underline{-9} \\ 18 \\ \underline{-18} \\ 0 \end{array}$$

Check: $12 \cdot 9 = 108$

3.
$$\begin{array}{r} 37 \\ 6\overline{)222} \\ \underline{-18} \\ 42 \\ \underline{-42} \\ 0 \end{array}$$

Check: $37 \cdot 6 = 222$

5.
$$\begin{array}{r} 338 \\ 3\overline{)1014} \\ \underline{-9} \\ 11 \\ \underline{-9} \\ 24 \\ \underline{-24} \\ 0 \end{array}$$

Check: $338 \cdot 3 = 1014$

7.
$$\begin{array}{r} 16 \text{ R } 2 \\ 6\overline{)98} \\ \underline{-6} \\ 38 \\ \underline{-36} \\ 2 \end{array}$$

Check: $16 \cdot 6 + 2 = 98$

9.
$$\begin{array}{r} 563 \text{ R } 1 \\ 2\overline{)1127} \\ \underline{-10} \\ 12 \\ \underline{-12} \\ 07 \\ \underline{-6} \\ 1 \end{array}$$

Check: $563 \cdot 2 + 1 = 1127$

11.
$$\begin{array}{r} 37 \text{ R } 1 \\ 5\overline{)186} \\ \underline{-15} \\ 36 \\ \underline{-35} \\ 1 \end{array}$$

Check: $37 \cdot 5 + 1 = 186$

13.
$$\begin{array}{r} 265 \text{ R } 1 \\ 8\overline{)2121} \\ \underline{-16} \\ 52 \\ \underline{-48} \\ 41 \\ \underline{-40} \\ 1 \end{array}$$

Check: $265 \cdot 8 + 1 = 2121$

15.
$$\begin{array}{r} 49 \\ 23\overline{)1127} \\ \underline{-92} \\ 207 \\ \underline{-207} \\ 0 \end{array}$$

Check: $49 \cdot 23 = 1127$

17.
$$\begin{array}{r} 13 \\ 55\overline{)\ 715} \\ \underline{-55} \\ 165 \\ \underline{-165} \\ 0 \end{array}$$

Check: $13 \cdot 55 = 715$

19.
$$\begin{array}{r} 97 \text{ R } 40 \\ 97\overline{)\ 9449} \\ \underline{-873} \\ 719 \\ \underline{-679} \\ 40 \end{array}$$

Check: $97 \cdot 97 + 40 = 9449$

21.
$$\begin{array}{r} 206 \\ 18\overline{)\ 3708} \\ \underline{-36} \\ 10 \\ \underline{-0} \\ 108 \\ \underline{-108} \\ 0 \end{array}$$

Check: $206 \cdot 18 = 3708$

23.
$$\begin{array}{r} 506 \\ 13\overline{)\ 6578} \\ \underline{-65} \\ 07 \\ \underline{-0} \\ 78 \\ \underline{-78} \\ 0 \end{array}$$

Check: $506 \cdot 13 = 6578$

25.
$$\begin{array}{r} 202 \text{ R } 7 \\ 46\overline{)\ 9299} \\ \underline{-92} \\ 09 \\ \underline{-0} \\ 99 \\ \underline{-92} \\ 7 \end{array}$$

Check: $202 \cdot 46 + 7 = 9299$

27.
$$\begin{array}{r} 45 \\ 236\overline{)10620} \\ \underline{-944} \\ 1180 \\ \underline{-1180} \\ 0 \end{array}$$

Check: $45 \cdot 236 = 10620$

29.
$$\begin{array}{r} 98 \text{ R } 100 \\ 103\overline{)10194} \\ \underline{-927} \\ 924 \\ \underline{-824} \\ 100 \end{array}$$

Check: $98 \cdot 103 + 100 = 10,194$

31.
$$\begin{array}{r} 202 \text{ R } 15 \\ 102\overline{)\ 20619} \\ \underline{-204} \\ 21 \\ \underline{-0} \\ 219 \\ \underline{-204} \\ 15 \end{array}$$

Check: $202 \cdot 102 + 15 = 20,619$

33.
$$\begin{array}{r} 202 \\ 223\overline{)\ 45046} \\ \underline{-446} \\ 44 \\ \underline{-0} \\ 446 \\ \underline{-446} \\ 0 \end{array}$$

Check: $202 \cdot 223 = 45,046$

35.
$$\begin{array}{r} 58 \\ 85\overline{)\,4930} \\ \underline{-425} \\ 680 \\ \underline{-680} \\ 0 \end{array}$$

There are 58 students in the group.

37.
$$\begin{array}{r} 252000 \\ 21\overline{)\,5292000} \\ \underline{-42} \\ 109 \\ \underline{-105} \\ 42 \\ \underline{-42} \\ 00 \\ \underline{-0} \\ 00 \\ \underline{-0} \\ 00 \\ \underline{-0} \\ 0 \end{array}$$

Each person receives $252,000.

39.
$$\begin{array}{r} 415 \\ 14\overline{)\,5810} \\ \underline{-56} \\ 21 \\ \underline{-14} \\ 70 \\ \underline{-70} \\ 0 \end{array}$$

The truck hauls 415 bushels each trip.

41.
$$\begin{array}{r} 88 \text{ R } 1 \\ 3\overline{)\,265} \\ \underline{-24} \\ 25 \\ \underline{-24} \\ 1 \end{array}$$

There are 88 bridges in 265 miles.

43.
$$\begin{array}{r} 23 \text{ R } 1 \\ 8\overline{)\,185} \\ \underline{-16} \\ 25 \\ \underline{-24} \\ 1 \end{array}$$

She has enough for 22 students, with another student's worth plus one more foot, or $8 + 1 = 9$ feet extra.

45.
$$\begin{array}{r} 17 \\ 6\overline{)\,102} \\ \underline{-6} \\ 42 \\ \underline{-42} \\ 0 \end{array}$$

He scored 17 touchdowns during 1999.

47.
$$\begin{array}{r} 1760 \\ 3\overline{)\,5280} \\ \underline{-3} \\ 22 \\ \underline{-21} \\ 18 \\ \underline{-18} \\ 00 \\ \underline{-0} \\ 0 \end{array}$$

There are 1760 yards in 1 mile.

49.
$$\begin{array}{r} 6,000 \\ 3,500,000\overline{)\,21,000,000,000} \\ \underline{-21,000,000} \\ 0\;0 \\ \underline{-0} \\ 00 \\ \underline{-0} \\ 0 \end{array}$$

A single home uses 6000 kilowatt-hours of electricity during one year on average.

51. There are six numbers.

$$\begin{array}{r} 14 \\ 22 \\ 45 \\ 18 \\ 30 \\ \underline{+\;27} \\ 156 \end{array} \qquad \begin{array}{r} 26 \\ 6\overline{)\,156} \\ \underline{-12} \\ 36 \\ \underline{-36} \\ 0 \end{array}$$

Average $= \frac{156}{6} = 26$

53. There are four numbers.

$$\begin{array}{r} 204 \\ 968 \\ 552 \\ +\ 268 \\ \hline 1992 \end{array} \qquad \begin{array}{r} 498 \\ 4\overline{)\ 1992} \\ \underline{-16} \\ 39 \\ \underline{-36} \\ 32 \\ \underline{-32} \\ 0 \end{array}$$

$\text{Average} = \frac{1992}{4} = 498$

55. There are five numbers.

$$\begin{array}{r} 86 \\ 79 \\ 81 \\ 69 \\ +\ 80 \\ \hline 395 \end{array} \qquad \begin{array}{r} 79 \\ 5\overline{)\ 395} \\ \underline{-35} \\ 45 \\ \underline{-45} \\ 0 \end{array}$$

$\text{Average} = \frac{395}{5} = 79$

57. Add those month's temperatures, then divide by 3, the number of temperatures.

$$\begin{array}{r} 18 \\ 12 \\ +\ 18 \\ \hline 48 \end{array} \qquad \begin{array}{r} 16 \\ 3\overline{)\ 48} \\ \underline{-3} \\ 18 \\ \underline{-18} \\ 0 \end{array}$$

The average temperature is $\frac{48}{3} = 16$ degrees.

59. Add the four greatest amounts, then divide by 4.

$$\begin{array}{r} 2,121,040,900 \\ 1,724,259,700 \\ 1,410,748,700 \\ +\ 1,264,353,200 \\ \hline 6,520,402,500 \end{array}$$

$$\begin{array}{r} 1630100625 \\ 4\overline{)\ 6520402500} \\ \underline{-4} \\ 25 \\ \underline{-24} \\ 12 \\ \underline{-12} \\ 00 \\ \underline{-0} \\ 04 \\ \underline{-4} \\ 00 \\ \underline{-0} \\ 02 \\ \underline{-0} \\ 25 \\ \underline{-24} \\ 10 \\ \underline{-8} \\ 20 \\ \underline{-20} \\ 0 \end{array}$$

The average amount spent by the top four companies is $1,630,100,625.

61. Increases, because 71 lies below the average, which is 83.

63. No, because all the numbers are greater than 86.

Integrated Review

1.

$$\begin{array}{r} {\scriptstyle 11} \\ 23 \\ 46 \\ +\ 79 \\ \hline 148 \end{array}$$

2.

$$\begin{array}{r} 7006 \\ -\ 451 \\ \hline 6555 \end{array}$$

3.

$$\begin{array}{r} 36 \\ \times\ 45 \\ \hline 180 \\ 1440 \\ \hline 1620 \end{array}$$

4. $$\begin{array}{r} 562 \\ 8\overline{)\ 4496} \\ \underline{-40} \\ 49 \\ \underline{-48} \\ 16 \\ \underline{-16} \\ 0 \end{array}$$

5. $1 \cdot 79 = 79$

6. $\frac{36}{0}$ is undefined.

7. $9 \div 1 = 9$

8. $9 \div 9 = 1$

9. $0 \cdot 13 = 0$

10. $7 \cdot 0 \cdot 8 = 0$

11. $0 \div 2 = 0$

12. $12 \div 4 = 3$

13. $$\begin{array}{r} 4219 \\ \underline{-\ 1786} \\ 2433 \end{array}$$

14. $$\begin{array}{r} {\scriptstyle 1\,1} \\ 1861 \\ \underline{+\ 7965} \\ 9826 \end{array}$$

15. $$\begin{array}{r} 213 \text{ R } 3 \\ 5\overline{)\ 1068} \\ \underline{-10} \\ 06 \\ \underline{-5} \\ 18 \\ \underline{-15} \\ 3 \end{array}$$

16. $$\begin{array}{r} 1259 \\ \underline{\times\ 63} \\ 3\ 777 \\ \underline{75,540} \\ 79,317 \end{array}$$

17. $3 \cdot 9 = 27$

18. $45 \div 5 = 9$

19. $$\begin{array}{r} 207 \\ \underline{-\ 69} \\ 138 \end{array}$$

20. $$\begin{array}{r} {\scriptstyle 1} \\ 207 \\ \underline{+\ 69} \\ 276 \end{array}$$

21. $$\begin{array}{r} 663 \text{ R } 6 \\ 32\overline{)\ 21222} \\ \underline{-192} \\ 202 \\ \underline{-192} \\ 102 \\ \underline{-96} \\ 6 \end{array}$$

22. $$\begin{array}{r} 1076 \text{ R } 60 \\ 65\overline{)\ 70000} \\ \underline{-65} \\ 50 \\ \underline{-0} \\ 500 \\ \underline{-455} \\ 450 \\ \underline{-390} \\ 60 \end{array}$$

23. The place value of the 3 in 732 is tens.

24. The place value of the 3 in 23,000 is thousands.

25. The place value of the 3 in 8003 is ones.

26. The place value of the 3 in 32,222 is ten-thousands.

27. 850 is written as eight hundred fifty.

28. 805 is written as eight hundred five.

29. 21,060 is written as twenty-one thousand, sixty.

30. 4044 is written as four thousand, forty-four.

31. Five thousand, six hundred twelve in standard form is 5612.

32. Seventy-three thousand, one in standard form is 73,001.

33. One hundred-thousand, three hundred six in standard from is 100,306.

34. Four million, twenty in standard form is 4,000,020.

35. To round 3265 to the nearest ten, observe that the digit in the ones place is 5. Since this digit is at least 5, we need to add 1 to the digit in the tens place. The number 3265 rounded to the nearest ten is 3270.
To round 3265 to the nearest hundred, observe that the digit in the tens place is 6. Since this digit is at least 5, we need to add 1 to the digit in the hundreds place. The number 3265 rounded to the nearest hundred is 3300.

36. To round 46,817 to the nearest ten, observe that the digit in the ones place is 7. Since this digit is at least 5, we need to add 1 to the digit in the tens place. The number 46,817 rounded to the nearest ten is 46,820. To round 46,817 to the nearest hundred, observe that the digit in the tens place is 1. Since this digit is less than 5, we do not add 1 to the digit in the hundreds place. The number 46,817 rounded to the nearest hundred is 46,800.

37. To round 60,505 to the nearest ten, observe that the digit in the ones place is 5. Since this digit is at least 5, we need to add 1 to the digit in the tens place. The number 60,505 rounded to the nearest ten is 60,510. To round 60,505 to the nearest hundred, observe that the digit in the tens place is 0. Since this digit is less than 5, we do not add 1 to the digit in the hundreds place. The number 60,505 rounded to the nearest hundred is 60,500.

38. To round 671 to the nearest ten, observe that the digit in the ones place is 1. Since this digit is less than 5, we do not need to add 1 to the digit in the tens place. The number 671 rounded to the nearest ten is 670. To round 671 to the nearest hundred, observe that the digit in the tens place is 7. Since this digit is at least 5, we need to add 1 to the digit in the hundreds place. The number 671 rounded to the nearest hundred is 700.

Section 1.7

Practice Problems

1. $2 \cdot 2 \cdot 2 = 2^3$

2. $3 \cdot 3 = 3^2$

3. $10 \cdot 10 \cdot 10 \cdot 10 \cdot 10 \cdot 10 = 10^6$

4. $5 \cdot 5 \cdot 4 \cdot 4 \cdot 4 = (5 \cdot 5)(4 \cdot 4 \cdot 4) = 5^2 \cdot 4^3$

5. $2^3 = 2 \cdot 2 \cdot 2 = 8$

6. $5^2 = 5 \cdot 5 = 25$

7. $10^1 = 10$

8. $4 \cdot 5^2 = 4(5 \cdot 5) = 4 \cdot 25 = 100$

9. $$\begin{aligned} 16 \div 4 - 2 &= 4 - 2 \\ &= 2 \end{aligned}$$

10. $$\begin{aligned} (9-8)^3 + 3 \cdot 2^4 &= 1^3 + 3 \cdot 2^4 \\ &= 1 + 3 \cdot 16 \\ &= 1 + 48 \\ &= 49 \end{aligned}$$

11. $$\begin{aligned} &24 \div [20 - (3 \cdot 4)] + 2^3 - 5 \\ &= 24 \div [20 - 12] + 2^3 - 5 \\ &= 24 \div 8 + 8 - 5 \\ &= 3 + 8 - 5 \\ &= 6 \end{aligned}$$

12. $\frac{60-5^2+1}{3(1+1)} = \frac{60-25+1}{3(2)}$
$= \frac{36}{6}$
$= 6$

13. Area of a square = $(\text{side})^2$
$= (11 \text{ centimeters})^2$
$= 121$ square centimeters

Exercise Set 1.7

1. $3 \cdot 3 \cdot 3 \cdot 3 = 3^4$

3. $7 \cdot 7 \cdot 7 \cdot 7 \cdot 7 \cdot 7 \cdot 7 \cdot 7 = 7^8$

5. $12 \cdot 12 \cdot 12 = 12^3$

7. $6 \cdot 6 \cdot 5 \cdot 5 \cdot 5 = (6 \cdot 6)(5 \cdot 5 \cdot 5) = 6^2 \cdot 5^3$

9. $9 \cdot 9 \cdot 9 \cdot 8 = (9 \cdot 9 \cdot 9)8 = 9^3 \cdot 8$

11. $3 \cdot 2 \cdot 2 \cdot 2 \cdot 2 \cdot 2 = 3(2 \cdot 2 \cdot 2 \cdot 2 \cdot 2) = 3 \cdot 2^5$

13. $3 \cdot 2 \cdot 2 \cdot 5 \cdot 5 \cdot 5 = 3(2 \cdot 2)(5 \cdot 5 \cdot 5) = 3 \cdot 2^2 \cdot 5^3$

15. $5^2 = 5 \cdot 5 = 25$

17. $5^3 = 5 \cdot 5 \cdot 5 = 125$

19. $2^6 = 2 \cdot 2 \cdot 2 \cdot 2 \cdot 2 \cdot 2 = 64$

21. $2^{10} = 2 \cdot 2 \cdot 2 \cdot 2 \cdot 2 \cdot 2 \cdot 2 \cdot 2 \cdot 2 \cdot 2 = 1024$

23. $7^1 = 7$

25. $3^5 = 3 \cdot 3 \cdot 3 \cdot 3 \cdot 3 = 243$

27. $2^8 = 2 \cdot 2 \cdot 2 \cdot 2 \cdot 2 \cdot 2 \cdot 2 \cdot 2 = 256$

29. $4^3 = 4 \cdot 4 \cdot 4 = 64$

31. $9^2 = 9 \cdot 9 = 81$

33. $9^3 = 9 \cdot 9 \cdot 9 = 729$

35. $10^2 = 10 \cdot 10 = 100$

37. $10^4 = 10 \cdot 10 \cdot 10 \cdot 10 = 10{,}000$

39. $10^1 = 10$

41. $1920^1 = 1920$

43. $3^6 = 3 \cdot 3 \cdot 3 \cdot 3 \cdot 3 \cdot 3 = 729$

45. $15 + 3 \cdot 2 = 15 + 6$
$= 21$

47. $20 - 4 \cdot 3 = 20 - 12$
$= 8$

49. $5 \cdot 9 - 16 = 45 - 16$
$= 29$

51. $28 \div 4 - 3 = 7 - 3$
$= 4$

53. $14 + \frac{24}{8} = 14 + 3$
$= 17$

55. $6 \cdot 5 + 8 \cdot 2 = 30 + 8 \cdot 2$
$= 30 + 16$
$= 46$

57. $0 \div 6 + 4 \cdot 7 = 0 + 4 \cdot 7$
$= 0 + 28$
$= 28$

59. $6 + 8 \div 2 = 6 + 4$
$= 10$

61. $(6 + 8) \div 2 = 14 \div 2$
$= 7$

63. $\left(6^2 - 4\right) \div 8 = (36 - 4) \div 8$
$= 32 \div 8$
$= 4$

65. $\left(3+5^2\right)\div 2=(3+25)\div 2$
$=28\div 2$
$=14$

67. $6^2\cdot(10-8)=36\cdot(10-8)$
$=36\cdot 2$
$=72$

69. $\frac{18+6}{2^4-4}=\frac{24}{16-4}$
$=\frac{24}{12}$
$=2$

71. $(2+5)\cdot(8-3)=7\cdot(8-3)$
$=7\cdot 5$
$=35$

73. $\frac{7(9-6)+3}{3^2-3}=\frac{7\cdot 3+3}{9-3}$
$=\frac{21+3}{6}$
$=\frac{24}{6}$
$=4$

75. $5\div 0+24$ is undefined because $5\div 0$ is undefined.

77. $3^4-[35-(12-6)]=3^4-[35-6]$
$=81-29$
$=52$

79. $(7\cdot 5)+[9\div(3\div 3)]=(7\cdot 5)+[9\div 1]$
$=35+9$
$=44$

81. $8\cdot[4+(6-1)\cdot 2]-50\cdot 2$
$=8\cdot[4+5\cdot 2]-50\cdot 2$
$=8\cdot[4+10]-100$
$=8\cdot 14-100$
$=112-100$
$=12$

83. $7^2-\left\{18-[40\div(4\cdot 2)+2]+5^2\right\}$
$=7^2-\left\{18-[40\div 8+2]+5^2\right\}$
$=49-\{18-[5+2]+25\}$
$=49-\{18-7+25\}$
$=49-36$
$=13$

85. Area of a square = $(\text{side})^2$
$=(20\text{ miles})^2$
$=400$ square miles

87. Area of a square = $(\text{side})^2$
$=(8\text{ centimeters})^2$
$=64$ square centimeters

89. Area of base = $(\text{side})^2$
$=(100\text{ meters})^2$
$=10{,}000$ square meters

91. $(2+3)\cdot 6-2=5\cdot 6-2$
$=30-2$
$=28$

93. $24\div(3\cdot 2)+2\cdot 5=24\div 6+2\cdot 5$
$=4+10$
$=14$

95. Missing side lengths are 60 – 40 = 20 feet and 30 – 12 = 18 feet.
Perimeter of one home = 12 + 60 + 30 + 40 + 18 + 20 = 180 feet.
Total perimeter = $7\times 180=1260$ feet.

97. $\left(7+2^4\right)^5-\left(3^5-2^4\right)^2$
$=(7+16)^5-\left(3^5-2^4\right)^2$
$=23^5-(243-16)^2$
$=6{,}436{,}343-227^2$
$=6{,}436{,}343-51{,}529$
$=6{,}384{,}814$

Section 1.8

Practice Problems

1. Replace x with 5 in the expression $x-2$.
   $x-2=5-2=3$

2. $y(x-3)=7(3-3)$
   $=7(0)$
   $=0$

3. $\frac{y+6}{x}=\frac{8+6}{2}$
   $=\frac{14}{2}$
   $=7$

4. $25-z^3+1=25-(2)^3+1$
   $=25-8+1$
   $=18$

5. $\frac{5(F-32)}{9}=\frac{5(41-32)}{9}$
   $=\frac{5(9)}{9}$
   $=5$

6. **a.** In words: twice a number
   Translate: $2\cdot x$ or $2x$

   b. In words: 8 increased by a number
   Translate: $x+8$

   c. In words: 10 minus a number
   Translate: $10-x$

   d. In words: 10 subtracted from a number
   Translate: $x-10$

   e. In words: the quotient of 6 and a number
   Translate: $6\div x$ or $\frac{6}{x}$

Exercise Set 1.8

1. $3+2z=3+2(3)$
   $=3+6$
   $=9$

3. $6xz-5x=6(2)(3)-5(2)$
   $=36-10$
   $=26$

5. $z-x+y=3-2+5$
   $=6$

7. $3x-z=3\cdot2-3$
   $=6-3$
   $=3$

9. $y^3-4x=5^3-4(2)$
   $=125-4(2)$
   $=125-8$
   $=117$

11. $2xy^2-6=2(2)(5)^2-6$
    $=2(2)(25)-6$
    $=100-6$
    $=94$

13. $8-(y-x)=8-(5-2)$
    $=8-3$
    $=5$

15. $y^4+(z-x)=5^4+(3-2)$
    $=5^4+1$
    $=625+1$
    $=626$

17. $\frac{6xy}{z}=\frac{6\cdot2\cdot5}{3}$
    $=\frac{60}{3}$
    $=20$

19. $\frac{2y-2}{x}=\frac{2\cdot5-2}{2}$
    $=\frac{10-2}{2}$
    $=\frac{8}{2}$
    $=4$

21. $\frac{x+2y}{z} = \frac{2+2\cdot 5}{3}$
$= \frac{2+10}{3}$
$= \frac{12}{3}$
$= 4$

23. $\frac{5x}{y} - \frac{10}{y} = \frac{5\cdot 2}{5} - \frac{10}{5}$
$= \frac{10}{5} - \frac{10}{5}$
$= 0$

25. $2y^2 - 4y + 3 = 2\cdot 5^2 - 4\cdot 5 + 3$
$= 2\cdot 25 - 4\cdot 5 + 3$
$= 50 - 20 + 3$
$= 33$

27. $(3y - 2x)^2 = (3\cdot 5 - 2\cdot 2)^2$
$= (15 - 4)^2$
$= (11)^2$
$= 121$

29. $(xy + 1)^2 = (2\cdot 5 + 1)^2$
$= (10 + 1)^2$
$= (11)^2$
$= 121$

31. $2y(4z - x) = 2\cdot 5(4\cdot 3 - 2)$
$= 2\cdot 5(12 - 2)$
$= 2\cdot 5(10)$
$= 10(10)$
$= 100$

33. $xy(5 + z - x) = 2\cdot 5(5 + 3 - 2)$
$= 2\cdot 5(8 - 2)$
$= 2\cdot 5(6)$
$= 10(6)$
$= 60$

35. $\frac{7x+2y}{3x} = \frac{7\cdot 2+2\cdot 5}{3\cdot 2}$
$= \frac{14+10}{6}$
$= \frac{24}{6}$
$= 4$

37.

t	1	2	3	4
$16t^2$	$16(1)^2$	$16(2)^2$	$16(3)^2$	$16(4)^2$
	$16\cdot 1$	$16\cdot 4$	$16\cdot 9$	$16\cdot 16$
	16	64	144	256

39. $x+5$

41. $x+8$

43. $20-x$

45. $512x$

47. $x \div 2$ or $\frac{x}{2}$

49. $5x+(17+x)$

51. $5x$

53. $11-x$

55. $50-8x$

57. $x^4 - y^2 = (23)^4 - (72)^2$ Substitute
$= 279,841 - 5184$ Exponentiate
$= 274,657$ Subtract

59. $x^2 + 5y - 112 = (23)^2 + 5(72) - 112$ Substitute
$= 529 + 360 - 112$ Exponentiate
$= 777$ Add and subtract

61. Compare expressions:
$\frac{x}{3} = \left(\frac{1}{3}\right)x$
$\left(\frac{1}{3}\right)x < 2x < 5x$
$5x$ is the largest.

63.

t	1	2	3	4
$16t^2$	16	64	144	256

As t gets larger, $16t^2$ gets larger.

Chapter 1 Review

1. 5<u>4</u>80
↓
hundreds

2. <u>4</u>6,200,120
↓
ten millions

3. 5480 = five thousand, four hundred eighty

4. 46,200,120 = forty-six million, two hundred thousand, one hundred twenty

5. 6279 = 6000 + 200 + 70 + 9

6. 403,225,000 = 400,000,000 + 3,000,000 + 200,000 + 20,000 + 5000

7. Fifty-nine thousand, eight hundred = 59,800

8. Six billion, three hundred four million = 6,304,000,000

9. In 1980, Houston had a population of 1,595,138 people.

10. In 1970, Los Angeles had a population of 2,811,801 people.

11. $$\begin{array}{r} 983,403 \\ -\ 789,704 \\ \hline 193,699 \end{array}$$

The population of Phoenix increased by 193,699 people from 1980 to 1990.

12. $$\begin{array}{r} 7,322,564 \\ -\ 7,071,639 \\ \hline 250,925 \end{array}$$

The population of New York increased by 250,925 people from 1980 to 1990.

13. $7 + 6 = 13$

14. $8 + 9 = 17$

15. $3 + 0 = 3$

16. $0 + 10 = 10$

17. $$\begin{array}{r} {\scriptstyle 1} \\ 25 \\ 8 \\ +\ 5 \\ \hline 38 \end{array}$$

18. $$\begin{array}{r} 27 \\ +\ 41 \\ \hline 68 \end{array}$$

19. $$\begin{array}{r} 32 \\ +\ 24 \\ \hline 56 \end{array}$$

20. $$\begin{array}{r} {\scriptstyle 1} \\ 19 \\ +\ 21 \\ \hline 40 \end{array}$$

21. $$\begin{array}{r} {\scriptstyle 11} \\ 47 \\ +\ 63 \\ \hline 110 \end{array}$$

22. $$\begin{array}{r} {\scriptstyle 11} \\ 77 \\ +\ 43 \\ \hline 120 \end{array}$$

23. $$\begin{array}{r} {\scriptstyle 11} \\ 567 \\ +\ 383 \\ \hline 950 \end{array}$$

24. $$\begin{array}{r} {\scriptstyle 111} \\ 463 \\ +\ 787 \\ \hline 1250 \end{array}$$

25. $$\begin{array}{r} {\scriptstyle 121} \\ 591 \\ 623 \\ +\ 497 \\ \hline 1711 \end{array}$$

26. $$\begin{array}{r} {\scriptstyle 111} \\ 5982 \\ 1647 \\ +\ 2238 \\ \hline 9867 \end{array}$$

27. $$\begin{array}{r} {\scriptstyle 111\ 1} \\ 62,589 \\ 65,340 \\ +\ 69,770 \\ \hline 197,699 \end{array}$$

Sean Cruise's total earnings during the three years is $197,699.

28. $$\begin{array}{r} {\scriptstyle 1\ 1} \\ 714 \\ +\ 7318 \\ \hline 8032 \end{array}$$

The total distance from Chicago to New Delhi via New York City is 8032 miles.

29. $$\begin{array}{r} {\scriptstyle 2} \\ 50 \\ 72 \\ 72 \\ +\ 82 \\ \hline 276 \end{array}$$

The perimeter is 276 feet.

30. $$\begin{array}{r} 11 \\ 20 \\ +\ 35 \\ \hline 66 \end{array}$$

The perimeter is 66 kilometers.

31. $$\begin{array}{r} 42 \\ -\ 9 \\ \hline 33 \end{array}$$

Check:
$$\begin{array}{r} {\scriptstyle 1} \\ 33 \\ +\ 9 \\ \hline 42 \end{array}$$

32. $$\begin{array}{r} 67 \\ -\ 24 \\ \hline 43 \end{array}$$

Check:
$$\begin{array}{r} 43 \\ +\ 24 \\ \hline 67 \end{array}$$

33. $\begin{array}{r} 93 \\ -\ 79 \\ \hline 14 \end{array}$

Check:
$\begin{array}{r} \scriptstyle 1 \\ 14 \\ +\ 79 \\ \hline 93 \end{array}$

34. $\begin{array}{r} 60 \\ -\ 27 \\ \hline 33 \end{array}$

Check:
$\begin{array}{r} \scriptstyle 1 \\ 33 \\ +\ 27 \\ \hline 60 \end{array}$

35. $\begin{array}{r} 599 \\ -\ 237 \\ \hline 362 \end{array}$

Check:
$\begin{array}{r} 362 \\ +\ 237 \\ \hline 599 \end{array}$

36. $\begin{array}{r} 462 \\ -\ 397 \\ \hline 65 \end{array}$

Check:
$\begin{array}{r} \scriptstyle 11 \\ 65 \\ +\ 397 \\ \hline 462 \end{array}$

37. $\begin{array}{r} 583 \\ -\ 279 \\ \hline 304 \end{array}$

Check:
$\begin{array}{r} \scriptstyle 1 \\ 304 \\ +\ 279 \\ \hline 583 \end{array}$

38. $\begin{array}{r} 600 \\ -\ 124 \\ \hline 476 \end{array}$

Check:
$\begin{array}{r} \scriptstyle 11 \\ 476 \\ +\ 124 \\ \hline 600 \end{array}$

39. $\begin{array}{r} 4000 \\ -\ 1886 \\ \hline 2114 \end{array}$

Check:
$\begin{array}{r} \scriptstyle 111 \\ 2114 \\ +\ 1886 \\ \hline 4000 \end{array}$

40. $\begin{array}{r} 4268 \\ -\ 3947 \\ \hline 321 \end{array}$

Check:
$\begin{array}{r} \scriptstyle 1 \\ 321 \\ +\ 3947 \\ \hline 4268 \end{array}$

41. $\begin{array}{r} 18,425 \\ -\ 1\ 599 \\ \hline 16,826 \end{array}$ $\begin{array}{r} 16,826 \\ -\ 1\ 200 \\ \hline 15,626 \end{array}$

Check:
$\begin{array}{r} \scriptstyle 1\ 11 \\ 15,626 \\ 1\ 200 \\ +\ 1\ 599 \\ \hline 18,425 \end{array}$

Shelly Winters paid $15,626 for the car.

42. $\begin{array}{r} 712 \\ -\ 315 \\ \hline 397 \end{array}$

Check:
$\begin{array}{r} \scriptstyle 11 \\ 397 \\ +\ 315 \\ \hline 712 \end{array}$

Bob Roma has 397 pages left.

43. During May, when it was $100.

44. During August, when it was $490.

45. During July, August, and September.

46. During April and May.

47. To round 93 to the nearest ten, observe that the digit in the ones place is 3. Since this digit is less than 5, we do not add 1 to the digit in the tens place. The number 93 rounded to the nearest ten is 90.

48. To round 45 to the nearest ten, observe that the digit in the ones place is 5. Since this digit is at least 5, we need to add 1 to the digit in the tens place. The number 45 rounded to the nearest ten is 50.

49. To round 467 to the nearest ten, observe that the digit in the ones place is 7. Since this digit is at least 5, we need to add 1 to the digit in the tens place. The number 467 rounded to the nearest ten is 470.

50. To round 493 to the nearest hundred, observe that the digit in the tens place is 9. Since this digit is at least 5, we need to add 1 to the digit in the hundreds place. The number 493 rounded to the nearest hundred it 500.

51. To round 4832 to the nearest hundred, observe that the digit in the tens place is 3. Since this digit is less than 5, we do not add 1 to the digit in the hundreds place. The number 4832 rounded to the nearest hundred is 4800.

52. To round 57,534 to the nearest thousand, observe that the digit in the hundreds place is 5. Since this digit is at least 5, we need to add 1 to the digit in the thousands place. The number 57,534 rounded to the nearest thousand is 58,000.

53. To round 49,683,712 to the nearest million, observe that the digit in the hundred-thousands place is 6. Since this digit is at least 5, we need to add 1 to the digit in the millions place. The number 49,683,712 rounded to the nearest million is 50,000,000.

54. To round 768,542 to the nearest hundred-thousand, observe that the digit in the ten-thousands place is 6. Since this digit is at least 5, we need to add 1 to the digit in the hundred-thousands place. The number 768,542 rounded to the nearest hundred-thousand is 800,000.

55.

4892	rounds to	4900
647	rounds to	600
+ 1867	rounds to	+ 1900
		7400

4892 + 647 + 1867 is approximately 7400.

56.

5925	rounds to	5900
− 1787	rounds to	− 1800
		4100

5925 − 1787 is approximately 4100.

57. To round 10,506 to the nearest thousand, observe that the digit in the hundreds place is 5. Since this digit is at least 5, we need to add 1 to the digit in the thousands place. The number 10,506 rounded to the nearest hundred is 11,000.

58. To round 1,752,693 to the nearest ten-thousand, observe that the digit in the thousands place is 2. Since this digit is less than 5, we do not need to add 1 to the ten-thousands place. The number 1,752,693 rounded to the nearest ten-thousand is 1,750,000.

59. $6 \cdot 7 = 42$

60. $8 \cdot 3 = 24$

61. $5(0) = 0$

62. $0(9) = 0$

63.
$$\begin{array}{r} 47 \\ \times\ 30 \\ \hline 0 \\ 1410 \\ \hline 1410 \end{array}$$

64. $$\begin{array}{r} 69 \\ \times\ 42 \\ \hline 138 \\ 2760 \\ \hline 2898 \end{array}$$

65. $20(8)(5) = 20 \cdot 40$
$= 800$

66. $25(9 \times 4) = 25 \cdot 36$
$= 900$

67. $$\begin{array}{r} 48 \\ \times\ 77 \\ \hline 336 \\ 3360 \\ \hline 3696 \end{array}$$

68. $$\begin{array}{r} 77 \\ \times\ 22 \\ \hline 154 \\ 1540 \\ \hline 1694 \end{array}$$

69. $49 \cdot 49 \cdot 0 = 0$,
because anything times zero is zero.

70. $62 \cdot 88 \cdot 0 = 0$,
because anything times zero is zero.

71. $$\begin{array}{r} 586 \\ \times\ 29 \\ \hline 5\,274 \\ 11,720 \\ \hline 16,994 \end{array}$$

72. $$\begin{array}{r} 242 \\ \times\ 37 \\ \hline 1694 \\ 7260 \\ \hline 8954 \end{array}$$

73. $$\begin{array}{r} 642 \\ \times\ 177 \\ \hline 4\,494 \\ 44,940 \\ 64,200 \\ \hline 113,634 \end{array}$$

74. $$\begin{array}{r} 347 \\ \times\ 129 \\ \hline 3\,123 \\ 6\,940 \\ 34,700 \\ \hline 44,763 \end{array}$$

75. $$\begin{array}{r} 1026 \\ \times\ 401 \\ \hline 1\,026 \\ 00 \\ 410,400 \\ \hline 411,426 \end{array}$$

76. $$\begin{array}{r} 2107 \\ \times\ 302 \\ \hline 4\,214 \\ 00 \\ 632,100 \\ \hline 636,314 \end{array}$$

77. $$\begin{array}{rcr} 49 & \text{rounds to} & 50 \\ \times\ 32 & \text{rounds to} & \times\ 30 \\ \cline{3-3} & & 1500 \end{array}$$

$49 \cdot 32$ is approximately 1500.

78. $$\begin{array}{rcr} 586 & \text{rounds to} & 600 \\ \times\ 357 & \text{rounds to} & \times\ 400 \\ \cline{3-3} & & 240,000 \end{array}$$

$586 \cdot 357$ is approximately 240,000.

79. $$\begin{array}{rcr} 5231 & \text{rounds to} & 5200 \\ \times\ 243 & \text{rounds to} & \times\ 200 \\ \cline{3-3} & & 1,040,000 \end{array}$$

$5231 \cdot 243$ is approximately 1,040,000.

80. $$\begin{array}{rcr} 7836 & \text{rounds to} & 7800 \\ \times\ 912 & \text{rounds to} & \times\ 900 \\ \cline{3-3} & & 7,020,000 \end{array}$$

$7836 \cdot 912$ is approximately 7,020,000.

81. $$\begin{array}{r} 5283 \\ \times\ 927 \\ \hline 36,981 \\ 105,660 \\ 4,754,700 \\ \hline 4,897,341 \end{array}$$

The tuition collected totaled \$4,897,341.

82. $$\begin{array}{r} 490 \\ \times\ 8 \\ \hline 3920 \end{array}$$

The whole cake contains 3920 calories.

83. $$\begin{array}{r} 12 \\ \times\ 5 \\ \hline 60 \end{array}$$

The area is 60 square miles.

84. $$\begin{array}{r} 20 \\ \times\ 25 \\ \hline 100 \\ 400 \\ \hline 500 \end{array}$$

The area is 500 square centimeters.

85. $$\begin{array}{r} 3 \\ 6\overline{)\ 18} \\ \underline{-18} \\ 0 \end{array}$$

Check: $3 \cdot 6 = 18$

86. $$\begin{array}{r} 4 \\ 9\overline{)\ 36} \\ \underline{-36} \\ 0 \end{array}$$

Check: $4 \cdot 9 = 36$

87. $$\begin{array}{r} 6 \\ 7\overline{)\ 42} \\ \underline{-42} \\ 0 \end{array}$$

Check: $6 \cdot 7 = 42$

88. $$\begin{array}{r} 5 \\ 5\overline{)\ 25} \\ \underline{-25} \\ 0 \end{array}$$

Check: $5 \cdot 5 = 25$

89. $$\begin{array}{r} 5\ \text{R}\ 2 \\ 5\overline{)\ 27} \\ \underline{-25} \\ 2 \end{array}$$

Check: $5 \cdot 5 + 2 = 27$

90. $$\begin{array}{r} 4\ \text{R}\ 2 \\ 4\overline{)\ 18} \\ \underline{-16} \\ 2 \end{array}$$

Check: $4 \cdot 4 + 2 = 18$

91. $16 \div 0$ is undefined

92. $0 \div 8 = 0$
Check: $0 \cdot 8 = 0$

93. $9 \div 9 = 1$
Check: $9 \cdot 1 = 9$

94. $10 \div 1 = 10$
Check: $10 \cdot 1 = 10$

95. $918 \div 0$ is undefined

96. $0 \div 668 = 0$
Check: $668 \cdot 0 = 0$

97. $$\begin{array}{r} 15 \\ 5\overline{)\ 75} \\ \underline{-5\ \ } \\ 25 \\ \underline{-25} \\ 0 \end{array}$$

Check: $15 \cdot 5 = 75$

98. $$\begin{array}{r} 19\ \text{R}\ 7 \\ 8\overline{)159} \\ \underline{-8\ \ } \\ 79 \\ \underline{-72} \\ 7 \end{array}$$

Check: $19 \cdot 8 + 7 = 159$

99. $$\begin{array}{r} 24\ \text{R}\ 2 \\ 26\overline{)\ 626} \\ \underline{-52\ \ } \\ 106 \\ \underline{-104} \\ 2 \end{array}$$

Check: $24 \cdot 26 + 2 = 626$

100. $$\begin{array}{r} 56 \\ 6\overline{)\,336} \\ \underline{-30} \\ 36 \\ \underline{-36} \\ 0 \end{array}$$

Check: $56 \cdot 6 = 336$

101. $$\begin{array}{r} 1 \text{ R } 17 \\ 32\overline{)\,49} \\ \underline{-32} \\ 17 \end{array}$$

Check: $1 \cdot 32 + 17 = 49$

102. $$\begin{array}{r} 35 \text{ R } 15 \\ 19\overline{)\,680} \\ \underline{-57} \\ 110 \\ \underline{-95} \\ 15 \end{array}$$

Check: $35 \cdot 19 + 15 = 680$

103. $$\begin{array}{r} 500 \\ 20\overline{)\,10000} \\ \underline{-100} \\ 00 \\ \underline{-0} \\ 00 \\ \underline{-0} \\ 0 \end{array}$$

Check: $500 \cdot 20 = 10{,}000$

104. $$\begin{array}{r} 21 \text{ R } 6 \\ 43\overline{)\,909} \\ \underline{-86} \\ 49 \\ \underline{-43} \\ 6 \end{array}$$

Check: $21 \cdot 43 + 6 = 909$

105. $$\begin{array}{r} 506 \\ 47\overline{)\,23782} \\ \underline{-235} \\ 28 \\ \underline{-0} \\ 282 \\ \underline{-282} \\ 0 \end{array}$$

Check: $506 \cdot 47 = 23{,}782$

106. $$\begin{array}{r} 16 \\ 30\overline{)\,480} \\ \underline{-30} \\ 180 \\ \underline{-180} \\ 0 \end{array}$$

Check: $16 \cdot 30 = 480$

107. $$\begin{array}{r} 199 \text{ R } 8 \\ 16\overline{)\,3192} \\ \underline{-16} \\ 159 \\ \underline{-144} \\ 152 \\ \underline{-144} \\ 8 \end{array}$$

Check: $199 \cdot 16 + 8 = 3192$

108. $$\begin{array}{r} 200 \\ 25\overline{)\,5000} \\ \underline{-50} \\ 00 \\ \underline{-0} \\ 00 \\ \underline{-0} \\ 0 \end{array}$$

Check: $200 \cdot 25 = 5000$

109. $$\begin{array}{r} 458 \\ 12\overline{)\,5496} \\ \underline{-48} \\ 69 \\ \underline{-60} \\ 96 \\ \underline{-96} \\ 0 \end{array}$$

Check: $458 \cdot 12 = 5496$

There are 458 feet in 5496 inches.

110. Add the numbers, then divide by four.

$$\begin{array}{r} 76 \\ 49 \\ 32 \\ \underline{+\ 47} \\ 204 \end{array} \qquad \begin{array}{r} 51 \\ 4\overline{)\,204} \\ \underline{-20} \\ 04 \\ \underline{-4} \\ 0 \end{array}$$

The average is $\frac{204}{4} = 51$.

111. $7 \cdot 7 \cdot 7 \cdot 7 = 7^4$

112. $3 \cdot 3 \cdot 3 = 3^3$

113. $4 \cdot 2 \cdot 2 \cdot 2 = 4 \cdot 2^3$

114. $5 \cdot 5 \cdot 7 \cdot 7 \cdot 7 = 5^2 \cdot 7^3$

115. $7^2 = 7 \cdot 7 = 49$

116. $2^6 = 2 \cdot 2 \cdot 2 \cdot 2 \cdot 2 \cdot 2 = 64$

117. $5^3 \cdot 3^2 = 5 \cdot 5 \cdot 5 \cdot 3 \cdot 3 = 1125$

118. $4^1 \cdot 10^2 \cdot 7^2 = 4 \cdot 10 \cdot 10 \cdot 7 \cdot 7 = 19,600$

119. $$\begin{aligned} 18 \div 3 + 7 &= 6 + 7 \\ &= 13 \end{aligned}$$

120. $$\begin{aligned} 12 - 8 \div 4 &= 12 - 2 \\ &= 10 \end{aligned}$$

121. $$\begin{aligned} \frac{6^2 - 3}{3^2 + 2} &= \frac{36 - 3}{9 + 2} \\ &= \frac{33}{11} \\ &= 3 \end{aligned}$$

122. $$\begin{aligned} \frac{16 - 8}{2^3} &= \frac{8}{8} \\ &= 1 \end{aligned}$$

123. $$\begin{aligned} 2 + 3[1 + (20 - 17) \cdot 3] &= 2 + 3[1 + 3 \cdot 3] \\ &= 2 + 3[1 + 9] \\ &= 2 + 3[10] \\ &= 2 + 30 \\ &= 32 \end{aligned}$$

124. $$\begin{aligned} &21 - \left[2^4 - (7 - 5) - 10\right] + 8 \cdot 2 \\ &= 21 - \left[2^4 - 2 - 10\right] + 8 \cdot 2 \\ &= 21 - [16 - 2 - 10] + 16 \\ &= 21 - 4 + 16 \\ &= 33 \end{aligned}$$

125. $$\begin{aligned} \text{Area} &= (\text{side})^2 \\ &= (7 \text{ meters})^2 \\ &= 49 \text{ square meters} \end{aligned}$$

126. $$\begin{aligned} \text{Area} &= (\text{side})^2 \\ &= (3 \text{ inches})^2 \\ &= 9 \text{ square inches} \end{aligned}$$

127. $\frac{2 \cdot 5}{2} = \frac{10}{2} = 5$

128 $4(5) - 3 = 20 - 3 = 17$

129. $\frac{5 + 7}{0} = \frac{12}{0}$ is undefined

130. $\frac{0}{5(5)} = 0$

131. $5^3 - 2 \cdot 2 = 125 - 4 = 121$

132. $\frac{7 + 5}{3(2)} = \frac{12}{6} = 2$

133. $(0 + 2)^2 = 2^2 = 4$

134. $\frac{100}{5} + \frac{0}{3} = 20 + 0 = 20$

135. $x - 5$

136. $x + 7$

137. $10 \div (x + 1)$

138. $5(x + 3)$

139.

x	0	1	2	3
$8x^2$	$8(0)^2$	$8(1)^2$	$8(2)^2$	$8(3)^2$
	0	8	32	72

Chapter 1 Test

1. $59 + 82 = 141$

2. $600 - 487 = 113$

3.
$$\begin{array}{r} 496 \\ \times\ \ 30 \\ \hline 000 \\ 1488 \\ \hline 14,880 \end{array}$$

4.
$$\begin{array}{r} 766 \text{ R } 42 \\ 69\overline{)\ 52896} \\ -483 \\ \hline 459 \\ -414 \\ \hline 456 \\ -414 \\ \hline 42 \end{array}$$

$$\begin{array}{r} 7666 \\ \times\ \ 69 \\ \hline 52854 \\ +\ \ 42 \\ \hline 52896 \end{array}$$

5. $2^3 \cdot 5^2 = 8 \cdot 25 = 200$

6. $6^1 \cdot 2^3 = 6 \cdot 2 \cdot 2 \cdot 2 = 48$

7. $98 \div 1 = 98$

8. $0 \div 49 = 0$

9. $62 \div 0$ is undefined.

10.
$$\begin{aligned}(2^4 - 5) &= (16 - 5) \cdot 3 \\ &= 11 \cdot 3 \\ &= 33\end{aligned}$$

11. $16 + 9 \div 3 \cdot 4 - 7 = 16 + 12 - 7 = 21$

12.
$$\begin{aligned}&2[(6-4)^2 + (22-19)^2] + 10 \\ &= 2[2^2 + 3^2] + 10 \\ &= 2[4 + 9] + 10 \\ &= 2(13) + 10 \\ &= 26 + 10 \\ &= 36\end{aligned}$$

13. 52,369 rounded to the nearest thousand is 52,000.

14.
$$\begin{array}{r} 6289 \\ 5403 \\ +\ 1957 \\ \hline \end{array} \qquad \begin{array}{r} 6300 \\ 5400 \\ +\ 2000 \\ \hline 13,700 \end{array}$$

15.
$$\begin{array}{r} 4267 \\ -2738 \\ \hline \end{array} \qquad \begin{array}{r} 4300 \\ -\ 2700 \\ \hline 1600 \end{array}$$

16.
$$\begin{array}{r} \$17 \\ 29\overline{)\ 493} \\ -29 \\ \hline 203 \\ -203 \\ \hline 0 \end{array}$$

$$\begin{array}{r} 17 \\ \times\ 29 \\ \hline \$493 \end{array}$$

17.
$$\begin{array}{r} 17 \\ \times\ 7 \\ \hline \$119 \end{array}$$

18.
$$\begin{array}{r} 725 \\ -\ 599 \\ \hline \$126 \end{array}$$ more expensive

19.
$$\begin{array}{r} 11 \\ -\ 4 \\ \hline 7 \end{array}$$
7 billion more tablets per year are taken for heart disease.

20.
$$\begin{array}{r} 34,000 \\ \times\qquad 5 \\ \hline 170,000 \end{array}$$
During a five-day work week the FBI receives over 170,000 fingerprint cards.

21. $5[(2)^3 - 2] = 5[8 - 2] = 5 \cdot 6 = 30$

22. $\dfrac{3(7) - 5}{2(8)} = \dfrac{21 - 5}{16} = 1$

23. a. $17x$

b. $20 - 2x$

24. Area = $(\text{side})^2 = (5\text{ cm})^2 = 25$ sq cm
Perimeter = 4(side) = 4(5 cm) = 20 cm

25. Perimeter = length + width + length + width
= 20 yards + 10 yards + 20 yards + 10 yards
= 60 yards

Area = length · width
= (20 yards)(10 yards)
= 200 square yards

Chapter 2

Pretest

1. If 0 represents the line of scrimmage, a loss of
22 yards is –22.

2. $-31 > -36$

3. $|-8| = 8$, because –8 is 8 units from 0.

4. The opposite of –12 is $-(-12) = 12$.

5. $-19 + 8$
We are adding two numbers with different signs.
$|-19| = 19$, $|8| = 8$
$19 - 8 = 11$
–19 has the larger absolute value and its sign is –.
$-19 + 8 = -11$

6. $x + y = -6 + (-11)$
We are adding two numbers with the same sign.
$|-6| = 6$, $|-11| = 11$
$6 + 11 = 17$
Their common sign is negative, so the sum is negative.
$-6 + (-11) = -17$

7. $-9 + 5 + 7 = -4 + 7 = 3$
The temperature was 3°F at 9 a.m.

8. $-9 - (-14) = -9 + 14 = 5$

9. $$\begin{aligned} 4 - 6 - (-2) + 8 &= 4 + (-6) + 2 + 8 \\ &= -2 + 2 + 8 \\ &= 0 + 8 \\ &= 8 \end{aligned}$$

10. $m - n = -5 - 10 = -5 + (-10) = -15$

11. $$\begin{aligned} 88 + 35 - 72 - 55 &= 88 + 35 + (-72) + (-55) \\ &= 123 + (-72) + (-55) \\ &= 51 + (-55) \\ &= -4 \end{aligned}$$

12. $(-8)(13) = -104$

13. $\dfrac{-36}{-4} = 9$

14. $\dfrac{x}{y} = \dfrac{-96}{-2} = 48$

15. $xy = 4 \cdot (-5) = -20$

16. $(-20) \cdot 3 = -60$

17. $-5^2 = -(5)(5) = -25$

18. $\dfrac{9-17}{-6+2} = \dfrac{9+(-17)}{-6+2} = \dfrac{-8}{-4} = 2$

19. $$\begin{aligned} (-2) \cdot |-7| - (-6) &= (-2) \cdot 7 + 6 \\ &= -14 + 6 \\ &= -8 \end{aligned}$$

20. $$\begin{aligned} x - y^2 &= 2 - (-6)^2 \\ &= 2 - 36 \\ &= 2 + (-36) \\ &= -34 \end{aligned}$$

Section 2.1

Practice Problems

1. a. If 0 represents the surface of the earth, then 800 feet below the surface can be represented by –800 feet.

   b. If 0 represents a balance of \$0, then \$2 million loss can be represented by –2 million dollars.

2. 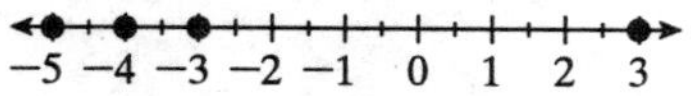

3. a. 0 is to the right of –3, so $0 > -3$.

   b. –5 is to the left of 5, so $-5 < 5$.

   c. –8 is to the right of –12, so $-8 > -12$.

4. a. $|-4| = 4$, because -4 is 4 units from 0.

   b. $|2| = 2$, because 2 is 2 units from 0.

   c. $|-8| = 8$, because 8 is 8 units from 0.

5. a. The opposite 9 of 7 is -7.

   b. The opposite of -17 is $-(-17)$ or 17.

6. a. $-|-2| = -2$

   b. $-|5| = -5$

   c. $-(-11) = 11$

Exercise Set 2.1

1. If 0 represents ground level, then 1445 feet underground is -1445.

3. If 0 represents sea level, then 14,494 above sea level is 14,494.

5. If 0 represents the line of scrimmage, a loss of 15 yards is -15.

7. Represent a fall in the stock market by a negative number: -317 points.

9. $-5{,}049$ thousand

11. $-10° < -5°$ on the Celsius scale so -10°C is cooler than -5°C.

13. If 0 represents 0%, a loss of 16% is -16.

15. 0 1 2 3 4 5 6 7 8

17. -7 -6 -5 -4 -3 -2 -1 0 1

19. 0 1 2 3 4 5 6 7 8

21. -7 -6 -5 -4 -3 -2 -1 0 1

23. $4 > 0$

25. $-7 < -5$

27. $0 > -3$

29. $-26 < 26$

31. $|5| = 5$, because 5 is 5 units from 0.

33. $|-8| = 8$, because -8 is 8 units from 0.

35. $|0| = 0$, because 0 is 0 units from 0.

37. $|-5| = 5$, because -5 is 5 units from 0.

39. The opposite of 5 is -5.

41. The opposite of -4 is $-(-4) = 4$.

43. The opposite of 23 is -23.

45. The opposite of -10 is $-(-10) = 10$.

47. $|-7| = 7$, because -7 is 7 units from 0.

49. $-|20| = -20$
    The opposite of the absolute value of 20 is the opposite of 20.

51. $-|-3| = -3$
    The opposite of the absolute value of -3 is the opposite of 3.

53. $-(-8) = 8$
    The opposite of negative 8 is 8.

55. $|-14| = 14$, because -14 is 14 units from 0.

57. $-(-29) = 29$
    The opposite of negative 29 is 29.

59. $-3 > -5$

61. $|-9| \ ? \ |-14|$
    $9 \ ? \ 14$
    $9 < 14$

63. $|-33| \ ? \ -(-33)$
    $33 \ ? \ 33$
    $33 = 33$

65. $-|-10| \ ? \ -(-10)$
$-10 \ ? \ 10$
$-10 < 10$

67. $0 \ ? -9$
$0 > -9$

69. $|0| \ ? \ |-9|$
$0 \ ? \ 9$
$0 < 9$

71. $-|-2| \ ? \ -|-10|$
$-2 > -10$

73. $-(-12) \ ? \ -(-18)$
$12 < 18$

75. $0 + 13 = 13$

77.
$$\begin{array}{r} 15 \\ +\ 20 \\ \hline 35 \end{array}$$

79.
$$\begin{array}{r} 47 \\ 236 \\ +\ 77 \\ \hline 360 \end{array}$$

81. $-12 < -8$
D

83. False; consider $-1 > -2$.

85. True; consider the values on a number line.

87. False; consider $a = -3$, then $-a = -(-3) = 3$.

89. Answers will vary.

Section 2.2

Practice Problems

1. $5 + (-1) = 4$

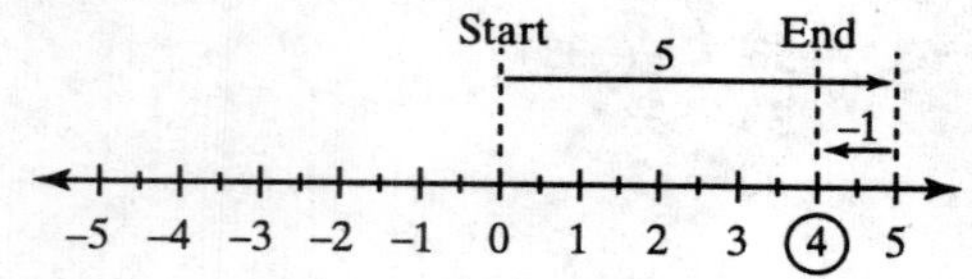

2. $-6 + (-2) = -8$

−2 −6
−9 −8 −7 −6 −5 −4 −3 −2 −1 0 1 2 3 4

3. $|-3| = 3$, $|-9| = 9$, and $3 + 9 = 12$
Their common sign is negative, so the sum is negative.
$(-3) + (-9) = -12$

4. $-12 + (-3) = -15$

5. $9 + 5 = 14$

6. $|-3| = 3$, $|9| = 9$, and $9 - 3 = 6$
9 has the larger absolute value and its sign is an understood +: $-3 + 9 = 6$.

7. $|2| = 2$, $|-8| = 8$, and $8 - 2 = 6$
-8 has the larger absolute value and its sign is $-$: $2 + (-8) = -6$

8. $-46 + 20 = -26$

9. $8 + (-6) = 2$

10. $-2 + 0 = -2$

11.
$$\begin{aligned} 8 + (-3) + (-13) &= 5 + (-13) \\ &= -8 \end{aligned}$$

12.
$$\begin{aligned} 5 + (-3) + 12 + (-14) &= 2 + 12 + (-14) \\ &= 14 + (-14) \\ &= 0 \end{aligned}$$

13. $x + y = -4 + 1 = -3$

14. $x + y = -11 + (-6) = -17$

15. $-8 + (4) + (7) = 3$
The temperature at 8 a.m. was 3° Fahrenheit.

Mental Math

1. $5 + 0 = 5$

2. $(-2) + 0 = -2$

3. $0 + (-35) = -35$

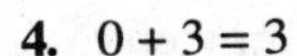

4. $0 + 3 = 3$

Exercise Set 2.2

1. $8 + 2 = 10$

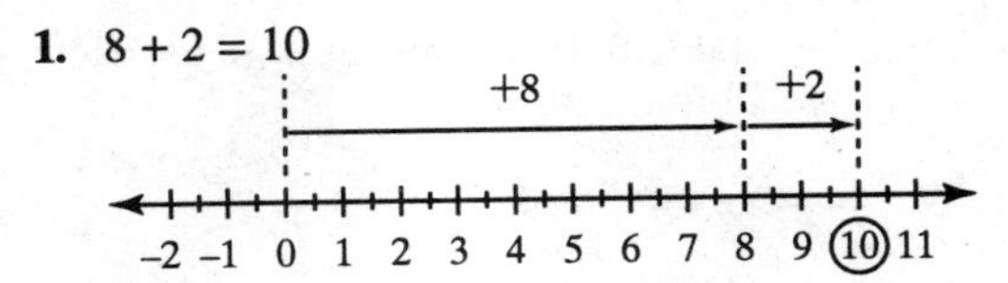

3. $-4 + 7 = 3$

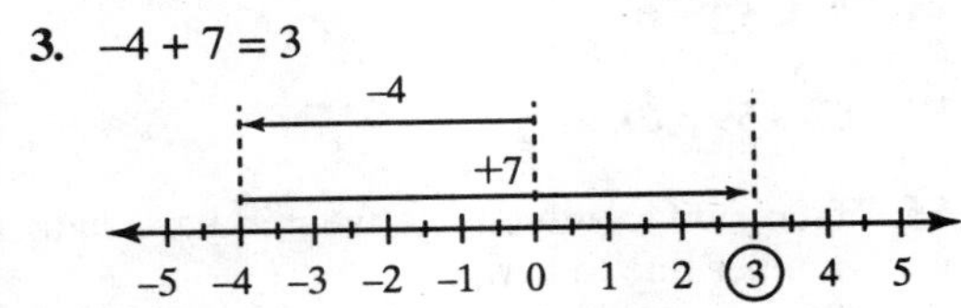

5. $-13 + 7 = -6$

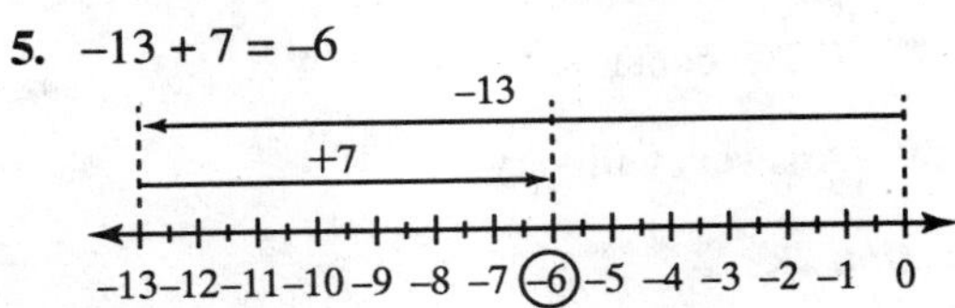

7. $|23| + |12| = 35$
Common sign is positive, so answer is +35.

9. $|-6| + |-2| = 8$
Common sign is negative, so answer is −8.

11. $|-43| - |43| = 0$
Sign does not matter.

13. $|6| - |-2| = 4$
$6 > 2$ so answer is +4.

15. $|8| - |-6| = 2$
$8 > 6$ so answer is +2.

17. $|-5| - |3| = 2$
$5 > 3$ so answer is −2.

19. $|-2| + |-7| = 9$
Common sign is negative, so answer is −9.

21. $|-12| + |-12| = 24$
Common sign is negative, so answer is −24.

23. $|-25| + |-32| = 57$
Common sign is negative, so answer is −57.

25. $|-123| + |-100| = 223$
Common sign is negative, so answer is −223.

27. $|7| - |-7| = 0$
Sign does not matter.

29. $|12| - |-5| = 7$
$12 > 5$, so answer is +7.

31. $|-6| - |3| = 3$
$6 > 3$, so answer is −3.

33. $|-12| - |3| = 9$
$12 > 3$, so answer is −9.

35. $|56| - |-26| = 30$
$56 > 26$, so answer is +30.

37. $|57| - |-37| = 20$
$57 > 37$, so answer is +20.

39. $|93| - |-42| = 51$
$93 > 42$, so answer is +51.

41. $|-67| - |34| = 33$
$67 > 34$, so answer is −33.

43. $|-144| - |124| = 20$
$144 > 124$, so answer is −20.

45. $|-82| + |-43| = 125$
Common sign is negative, so −125.

47. $-4 + 2 + (-5)$ add from left to right
$= -2 + (-5)$
$= -7$

49. $-52 + (-77) + (-117)$ add from left to right
$= -129 + (-117)$ add from left to right
$= -246$

51. $12 + (-4) + (-4) + 12$ add from left to right
$= 8 + (-4) + 12$ add from left to right
$= 4 + 12$
$= 16$

53. $(-10) + 14 + 25 + (-16)$ add from left to right
$= 4 + 25 + (-16)$ add from left to right
$= 29 + (-16)$
$= 13$

55. $-8 + (-14) + (-11)$ add from left to right
$= -22 + (-11)$ add from left to right
$= -33$

57. $5 + (-1) + 17$ add from left to right
$= 4 + 17$ add from left to right
$= 21$

59. $13 + 14 + (-18)$ add from left to right
$= 27 + (-18)$ add from left to right
$= 9$

61. $-3 + (-8) + 12 + (-1)$ add from left to right
$= -11 + 12 + (-1)$ add from left to right
$= 1 + (-1)$
$= 0$

63. $x + y$ substitute values
$= (-2) + 3$ add from left to right
$= 1$

65. $x + y$ substitute values
$= (-20) + (-50)$ add from left to right
$= -70$

67. $x + y$ substitute values
$= 3 + (-30)$ add from left to right
$= -27$

69. $-10 + 12 = 2$
The temperature at 11 p.m. was 2°C.

71. $(-165) + (-16) = -181$
The diver's present depth is 181 feet below the surface.

73. Team 1 Total $= -2 + (-13) + 20 + 2$
$= 7$
Team 2 Total $= 5 + 11 + (-7) + (-3)$
$= 6$
$7 > 6$ so the winning team is Team 1.

75. $-69 + 21 = -48$
Maine's record low temperature is −48°F.

77. $-170 + (-181) + (-230)$ add from left to right
$= -351 + (-230)$
$= -581$
The total U.S. trade balance was −$581 billion.

79. $44 - 0 = 44$

81. $52 - 52 = 0$

83. $87 - 59 = 28$

85. True; add any two negative numbers on a number line to verify.

87. False; consider $5 + (-1) = 4$.

89. Answers will vary.

Section 2.3

Practice Problems

1. $12 - 7 = 12 + (-7) = 5$

2. $-6 - 4 = -6 + (-4) = -10$

3. $11 - (-14) = 11 + 14 = 25$

4. $-9 - (-1) = -9 + 1 = -8$

5. $5 - 9 = 5 + (-9) = -4$

6. $-12 - 4 = -12 + (-4) = -16$

7. $-2 - (-7) = -2 + 7 = 5$

8. $-10 - 5 = -10 + (-5) = -15$

9. $-4 - 3 - 7 - (-5) = -4 + (-3) + (-7) + 5$
$= -7 + (-7) + 5$
$= -14 + 5$
$= -9$

10. $3 + (-5) - 6 - (-4) = 3 + (-5) + (-6) + 4$
$= -2 + (-6) + 4$
$= -8 + 4$
$= -4$

11. $x - y = -2 - 14 = -2 + (-14) = -16$

12. $y - z = -3 - (-4) = -3 + 4 = 1$

13. $29,028 - (-1312)$
$= 29,028 + 1312$
$= 30,340$
Mount Everest is 30,340 feet higher than the Dead Sea.

Exercise Set 2.3

1. $5 - 5 = 5 + (-5) = 0$

3. $8 - 3 = 8 + (-3) = 5$

5. $3 - 8 = 3 + (-8) = -5$

7. $7 - (-7) = 7 + (7) = 14$

9. $-5 - (-8) = -5 + (8) = 3$

11. $-14 - 4 = -14 + (-4) = -18$

13. $2 - 16 = 2 + (-16) = -14$

15. $-10 - (-10) = -10 + 10 = 0$

17. $-15 - (-15) = -15 + 15 = 0$

19. $3 - 7 = 3 + (-7) = -4$

21. $30 - 45 = 30 + (-45) = -15$

23. $-4 - 10 = -4 + (-10) = -14$

25. $-230 - 870 = -230 + (-870) = -1100$

27. $4 - (-6) = 4 + 6 = 10$

29. $-7 - (-3) = -7 + 3 = -4$

31. $-16 - (-23) = -16 + 23 = 7$

33. $-20 - 18 = -20 + (-18) = -38$

35. $-20 - (-3) = -20 + (3) = -17$

37. $2 - (-11) = 2 + (11) = 13$

39. $7 - 3 - 2 = 7 + (-3) + (-2)$
$= 4 + (-2)$
$= 2$

41. $12 - 5 - 7 = 12 + (-5) + (-7)$
$= 7 + (-7)$
$= 0$

43. $-5 - 8 - (-12) = -5 + (-8) + 12$
$= -13 + 12$
$= -1$

45. $-10 + (-5) - 12 = -10 + (-5) + (-12)$
$= -15 + (-12)$
$= -27$

47. $12 - (-34) + (-6) = 12 + 34 + (-6)$
$= 46 + (-6)$
$= 40$

49. $-(-6) - 12 + (-16) = 6 + (-12) + (-16)$
$= -6 + (-16)$
$= -22$

51. $-9 - (-12) + (-7) - 4 = -9 + 12 + (-7) + (-4)$
$= 3 + (-7) + (-4)$
$= -4 + (-4)$
$= -8$

53. $-3 + 4 - (-23) - 10 = -3 + 4 + 23 + (-10)$
$= 1 + 23 + (-10)$
$= 24 + (-10)$
$= 14$

55. $x - y$
$= (-3) - (5)$ show substitution exactly
$= -3 + (-5)$ change subtraction to addition
$= -8$

57. $x - y$
$= 6 - (-30)$ show substitution exactly
$= 6 + (30)$ change subtraction to addition
$= 36$

59. $x - y$
$= (-4) - (-4)$ show substitution exactly
$= -4 + (4)$ change subtraction to addition
$= 0$

61. $x - y$
$= (1) - (-18)$ show substitution exactly
$= 1 + (18)$ change subtraction to addition
$= 19$

63. To find how many degrees warmer, subtract the coldest temperature from the warmest temperature.
$136 - (-129) = 136 + 129 = 265$
136°F is 265°F warmer than −129°F.

65. Subtract check amounts from the checking account, and add the deposit.
$125 - 117 + 45 - 69$
$= 125 + (-117) + 45 + (-69)$
$= 8 + 45 + (-69)$
$= 53 + (-69)$
$= -16$
Aaron has overdrawn his checking account by \$16.

67. $-6 + (-3) + 4 + (-7)$ add from left to right
$= -9 + 4 + (-7)$ add from left to right
$= -5 + (-7)$
$= -12$°C

69. $600 - (-52) = 600 + 52 = 652$ feet

71. $144 - 0 = 144$ feet

73. $167 - (-215) = 167 + 215 = 382$
The surface of Mercury is 382°C warmer than the surface of Neptune.

75. $-176 - (-215) = -176 + 215 = 39$
Saturn has the warmer average temperature because $-176 > -215$. It is 39°C warmer than Uranus.

77. $663 - 912 = 663 + (-912) = -249$
The U.S. trade balance in 1998 was −\$249 billion.

79. $8 \cdot 0 = 0$

81. $1 \cdot 8 = 8$

83.
$$\begin{array}{r} 23 \\ \times\ 46 \\ \hline 138 \\ +\ 92\ \ \\ \hline 1058 \end{array}$$

85. $x - y - z$ show substitution exactly
$= (-4) - (3) - (15)$ change subtraction to addition
$= -4 + (-3) + (-15)$ add left to right
$= -7 + (-15)$
$= -22$

87. $a + b - c$ show substitution exactly
$= (-16) + (14) - (-22)$ change subtraction to addition
$= -16 + 14 + 22$ add from left to right
$= -2 + 22$
$= 20$

89. $|-3| - |-7| = 3 - 7 = 3 + (-7) = -4$

91. $|-6| - |6| = 6 - 6 = 6 + (-6) = 0$

93. False. $|-8 - 3| = |-8 + (-3)| = |-11| = 11$

95. Answers may vary.

97. Answers may vary.

Integrated Review

1. If 0 represents an elevation of 0 feet, then 29,028 feet above sea level is 29,028.

2. If 0 represents an elevation of 0 feet, then 35,840 feet below sea level is −35,840.

3. 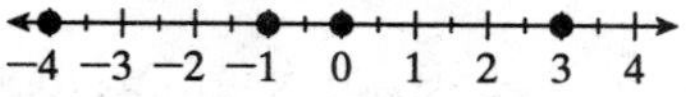

4. 0 is to the right of −3, so $0 > -3$.

5. −15 is to the left of −5, so $-15 < -5$.

6. −1 is to the left of 1, so $-1 < 1$.

7. −2 is to the right of −7, so $-2 > -7$.

8. $|-1| = 1$, because −1 is 1 unit from 0.

9. $-|-4| = -4$

10. $|-8| = 8$, because −8 is 8 units from 0.

11. $-(-5) = 5$

12. The opposite of 6 is –6.

13. The opposite of –3 is – (–3) = 3.

14. The opposite of 89 is –89.

15. The opposite of 0 is 0.

16. $-7 + 12 = 5$

17. $-9 + (-11) = -20$

18. $25 + (-35) = -10$

19. $1 - 3 = 1 + (-3) = -2$

20. $26 - (-26) = 26 + 26 = 52$

21. $-2 - 1 = -2 + (-1) = -3$

22. $-18 - (-102) = -18 + 102 = 84$

23. $-8 + (-6) + 20 = -14 + 20 = 6$

24. $-11 - 7 - (-19) = -11 + (-7) + 19$
$= -18 + 19 = 1$

25. $$\begin{aligned}-4+(-8)-16-(-9) &= -4+(-8)+(-16)+9\\ &= -12+(-16)+9\\ &= -28+9\\ &= -19\end{aligned}$$

26. $26 - 14 = 26 + (-14) = 12$

27. $-12 - (-8) = -12 + 8 = -4$

28. $-21 < -18$; D

29. $-5 < 0$; $-5 < 3$; $-5 < -1$; A, B, C

Section 2.4

Practice Problems

1. $-2 \cdot 6 = -12$

2. $-4(-3) = 12$

3. $0 \cdot (-10) = 0$

4. $5(-15) = -75$

5. a. $$\begin{aligned}(-2)(-1)(-2)(-1) &= 2(-2)(-1)\\ &= -4(-1)\\ &= 4\end{aligned}$$

b. $7(2)(-4) = 14(-4) = -56$

6. $$\begin{aligned}(-3)^4 &= (-3)(-3)(-3)(-3)\\ &= 9(-3)(-3)\\ &= -27(-3)\\ &= 81\end{aligned}$$

7. $\frac{28}{-7} = -4$

8. $-18 \div (-2) = 9$

9. $\frac{-60}{10} = -6$

10. $\frac{-1}{0}$ is undefined because there is no number that gives a product of –1 when multiplied by 0.

11. $\frac{0}{-2} = 0$ because $0 \cdot (-2) = 0$.

12. $xy = 5(-9) = -45$

13. $\frac{x}{y} = \frac{-9}{-3} = 3$

14. $4(-12) = -48$
The card player's total score was –48.

Exercise Set 2.4

1. $-2(-3) = 6$

3. $-4(9) = -36$

5. $8(-8) = -64$

7. $0(-14) = 0$

9. $6(-4)(2) = -24(2) = -48$

11. $-1(-2)(-4) = 2(-4) = -8$

13. $-4(4)(-5) = -16(-5) = 80$

15. $10(-5)(0) = -50(0) = 0$

17. $-5(3)(-1)(-1)$
$= -15(-1)(-1)$
$= 15(-1)$
$= -15$

19. $(-2)^2 = (-2)(-2) = 4$

21. $(-3)^3 = (-3)(-3)(-3) = 9(-3) = -27$

23. $(-5)^2 = (-5)(-5) = 25$

25. $(-2)^3 = (-2)(-2)(-2) = 4(-2) = -8$

27. $-24 \div 6 = -4$

29. $\frac{-30}{6} = -5$

31. $\frac{-88}{-11} = 8$

33. $\frac{0}{14} = 0$

35. $\frac{2}{0}$ is undefined.

37. $\frac{39}{-3} = -13$

39. $-12(0) = 0$

41. $-4(3) = -12$

43. $-9 \cdot 6 = -54$

45. $-7(-6) = 42$

47. $-3(-4)(-2) = 12(-2) = -24$

49. $(-4)^2 = (-4)(-4) = 16$

51. $-\frac{10}{5} = -2$

53. $-\frac{56}{8} = -7$

55. $-12 \div 3 = -4$

57. $4(-4)(-3) = -16(-3) = 48$

59. $-30(6)(-2)(-3) = -180(-2)(-3)$
$= 360(-3)$
$= -1080$

61. $3 \cdot (-2) \cdot 0 = (-6) \cdot 0 = 0$

63. $\frac{100}{-20} = -5$

65. $240 \div (-40) = -6$

67. $\frac{-12}{-4} = 3$

69. $(-1)^4 = (-1)(-1)(-1)(-1)$
$= 1 \cdot (-1)(-1)$
$= (-1) \cdot (-1)$
$= 1$

71. $(-3)^5 = (-3)(-3)(-3)(-3)(-3)$
$= 9 \cdot (-3)(-3)(-3)$
$= (-27)(-3)(-3)$
$= 81 \cdot (-3)$
$= -243$

73. $-2(3)(5)(-6) = (-6)(5)(-6)$
$= (-30)(-6)$
$= 180$

75. $(-1)^{32} = (-1)(-1)\cdots(-1)$ 32 factors
$= (-1)(-1)(-1)(-1)\cdots(-1)$ 30 factors
$= (1)(-1)(-1)\cdots(-1)$ 30 factors
$= (-1)(-1)\cdots(-1)$ 30 factors
Repeating the pattern 15 more times, we find that $(-1)^{32} = 1$.

77. $-2(-2)(-5) = 4(-5) = -20$

79.
$$\begin{array}{r} 42 \\ \times\ 23 \\ \hline 126 \\ 840 \\ \hline 966 \end{array}$$
$-42 \cdot 23 = -966$

81.
$$\begin{array}{r} 25 \\ \times\ 82 \\ \hline 50 \\ 2000 \\ \hline 2050 \end{array}$$
$25 \cdot (-82) = -2050$

83. $a \cdot b = (-4) \cdot (7)$ substitute
$= -28$

85. $a \cdot b = (3) \cdot (-2)$ substitute
$= -6$

87. $a \cdot b = (-5) \cdot (-5)$ substitute
$= 25$

89. $\frac{x}{y}$
$= \frac{5}{-5}$ show substitution exactly
$= -1$

91. $\frac{x}{y}$
$= \frac{-12}{0}$ show substitution exactly
undefined division by zero

93. $\frac{x}{y}$
$= \frac{-36}{-6}$ show substitution exactly
$= 6$

95. $x \cdot y$
$= (-4)(-2)$ show substitution exactly
$= 8$
$\frac{x}{y}$
$= \frac{-4}{-2}$ show substitution exactly
$= 2$

97. $x \cdot y$
$= (0)(-6)$ show substitution exactly
$= 0$ multiplication property of zero
$\frac{x}{y}$
$= \frac{0}{-6}$ show substitution exactly
$= 0$ zero division property

99. $(-4)(3) = -12$
A loss of 12 yards

101. $5(-20) = -100$
He is at a depth of 100 feet.

103. $\frac{1+0+(-4)+(-5)}{4} = \frac{-8}{4} = -2$
His average score per round was –2.

105. $-62 \cdot 3 = -186$
After three years the income would be –$186 million.

107. **a.** $27 - 35 = -8$
There was a change of –8 condors.

b. 1979 to 1987 is 8 years.
$\frac{-8}{8} = -1$
There was a change of –1 condor per year.

109. $(3 \cdot 5)^2 = (3 \cdot 5)(3 \cdot 5) = (15)(15) = 225$

111. $90 + 12^2 - 5^3 = 90 + 12 \cdot 12 - 5 \cdot 5 \cdot 5$
$= 90 + 144 - 125$
$= 109$

113. $12 \div 4 - 2 + 7 = 3 - 2 + 7 = 8$

115. False; The product of two negative numbers is always a positive number.

117. True

119. **a.** $231{,}786 - 250{,}810 = -19{,}024$
There was a change of –19,024 Saturns.

b. $2(-19{,}024) = -38{,}048$
The change will be –38,048 Saturns.

c. One year from 1997 is 1998.
$231{,}786 - 38{,}048 = 193{,}738$
Sales would be 193,738 Saturns.

121. False

123. False

125. Answers may vary.

Section 2.5

Practice Problems

1. $(-2)^4 = (-2)(-2)(-2)(-2) = 4(-2)(-2) = -8(-2) = 16$

2. $-2^4 = -(2)(2)(2)(2) = -4(2)(2) = -8(2) = -16$

3. $3 \cdot 6^2 = 3 \cdot (6 \cdot 6) = 3 \cdot 36 = 108$

4. $(-5)^3 = (-5)(-5)(-5) = 25(-5) = -125$

5. $-3^4 = -(3 \cdot 3 \cdot 3 \cdot 3) = -81$

3. $\left(-\frac{1}{3}\right)^3 = \left(-\frac{1}{3}\right)\left(-\frac{1}{3}\right)\left(-\frac{1}{3}\right) = \frac{1}{9}\left(-\frac{1}{3}\right) = -\frac{1}{27}$

6. $\frac{25}{5(-1)} = \frac{25}{-5} = -5$

7. $\frac{-18+6}{-3-1} = \frac{-12}{-4} = 3$

8. $20 + 50 + (-4)^3 = 20 + 50 + (-64) = 70 + (-64) = 6$

9. $-2^3 + (-4)^2 + 1^5 = -8 + 16 + 1 = 8 + 1 = 9$

10. $2(2-8) + (-12) - 3 = 2(-6) + (-12) - 3 = -12 + (-12) - 3 = -24 - 3 = -27$

11. $(-5) \cdot |-4| + (-3) + 2^3 = -5 \cdot 4 + (-3) + 2^3 = -5 \cdot 4 + (-3) + 8 = -20 + (-3) + 8 = -23 + 8 = -15$

12. $4(-6) \div \left[3(5-7)^2\right] = 4(-6) \div \left[3(-2)^2\right] = 4(-6) \div [3 \cdot (4)] = 4(-6) \div (12) = -24 \div 12 = -2$

13. $x^2 = (-12)^2 = (-12)(-12) = 144$
$-x^2 = -(-12)^2 = -(-12)(-12) = -144$

14. $5y^2 = 5(3)^2 = 5(9) = 45$
$5y^2 = 5(-3)^2 = 5(9) = 45$

15. $x^2 + y = (-5)^2 + (-2) = 25 + (-2) = 23$

16. $4 - x^2 = 4 - (-9)^2 = 4 - 81 = -77$

Mental Math

1. -3^2 base: 3; exponent: 2

2. $(-3)^2$ base: −3; exponent: 2

3. $4 \cdot 2^3$ base: 2; exponent: 3

4. $9 \cdot 5^6$ base: 5; exponent: 6

5. $(-7)^5$ base: −7; exponent: 5

6. -9^4 base: −9; exponent: 4

7. $5^7 \cdot 10$ base: 5; exponent: 7

8. $2^8 \cdot 11$ base: 2; exponent: 8

Exercise Set 2.5

1. $-1(-2)+1$ multiply before adding
$=2+1$
$=3$

3. $3-6+2$ change to addition
$=3+(-6)+2$ add from left to right
$=-3+2$
$=-1$

5. $9-12-4$ change to addition
$=9+(-12)+(-4)$ add from left to right
$=-3+(-4)$
$=-7$

7. $4+3(-6)$ multiply before adding
$=4+(-18)$
$=-14$

9. $5(-9)+2$ multiply before adding
$=-45+2$
$=-43$

11. $(-10)+4\div 2$ divide before adding
$=(-10)+2$
$=-8$

13. $25\div(-5)+12$ divide before adding
$=-5+12$
$=7$

15. $\frac{16-13}{-3}$ simplify numerator
$=\frac{3}{-3}$
$=-1$

17. $\frac{24}{10+(-4)}$ simplify the bottom of the fraction bar first
$=\frac{24}{6}$
$=4$

19. $5(-3)-(-12)$ multiply before adding
$=-15-(-12)$ change to addition
$=-15+12$
$=-3$

21. $(-19)-12(3)$ multiply before adding
$=(-19)-36$ change to addition
$=(-19)+(-36)$
$=-55$

23. $8+4^2$ exponents first
$=8+16$
$=24$

25. $(8+(-4))^2$ grouping symbols first
$=[4]^2$
$=16$

27. 3^3-12 exponents first
$=27-12$ change to addition
$=27+(-12)$
$=15$

29. $16-(-3)^4$ exponents first
$=16-(81)$ change to addition
$=16+(-81)$
$=-65$

31. $|5+3|\cdot 2^3$ grouping symbols first
$=|8|\cdot 2^3$ exponents before multiplication
$=8\cdot 8$
$=64$

33. $7\cdot 8^2+4$ exponents first
$=7\cdot 64+4$ multiply before adding
$=448+4$
$=452$

35. $5^3-(4-2^3)$ exponents inside grouping symbols first
$=5^3-(4-8)$ grouping symbols next
$=5^3-(-4)$ exponents before subtraction
$=125-(-4)$ change subtraction to addition
$=125+4$
$=129$

37. $(3-12)\div 3$ grouping symbols first
$=(-9)\div 3$
$=-3$

39. $5+2^3-4^2$ exponents first
$= 5+8-16$ change to addition
$= 5+8+(-16)$ add from left to right
$= -3$

41. $(5-9)^2 \div (4-2)^2$ grouping symbols first
$= (-4)^2 \div (2)^2$ exponents next
$= 16 \div 4$
$= 4$

43. $|8-24| \cdot (-2) \div (-2)$ grouping symbol first
$= |-16| \cdot (-2) \div (-2)$ absolute value
$= 16 \cdot (-2) \div (-2)$ multiply and divide from left to right
$= -32 \div (-2)$
$= 16$

45. $(-12-20) \div 16 - 25$ grouping symbols first
$= (-32) \div 16 - 25$ divide
$= -2-25$
$= -27$

47. $5(5-2)+(-5)^2-6$ grouping symbols first
$= 5(3)+5^2-6$ exponents next
$= 5(3)+25-6$ multiply before adding
$= 15+25-6$
$= 34$

49. $(2-7) \cdot (6-19)$ grouping symbols first
$= (-5) \cdot (-13)$
$= 65$

51. $2-7 \cdot 6-19$ multiply
$= 2-42-19$ subtract from left to right
$= -59$

53. $(-36 \div 6)-(4 \div 4)$ divide
$= -6-1$
$= -7$

55. -5^2-6^2 exponents first (watch the signs)
$= -25-36$
$= -61$

57. $(-5)^2-6^2$ exponents first (watch the signs)
$= 25-36$
$= -11$

59. $(10-4^2)^2 = (10-16)^2$
$= (-6)^2$
$= (-6)(-6)$
$= 36$

61. $2(8-10)^2-5(1-6)^2$
$= 2(-2)^2-5(-5)^2$
$= 2(-2)(-2)-5(-5)(-5)$
$= 2(-2)(-2)+(-5)(-5)(-5)$
$= (-4)(-2)+(25)(-5)$
$= 8+(-125)$
$= -117$

63. $3(-10) \div (5(-3)-7(-2))$
$= 3(-10) \div (5(-3)+(-7)(-2))$
$= 3(-10) \div (-15+14)$
$= 3(-10) \div (-1)$
$= -30 \div -1$
$= 30$

65. $\dfrac{(-7)(-3)-(4)(3)}{3[7 \div (3-10)]} = \dfrac{(-7)(-3)+(-4)(3)}{3(7 \div (3+(-10)))}$
$= \dfrac{21+(-12)}{3(7 \div (-7))}$
$= \dfrac{9}{3(-1)}$
$= \dfrac{9}{-3}$
$= -3$

67. $x+y+z$ show substitution exactly
$= (-2)+4+(-1)$ add from left to right
$= 2+(-1)$
$= 1$

69. $2x-y^2$ show substitution exactly
$= 2(-2)-(4)^2$ exponents first
$= 2(-2)-16$ multiplication before subtraction
$= -4-16$
$= -20$

71. x^2-y show substitution exactly
$= (-2)^2-4$ exponents first
$= 4-4$
$= 0$

73. $\frac{5y}{z}$ show substitution exactly

$=\frac{5\cdot 4}{-1}$ simplify numerator

$=\frac{20}{-1}$

$=-20$

75. $x^2=(-3)^2=9$

77. $-z^2=-(-4)^2=-16$

79. $10-(-3)^2=10-9=1$

81. $2x^3=2(-3)^3=2(-27)=-54$

83. $45\cdot 90=4050$

85. $90-45=45$

87. $p=4(8)=32$ in.

89. $p=2(9)+2(6)$
$=18+12$
$=30$ feet

91. $2\cdot(7-5)\cdot 3=2\cdot(2)\cdot 3=4\cdot 3=12$

93. $-6\cdot(10-4)=-6\cdot 6=-36$

95. $(-12)^4=(-12)(-12)(-12)(-12)$
$=144(-12)(-12)$
$=-1728(-12)$
$=20,736$

97. x^3-y^2
$=21^3-(-19)^2$ substitute
$=21\cdot 21\cdot 21-(-19)(-19)$
$=21\cdot 21\cdot 21+(19)(-19)$
$=441\cdot 21+(-361)$
$=9261+(-361)$
$=8900$

99. $(xy+z)^x$
$=[2\cdot(-5)+7]^2$ substitute
$=(-3)^2$
$=(-3)(-3)$
$=9$

101. Answers may vary.

Chapter 2 Review

1. 1435 feet down in a mine is represented as −1435.

2. 7562 meters above sea level is represented as 7562.

3. 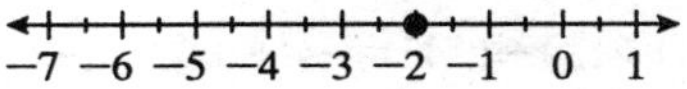

4. 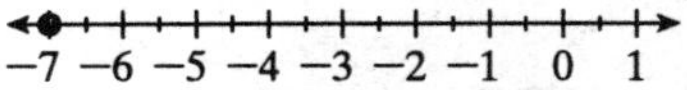

5. $-18>-20$

6. $-5<5$

7. $123>-198$

8. $|-12|=12$

9. $|0|=0$

10. $-|6|=-6$

11. $-(-12)=12$

12. $-(-(-3))=-3$

13. False

14. True

15. True

16. True

17. $5+(-3)=2$

18. $18+(-4)=14$

19. $-12+16=4$

20. $-23 + 40 = 17$

21. $-8 + (-15) = -23$

22. $-5 + (-17) = -22$

23. $-24 + 3 = -21$

24. $-89 + 19 = -70$

25. $15 + (-15) = 0$

26. $-24 + 24 = 0$

27. $-43 + (-108) = -151$

28. $-100 + (-506) = -606$

29. $$\begin{aligned} -7+3+(-2)+(-1) &= -4+(-2)+(-1) \\ &= -6+(-1) \\ &= -7 \end{aligned}$$
Her total score was –7.

30. $-8 + (-1) = -9$
She had a score of –9.

31. $-15 - 5 = -20$
The temperature was –20°C at 6 a.m.

32. $-127 + (-23) = -150$
The divers current depth is –150 feet.

33. $12 - 4 = 12 + (-4) = 8$

34. $-12 - 4 = -12 + (-4) = -16$

35. $8 - 19 = 8 + (-19) = -11$

36. $-8 - 19 = -8 + (-19) = -27$

37. $7 - (-13) = 7 + 13 = 20$

38. $-6 - (-14) = -6 + 14 = 8$

39. $16 - 16 = 16 + (-16) = 0$

40. $-16 - 16 = -16 + (-16) = -32$

41. $-12 - (-12) = -12 + 12 = 0$

42. $|-5| - |-12| = 5 - 12 = 5 + (-12) = -7$

43. $-(-5) - 12 + (-3) = 5 + (-12) + (-3) = -10$

44. $$\begin{aligned} -8+|-12|-10-|-3| &= -8+12+(-10)-3 \\ &= -8+12+(-10)+(-3) \\ &= 4+(-10)+(-3) \\ &= -6+(-3) \\ &= -9 \end{aligned}$$

45. $142-125+43-85$
$= 142+(-125)+43+(-85)$
$= 17+43+(-85)$
$= 60+(-85)$
$= -25$
The balance of his account represented as an integer is –25.

46. $600 - (-92) = 600 + 92 = 692$ feet

47. $|-5|-|-6| \stackrel{?}{=} 5-6$
$5-6 \stackrel{?}{=} 5-6$ True

48. $|-5-(-6)| \stackrel{?}{=} 5+6$
$|-5+6| \stackrel{?}{=} 11$
$|1| \stackrel{?}{=} 11$
$1 \stackrel{?}{=} 11$ False

49. True

50. True

51. $(-3) \cdot (-7) = 21$

52. $(-6) \cdot (3) = -18$

53. $-4 \cdot 16 = -64$

54. $(-5) \cdot (-12) = 60$

55. $-15 \div 3 = -5$

56. $\frac{-24}{-8} = 3$

57. $0 \div (-3) = 0$

58. $-2 \div 0$ is undefined

59. $\frac{-38}{-1} = 38$

60. $45 \div (-9) = -5$

61. $(-5)(2) = -10$

62. $(-50)(4) = -200$

63. $(-7)^2 = (-7)(-7) = 49$

64. $-7^2 = -(7)(7) = -49$

65. $-2^5 = -(2)(2)(2)(2)(2) = -32$

66. $(-2)^5 = (-2)(-2)(-2)(-2)(-2) = -32$

67. $5 - 8 + 3 = 0$

68. $$\begin{aligned} -3+12+(-7)-10 &= -3+12+(-7)+(-10) \\ &= -8 \end{aligned}$$

69. $-10 + 3 \cdot (-2) = -10 + (-6) = -16$

70. $5 - 10 \cdot (-3) = 5 - (-30) = 5 + 30 = 35$

71. $16 \cdot (-2) + 4 = -32 + 4 = -28$

72. $3 \cdot (-12) - 8 = -36 + (-8) = -44$

73. $5 + 6 \div (-3) = 5 - 2 = 3$

74. $-6 + (-10) \div (-2) = -6 + 5 = -1$

75. $$\begin{aligned} 16+(-3)\cdot 12 \div 4 &= 16+(-36)\div 4 \\ &= 16+(-9) \\ &= 7 \end{aligned}$$

76. $$\begin{aligned} (-12)+25\cdot 1 \div (-5) &= (-12)+25\div(-5) \\ &= (-12)+(-5) \\ &= -17 \end{aligned}$$

77. $4^3 - (8-3)^2 = 4^3 - 5^2 = 64 - 25 = 39$

78. $4^3 - 90 = 64 + (-90) = -26$

79. $$\begin{aligned} -(-4)\cdot|-3|-5 &= 4\cdot 3-5 \\ &= 12-5 \\ &= 12+(-5) \\ &= 7 \end{aligned}$$

80. $$\begin{aligned} |5-1|^2\cdot(-5) &= |4|^2\cdot(-5) \\ &= 4^2\cdot(-5) \\ &= 16\cdot(-5) \\ &= -80 \end{aligned}$$

81. $$\begin{aligned} \frac{(-4)(-3)-(-2)(-1)}{-10+5} &= \frac{12-2}{-10+5} \\ &= \frac{12+(-2)}{-10+5} \\ &= \frac{10}{-5} \\ &= -2 \end{aligned}$$

82. $$\begin{aligned} \frac{4(12-18)}{-10\div(-2-3)} &= \frac{4[12+(-18)]}{-10\div[-2+(-3)]} \\ &= \frac{4(-6)}{-10\div(-5)} \\ &= \frac{-24}{2} \\ &= -12 \end{aligned}$$

83. $2(-2) - 1 = -4 - 1 = -5$

84. $1^2 + (-2)^2 = 1 + 4 = 5$

85. $\frac{3(-2)}{6} = \frac{-6}{6} = -1$

86. $\frac{5(1)-(-2)}{-1} = \frac{5+2}{-1} = \frac{7}{-1} = -7$

87. $x^2 = (-2)^2 = 4$

88. $-x^2 = -(-2)^2 = -4$

89. $$\begin{aligned} 7-x^2 &= 7-(-2)^2 \\ &= 7-4 \\ &= 3 \end{aligned}$$

90. $$\begin{aligned} 100-x^3 &= 100-(-2)^3 \\ &= 100-(-8) \\ &= 100+8 \\ &= 108 \end{aligned}$$

Chapter 2 Test

1. $-5 + 8 = 3$

2. $18 - 24 = 18 + (-24) = -6$

3. $5 \cdot (-20) = -100$

4. $(-16) \div (-4) = 4$

5. $(-18) + (-12) = -30$

6. $-7 - (-19) = -7 + 19 = 12$

7. $(-5) \cdot (-13) = 65$

8. $\frac{-25}{-5} = 5$

9. $|-25| + (-13) = 25 + (-13) = 12$

10. $$\begin{aligned} 14 - |-20| &= 14 - 20 \\ &= 14 + (-20) \\ &= -6 \end{aligned}$$

11. $|5| \cdot |-10| = 5 \cdot 10 = 50$

12. $\frac{|-10|}{-|-5|} = \frac{10}{-5} = -2$

13. $(-8) + 9 \div (-3) = -8 + (-3) = -11$

14. $$\begin{aligned} -7 + (-32) - 12 + 5 &= -7 + (-32) + (-12) + 5 \\ &= -51 + 5 \\ &= -46 \end{aligned}$$

15. $$\begin{aligned} (-5)^3 - 24 \div (-3) &= (-125) - 24 \div (-3) \\ &= (-125) - (-8) \\ &= -125 + 8 \\ &= -117 \end{aligned}$$

16. $$\begin{aligned} (5-9)^2 \cdot (8-2)^3 &= (-4)^2 \cdot (6)^3 \\ &= 16 \cdot 216 \\ &= 3456 \end{aligned}$$

17. $$\begin{aligned} -(-7)^2 \div 7 \cdot (-4) &= -49 \div 7 \cdot (-4) \\ &= -7 \cdot (-4) \\ &= 28 \end{aligned}$$

18. $3 - (8-2)^3 = 3 - (6)^3 = 3 - 216 = -213$

19. $-6 + (-15) \div (-3) = -6 + 5 = -1$

20. $\frac{4}{2} - \frac{8^2}{16} = \frac{4}{2} - \frac{64}{16} = 2 - 4 = -2$

21. $\frac{(-3)(-2) + 12}{-1(-4-5)} = \frac{6+12}{-1(-9)} = \frac{18}{9} = 2$

22. $$\begin{aligned} \frac{|25-30|^2}{2(-6)+7} &= \frac{|-5|^2}{2(-6)+7} \\ &= \frac{5^2}{-12+7} \\ &= \frac{25}{-5} \\ &= -5 \end{aligned}$$

23. $3(0) + (-3) = 0 + (-3) = -3$

24. $|-3| + |0| + |2| = 3 + 0 + 2 = 5$

25. $\frac{3(2)}{2(-3)} = \frac{6}{-6} = -1$

26. $5 - (-3) = 5 + 3 = 8$

27. $14{,}893 - 147 = 14{,}893 + (-147) = 14{,}746$
His final elevation is 14,746 feet.

28. $$\begin{aligned} 237 - 157 - 77 + 38 &= 80 - 77 + 38 \\ &= 3 + 38 \\ &= 41 \end{aligned}$$
Jane has $41 in her account.

29. $6288 - (-25{,}354) = 6288 + 25{,}354 = 31{,}642$
The difference in elevation is 31,642 feet.

30. $1495 + (-5315) = -3820$
The elevation of the lake's deepest point is -3820 feet or 3820 feet below sea level.

Chapter 3

Pretest

1. $$\begin{aligned}9x-4+6x+8&=9x+6x-4+8\\&=(9+6)x+4\\&=15x+4\end{aligned}$$

2. $$\begin{aligned}&3(2x-1)-(x-8)\\&=3(2x-1)+(-1)(x-8)\\&=3\cdot 2x-3\cdot 1+(-1)(x)+(-1)(-8)\\&=6x-3-x+8\\&=6x-x-3+8\\&=5x+5\end{aligned}$$

3. $$\begin{aligned}8(7b)&=(8\cdot 7)b\\&=56b\end{aligned}$$

4. $$\begin{aligned}-5(2y-7)&=-5\cdot 2y+(-5)(-7)\\&=-10y+35\end{aligned}$$

5. $$\begin{aligned}\text{perimeter}&=4a+17+9a\\&=13a+17\end{aligned}$$
   The perimeter is $(13a + 17)$ inches.

6. $$\begin{aligned}\text{area}&=\text{length}\cdot\text{width}\\&=(4x-1)\cdot 3\\&=3(4x-1)\\&=3\cdot 4x-3\cdot 1\\&=12x-3\end{aligned}$$
   The area is $(12x-3)$ square feet.

7. $$\begin{aligned}2y-3&=-5\\2(-4)-3&\stackrel{?}{=}-5\\-8-3&\stackrel{?}{=}-5\\-11&=-5\quad\text{False}\end{aligned}$$
   No, -4 is not a solution.

8. $$\begin{aligned}-7&=x+5\\-7-5&=x+5-5\\-12&=x\end{aligned}$$

9. $$\begin{aligned}9n+3-8n&=8-14\\n+3&=-6\\n+3-3&=-6-3\\n&=-9\end{aligned}$$

10. $$\begin{aligned}-6x&=42\\\frac{-6x}{-6}&=\frac{42}{-6}\\x&=-7\end{aligned}$$

11. $$\begin{aligned}2m-9m&=-77\\-7m&=-77\\\frac{-7m}{-7}&=\frac{-77}{-7}\\m&=11\end{aligned}$$

12. $2(x-12)$

13. $3x$

14. $$\begin{aligned}18-3x&=-9\\18-3x-18&=-9-18\\-3x&=-27\\\frac{-3x}{-3}&=\frac{-27}{-3}\\x&=9\end{aligned}$$

15. $$\begin{aligned}8a-5&=2a+7\\8a-5+5&=2a+7+5\\8a&=2a+12\\8a-2a&=2a+12-2a\\6a&=12\\\frac{6a}{6}&=\frac{12}{6}\\a&=2\end{aligned}$$

16. $$\begin{aligned}4(x+1)&=9x-1\\4x+4&=9x-1\\4x+4-4&=9x-1-4\\4x&=9x-5\\4x-9x&=9x-5-9x\\-5x&=-5\\\frac{-5x}{-5}&=\frac{-5}{-5}\\x&=1\end{aligned}$$

17.
$$\begin{aligned} -2(3x+4)-10&=0\\ -2\cdot 3x+(-2)(4)-10&=0\\ -6x-8-10&=0\\ -6x-18&=0\\ -6x-18+18&=0+18\\ -6x&=18\\ \frac{-6x}{-6}&=\frac{18}{-6}\\ x&=-3 \end{aligned}$$

18. $\frac{54}{-6}=-9$

19. $x-5=12$

20.
$$\begin{aligned} 68&=9x+5\\ 68-5&=9x+5-5\\ 63&=9x\\ \frac{63}{9}&=\frac{9x}{9}\\ 7&=x \end{aligned}$$

Section 3.1

Practice Problems

1. a.
$$\begin{aligned} 8m-11m&=(8-11)m\\ &=-3m \end{aligned}$$

b.
$$\begin{aligned} 5a+a&=5a+1a\\ &=(5+1)a\\ &=6a \end{aligned}$$

c.
$$\begin{aligned} -y^2+3y^2+7&=-1y^2+3y^2+7\\ &=(-1+3)y^2+7\\ &=2y^2+7 \end{aligned}$$

2.
$$\begin{aligned} 8m+5+m-4&=8m+5+m+(-4)\\ &=8m+m+5+(-4)\\ &=(8+1)m+5+(-4)\\ &=9m+1 \end{aligned}$$

3. $7y+11y-8=18y-8$

4.
$$\begin{aligned} 2y-6+y+7&=2y+y-6+7\\ &=3y+1 \end{aligned}$$

5.
$$\begin{aligned} &-9y+2-4y-8x+12-x\\ &=-9y+2+(-4y)+(-8x)+12+(-x)\\ &=-9y+(-4y)+(-8x)+(-x)+2+12\\ &=-13y-9x+14 \end{aligned}$$

6.
$$\begin{aligned} 7(8a)&=(7\cdot 8)a\\ &=56a \end{aligned}$$

7.
$$\begin{aligned} -5(9x)&=(-5\cdot 9)x\\ &=-45x \end{aligned}$$

8.
$$\begin{aligned} 7(y+2)&=7\cdot y+7\cdot 2\\ &=7y+14 \end{aligned}$$

9.
$$\begin{aligned} 4(7a-5)&=4\cdot 7a+4\cdot(-5)\\ &=28a-20 \end{aligned}$$

10. $6(5-y)=6\cdot 5-6\cdot y=30-6y$

11.
$$\begin{aligned} 5(y-3)-8+y&=5\cdot y-5\cdot 3-8+y\\ &=5y-15-8+y\\ &=6y-23 \end{aligned}$$

12.
$$\begin{aligned} &5(2x-3)+7(x-1)\\ &=5\cdot 2x-5\cdot 3+7\cdot x-7\cdot 1\\ &=10x-15+7x-7\\ &=10x+(-15)+7x+(-7)\\ &=17x-22 \end{aligned}$$

13.
$$\begin{aligned} &-(y+1)+3y-12\\ &=-1(y+1)+3y+(-12)\\ &=-1\cdot y+(-1)(1)+3y+(-12)\\ &=-y+(-1)+3y+(-12)\\ &=2y-13 \end{aligned}$$

14. Find the sum of the lengths of the sides. A square has 4 sides of equal length.
$$\begin{aligned} \text{perimeter}&=2x+2x+2x+2x\\ &=8x \end{aligned}$$
The perimeter is $8x$ centimeters.

15.
$$\begin{aligned} A&=\text{length}\cdot\text{width}\\ &=(12y+9)\cdot 3\\ &=3(12y+9)\\ &=3\cdot 12y+3\cdot 9\\ &=36y+27 \end{aligned}$$
The area is $(36y+27)$ square yards.

Mental Math

1. The numerical coefficient of $5y$ is 5.

2. The numerical coefficient of $-2z$ is -2.

3. The numerical coefficient of z is 1.

4. The numerical coefficient of $3xy^2$ is 3.

5. The numerical coefficient of $11a$ is 11.

6. The numerical coefficient of $-x$ is -1.

Exercise Set 3.1

1. $3x + 5x = (3 + 5)x = 8x$

3. $5n - 9n = (5 - 9)n = -4n$

5. $4c + c - 7c = (4 + 1 - 7)c = -2c$

7. $5x - 7x + x - 3x = (5 - 7 + 1 - 3)x = -4x$

9. $4a + 3a + 6a - 8 = (4 + 3 + 6)a - 8 = 13a - 8$

11. $6(5x) = (6 \cdot 5)x = 30x$

13. $-2(11y) = (-2 \cdot 11)y = -22y$

15. $12(6a) = (12 \cdot 6)a = 72a$

17. $2(y + 2) = 2 \cdot y + 2 \cdot 2 = 2y + 4$

19. $5(a - 8) = 5 \cdot a + 5 \cdot (-8) = 5a - 40$

21. $$\begin{aligned} -4(3x+7) &= -4 \cdot 3x + (-4) \cdot 7 \\ &= -12x - 28 \end{aligned}$$

23. $$\begin{aligned} 2(x+4)+7 &= 2 \cdot x + 2 \cdot 4 + 7 \\ &= 2x + 8 + 7 \\ &= 2x + 15 \end{aligned}$$

25. $$\begin{aligned} -4(6n-5)+3n &= -4 \cdot 6n + (-4) \cdot (-5) + 3n \\ &= -24n + 20 + 3n \\ &= -21n + 20 \end{aligned}$$

27. $$\begin{aligned} 5(3c-1)+8 &= 5 \cdot 3c - 5 \cdot 1 + 8 \\ &= 15c - 5 + 8 \\ &= 15c + 3 \end{aligned}$$

29. $$\begin{aligned} 3+6(w+2)+w &= 3 + 6 \cdot w + 6 \cdot 2 + w \\ &= 3 + 6w + 12 + w \\ &= 6w + w + 3 + 12 \\ &= 7w + 15 \end{aligned}$$

31. $$\begin{aligned} &2(3x + 1) + 5(x - 2) \\ &= 2(3x) + 2(1) + 5(x) - 5(2) \\ &= 6x + 2 + 5x - 10 \\ &= 6x + 5x + 2 + (-10) \\ &= 11x - 8 \end{aligned}$$

33. $$\begin{aligned} -(5x-1)-10 &= -1(5x-1) - 10 \\ &= -1 \cdot 5x + (-1) \cdot (-1) - 10 \\ &= -5x + 1 - 10 \\ &= -5x - 9 \end{aligned}$$

35. $18y - 20y = (18 - 20)y = -2y$

37. $z - 8z = (1 - 8)z = -7z$

39. $9d - 3c - d = 9d - d - 3c = 8d - 3c$

41. $$\begin{aligned} 2y - 6 + 4y - 8 &= 2y + 4y - 6 - 8 \\ &= 6y - 14 \end{aligned}$$

43. $$\begin{aligned} 5q + p - 6q - p &= p - p + 5q - 6q \\ &= -q \end{aligned}$$

45. $$\begin{aligned} 2(x+1)+20 &= 2 \cdot x + 2 \cdot 1 + 20 \\ &= 2x + 2 + 20 \\ &= 2x + 22 \end{aligned}$$

47. $$\begin{aligned} 5(x-7)-8x &= 5 \cdot x - 5 \cdot 7 - 8x \\ &= 5x - 35 - 8x \\ &= 5x - 8x - 35 \\ &= -3x - 35 \end{aligned}$$

49. $$\begin{aligned} -5(z+3)+2z &= -5 \cdot z + (-5) \cdot 3 + 2z \\ &= -5z - 15 + 2z \\ &= -5z + 2z - 15 \\ &= -3z - 15 \end{aligned}$$

51. $$\begin{aligned} &8 - x + 4x - 2 - 9x \\ &= -x + 4x - 9x + 8 - 2 \\ &= -6x + 6 \end{aligned}$$

53. $-7(x+5)+5(2x+1)$
$=-7\cdot x+(-7)\cdot 5+5\cdot 2x+5\cdot 1$
$=-7x-35+10x+5$
$=-7x+10x-35+5$
$=3x-30$

55. $3r-5r+8+r=3r-5r+r+8$
$=-r+8$

57. $-3(n-1)-4n=-3\cdot n-(-3)\cdot 1-4n$
$=-3n+3-4n$
$=-3n-4n+3$
$=-7n+3$

59. $4(z-3)+5z-2=4\cdot z-4\cdot 3+5z-2$
$=4z-12+5z-2$
$=4z+5z-12-2$
$=9z-14$

61. $6(2x-1)-12x=6\cdot 2x-6\cdot 1-12x$
$=12x-6-12x$
$=12x-12x-6$
$=-6$

63. $-(4x-5)+5=-1(4x-5)+5$
$=-1\cdot 4x-(-1)\cdot 5+5$
$=-4x+5+5$
$=-4x+10$

65. $-(4xy-10)+2(3xy+5)$
$=-1(4xy-10)+2\cdot 3xy+2\cdot 5$
$=-1\cdot 4xy-(-1)\cdot 10+6xy+10$
$=-4xy+10+6xy+10$
$=-4xy+6xy+10+10$
$=2xy+20$

67. $3a+4(a+3)=3a+4\cdot a+4\cdot 3$
$=3a+4a+12$
$=7a+12$

69. $5y-2(y-1)+3=5y-2y-2(-1)+3$
$=3y+2+3$
$=3y+5$

71. There are five sides, each of length $-5x+11$ inches, so the perimeter is:
$5(-5x+11)=5(-5x)+5(11)$
$=-(25x+55)$ inches

73. $5y+16+3y+4y+2y+6$
$=5y+3y+4y+2y+16+6$
$=(14y+22)$ meters

75. There are three sides of length $2a$ feet, two sides of length 6 feet, and one side of length $5a$ feet, so the perimeter is:
$3(2a)+2\cdot 6+5a=6a+12+5a$
$=6a+5a+12$
$=(11a+12)$ feet

77. Area = $(\text{side})^2$
$=(4z)^2$
$=(4z)(4z)$
$=(4)(4)(z)(z)$
$=16z^2$ square centimeters

79. $-13+10=-3$

81. $-4-(-12)=-4+12=8$

83. $-4+4=0$

85. $9684q-686-4860q+12{,}960$
$=(9684-4860)q+(12{,}960-686)$
$=4824q+12{,}274$

87. Answers may vary.

89. Add the areas of the two rectangles.
Area = (length) · (width)
$\begin{pmatrix}\text{Area of left}\\ \text{rectangle}\end{pmatrix}+\begin{pmatrix}\text{Area of right}\\ \text{rectangle}\end{pmatrix}$
$=7(2x+1)+3(2x+3)$
$=7(2x)+7(1)+3(2x)+3(3)$
$=14x+7+6x+9$
$=(14+6)x+16$
$=(20x+16)$
The area is $(20x+16)$ square miles.

91. Answers will vary.

Section 3.2

Practice Problems

1. $3(y-6) = 6$
$3(4-6) \stackrel{?}{=} 6$
$3(-2) \stackrel{?}{=} 6$
$-6 = 6$ False
Since $-6 = 6$ is false, 4 is not a solution of the equation.

2. $-4x-3 = 5$
$-4(-2)-3 \stackrel{?}{=} 5$
$8-3 \stackrel{?}{=} 5$
$5 = 5$ True
Since $5 = 5$ is true, -2 is a solution of the equation.

3. $y-5=-3$
$y-5+5=-3+5$
$y=2$

4. $z+9=1$
$z+9-9=1-9$
$z=-8$

5. $10x=-2+9x$
$10x-9x=-2+9x-9x$
$1x=-2$ or $x=-2$

6. $x+6=1-3$
$x+6=-2$
$x+6-6=-2-6$
$x=-8$

7. $-6y-1+7y=17$
$y-1=17$
$y-1+1=17+1$
$y=18$

8. $13x=4(3x-1)$
$13x=4\cdot 3x-4\cdot 1$
$13x=12x-4$
$13x-12x=12x-4-12x$
$1x=-4$ or $x=-4$

Exercise Set 3.2

1. $x-8=2$
$10-8 \stackrel{?}{=} 2$
$2 \stackrel{?}{=} 2$ True
Yes, 10 is a solution.

3. $z+8=14$
$6+8 \stackrel{?}{=} 14$
$14 \stackrel{?}{=} 14$ True
Yes, 6 is a solution.

5. $x+12=7$
$-5+12 \stackrel{?}{=} 7$
$7 \stackrel{?}{=} 7$ True
Yes, -5 is a solution.

7. $7f=64-f$
$7(8) \stackrel{?}{=} 64-8$
$56 \stackrel{?}{=} 64-8$
$56 \stackrel{?}{=} 56$ True
Yes, 8 is a solution.

9. $h-8=-8$
$0-8 \stackrel{?}{=} -8$
$-8 \stackrel{?}{=} -8$ True
Yes, 0 is a solution.

11. $4c+2-3c=-1+6$
$4(3)+2-3(3) \stackrel{?}{=} 5$
$12+2-9 \stackrel{?}{=} 5$
$14-9 \stackrel{?}{=} 5$
$5 \stackrel{?}{=} 5$ True
Yes, 3 is a solution.

13. $a+5=23$
$a+5-5=23-5$
$a=18$
Check: $a+5=23$
$18+5 \stackrel{?}{=} 23$
$23 \stackrel{?}{=} 23$ True
The solution is 18.

15.
$$d-9=17$$
$$d-9+9=17+9$$
$$d=26$$
Check: $d-9=17$
$$26-9 \stackrel{?}{=} 17$$
$$17 \stackrel{?}{=} 17 \text{ True}$$
The solution is 26.

17.
$$7=y-2$$
$$7+2=y-2+2$$
$$9=y$$
Check: $7=y-2$
$$7 \stackrel{?}{=} 9-2$$
$$7 \stackrel{?}{=} 7 \text{ True}$$
The solution is 9.

19.
$$-12=x+4$$
$$-12-4=x+4-4$$
$$-16=x$$
Check: $-12=x+4$
$$-12 \stackrel{?}{=} -16+4$$
$$-12 \stackrel{?}{=} -12 \text{ True}$$
The solution is –16.

21.
$$3x=2x+11$$
$$3x-2x=2x+11-2x$$
$$x=11$$
Check: $3x=2x+11$
$$3(11) \stackrel{?}{=} 2(11)+11$$
$$33 \stackrel{?}{=} 22+11$$
$$33 \stackrel{?}{=} 33 \text{ True}$$
The solution is 11.

23.
$$-4+y=2y$$
$$-4+y-y=2y-y$$
$$-4=y$$
Check: $-4+y=2y$
$$-4+(-4) \stackrel{?}{=} 2(-4)$$
$$-8 \stackrel{?}{=} -8 \text{ True}$$
The solution is –4.

25.
$$x-3=-1+4$$
$$x-3=3$$
$$x-3+3=3+3$$
$$x=6$$
Check: $x-3=-1+4$
$$6-3 \stackrel{?}{=} -1+4$$
$$3 \stackrel{?}{=} 3 \text{ True}$$
The solution is 6.

27.
$$y+1=-3+4$$
$$y+1=1$$
$$y+1-1=1-1$$
$$y=0$$
Check: $y+1=-3+4$
$$0+1 \stackrel{?}{=} -3+4$$
$$1 \stackrel{?}{=} 1 \text{ True}$$
The solution is 0.

29.
$$-7+10=m-5$$
$$3=m-5$$
$$3+5=m-5+5$$
$$8=m$$
Check: $-7+10=m-5$
$$-7+10 \stackrel{?}{=} 8-5$$
$$3 \stackrel{?}{=} 3 \text{ True}$$
The solution is 8.

31.
$$-2-3=-4+x$$
$$-5=-4+x$$
$$-5+4=-4+x+4$$
$$-1=x$$
Check: $-2-3=-4+x$
$$-2-3 \stackrel{?}{=} -4+(-1)$$
$$-5 \stackrel{?}{=} -5 \text{ True}$$
The solution is –1.

33.
$$2(5x-3)=11x$$
$$2 \cdot 5x-2 \cdot 3=11x$$
$$10x-6=11x$$
$$10x-6-10x=11x-10x$$
$$-6=x$$
Check: $2(5x-3)=11x$
$$2(5 \cdot (-6)-3) \stackrel{?}{=} 11 \cdot (-6)$$
$$2(-30-3) \stackrel{?}{=} -66$$
$$2(-33) \stackrel{?}{=} -66$$
$$-66 \stackrel{?}{=} -66 \text{ True}$$
The solution is –6.

35.
$$\begin{aligned} 3y &= 2(y+12) \\ 3y &= 2\cdot y + 2\cdot 12 \\ 3y &= 2y+24 \\ 3y-2y &= 2y+24-2y \\ y &= 24 \end{aligned}$$

Check:
$$\begin{aligned} 3y &= 2(y+12) \\ 3\cdot 24 &\stackrel{?}{=} 2(24+12) \\ 72 &\stackrel{?}{=} 2(36) \\ 72 &\stackrel{?}{=} 72 \text{ True} \end{aligned}$$

The solution is 24.

37.
$$\begin{aligned} -8x+4+9x &= -1+7 \\ x+4 &= 6 \\ x+4-4 &= 6-4 \\ x &= 2 \end{aligned}$$

Check:
$$\begin{aligned} -8x+4+9x &= -1+7 \\ -8(2)+4+9(2) &\stackrel{?}{=} -1+7 \\ -16+4+18 &\stackrel{?}{=} 6 \\ 6 &\stackrel{?}{=} 6 \text{ True} \end{aligned}$$

The solution is 2.

39.
$$\begin{aligned} 2-2 &= 5x-4x \\ 0 &= x \end{aligned}$$

Check:
$$\begin{aligned} 2-2 &= 5x-4x \\ 2-2 &\stackrel{?}{=} 5(0)-4(0) \\ 0 &\stackrel{?}{=} 0-0 \\ 0 &\stackrel{?}{=} 0 \text{ True} \end{aligned}$$

The solution is 0.

41.
$$\begin{aligned} 7x+14-6x &= -4-10 \\ x+14 &= -14 \\ x+14-14 &= -14-14 \\ x &= -28 \end{aligned}$$

Check:
$$\begin{aligned} 7x+14-6x &= -4-10 \\ 7(-28)+14-6(-28) &\stackrel{?}{=} -4-10 \\ -196+14+168 &\stackrel{?}{=} -14 \\ -14 &\stackrel{?}{=} -14 \text{ True} \end{aligned}$$

The solution is –28.

43.
$$\begin{aligned} 57 &= y-16 \\ 57+16 &= y-16+16 \\ 73 &= y \end{aligned}$$

45.
$$\begin{aligned} 67 &= z+67 \\ 67+(-67) &= z+67+(-67) \\ 0 &= z \end{aligned}$$

47.
$$\begin{aligned} x+5 &= 4-3 \\ x+5 &= 1 \\ x+5+(-5) &= 1+(-5) \\ x &= -4 \end{aligned}$$

49.
$$\begin{aligned} z-23 &= -88 \\ z-23+23 &= -88+23 \\ z &= -65 \end{aligned}$$

51.
$$\begin{aligned} 7a+7-6a &= 20 \\ a+7 &= 20 \\ a+7+(-7) &= 20+(-7) \\ a &= 13 \end{aligned}$$

53.
$$\begin{aligned} -12+x &= -15 \\ -12+x+12 &= -15+12 \\ x &= -3 \end{aligned}$$

55.
$$\begin{aligned} -8-9 &= 3x+5-2x \\ -17 &= x+5 \\ -17-5 &= x+5-5 \\ -22 &= x \end{aligned}$$

57.
$$\begin{aligned} 8(3x-2) &= 25x \\ 8\cdot 3x-8\cdot 2 &= 25x \\ 24x-16 &= 25x \\ 24x-16-24x &= 25x-24x \\ -16 &= x \end{aligned}$$

59.
$$\begin{aligned} 7x+7-6x &= 10 \\ x+7 &= 10 \\ x+7+(-7) &= 10+(-7) \\ x &= 3 \end{aligned}$$

61.
$$\begin{aligned} 50y &= 7(7y+4) \\ 50y &= 7\cdot 7y+7\cdot 4 \\ 50y &= 49y+28 \\ 50y-49y &= 49y+28-49y \\ y &= 28 \end{aligned}$$

63. 1900

65. In 1995 there were about 900 trumpeter swans. In 1985 there were about 200 trumpeter swans. Thus, there were about 900 – 200 = 700 more trumpeter swans in 1995 than in 1985.

67. $\frac{8}{8}=1$

69. $\frac{-3}{-3}=1$

71. Answers may vary.

73.
$$x-76,862=86,102$$
$$x-76,862+76,862=86,102+76,862$$
$$x=162,964$$

75.
$$T=P+R$$
$$5419=P+1519$$
$$5419-1519=P+1519-1519$$
$$3900=P$$
The total passing yardage was 3900 yards.

77.
$$I=R-E$$
$$3056=R-103,090$$
$$3056+103,090=R-103,090+103,090$$
$$106,146=R$$
The total revenues were $106,146 million.

Section 3.3

Practice Problems

1.
$$3y=-18$$
$$\frac{3y}{3}=\frac{-18}{3}$$
$$\frac{3}{3}\cdot y=\frac{-18}{3}$$
$$1y=-6 \text{ or } y=-6$$

2.
$$-16=8x$$
$$\frac{-16}{8}=\frac{8x}{8}$$
$$\frac{-16}{8}=\frac{8}{8}\cdot x$$
$$-2=x \text{ or } x=-2$$

3.
$$-3y=-27$$
$$\frac{-3y}{-3}=\frac{-27}{-3}$$
$$\frac{-3}{-3}\cdot y=\frac{-27}{-3}$$
$$y=9$$

4.
$$10=2m-4m$$
$$10=-2m$$
$$\frac{10}{-2}=\frac{-2m}{-2}$$
$$-5=m \text{ or } m=-5$$

5.
$$-8+6=-a$$
$$-2=-a$$
$$\frac{-2}{-1}=\frac{-1a}{-1}$$
$$2=a$$

6.
$$-4-10=4y+3y$$
$$-14=7y$$
$$\frac{-14}{7}=\frac{7y}{7}$$
$$-2=y$$

7. **a.** The product of 5 and a number is $5x$. The product of 5 and a number decreased by 25 is $5x-25$.

b. The sum of a number and 3 is $x+3$. Twice the sum of a number and 3 is $2(x+3)$.

c. Twice a number is $2x$. The quotient of 39 and twice a number is $\frac{39}{2x}$.

Exercise Set 3.3

1.
$$5x=20$$
$$\frac{5x}{5}=\frac{20}{5} \quad \text{divide by 5}$$
$$x=4$$

3.
$$-3z=12$$
$$\frac{-3z}{-3}=\frac{12}{-3} \quad \text{divide by } -3$$
$$z=-4$$

5.
$$4y=0$$
$$\frac{4y}{4}=\frac{0}{4} \quad \text{divide by 4}$$
$$y=0$$

7. $2z = -34$
$\frac{2z}{2} = \frac{-34}{2}$ divide by 2
$z = -17$

9. $-3x = -15$
$\frac{-3x}{-3} = \frac{-15}{-3}$ divide by -3
$x = 5$

11. $2w - 12w = 40$ combine like terms
$-10w = 40$
$\frac{-10w}{-10} = \frac{40}{-10}$ divide by -10
$w = -4$

13. $16 = 10t - 8t$ combine like terms
$16 = 2t$
$\frac{16}{2} = \frac{2t}{2}$ divide by 2
$8 = t$

15. $2z = 12 - 14$ combine like terms
$2z = -2$
$\frac{2z}{2} = \frac{-2}{2}$ divide by 2
$z = -1$

17. $4 - 10 = -3z$ combine like terms
$-6 = -3z$
$\frac{-6}{-3} = \frac{-3z}{-3}$ divide by -3
$2 = z$

19. $-3x - 3x = 50 - 2$ combine like terms
$-6x = 48$
$\frac{6x}{-6} = \frac{48}{-6}$ divide by -6
$x = -8$

21. $-10x = 10$
$\frac{-10x}{-10} = \frac{10}{-10}$ divide by -10
$x = -1$

23. $5x = -35$
$\frac{5x}{5} = \frac{-35}{5}$ divide by 5
$x = -7$

25. $0 = 3x$
$\frac{0}{3} = \frac{3x}{3}$ divide by 3
$0 = x$

27. $24 = t + 3t$ combine like terms
$24 = 4t$
$\frac{24}{4} = \frac{4t}{4}$ divide by 4
$6 = t$

29. $10z - 3z = -63$ combine like terms
$7z = -63$
$\frac{7z}{7} = \frac{-63}{7}$ divide by 7
$z = -9$

31. $3z - 10z = -63$ combine like terms
$-7z = -63$
$\frac{-7z}{-7} = \frac{-63}{-7}$ divide by -7
$z = 9$

33. $12 = 13y - 10y$ combine like terms
$12 = 3y$
$\frac{12}{3} = \frac{3y}{3}$ divide by 3
$4 = y$

35. $-4x = 20 - (-4)$ combine like terms
$-4x = 20 + 4$
$-4x = 24$
$\frac{-4x}{-4} = \frac{24}{-4}$ divide by -4
$x = -6$

37. $18 - 11 = 7x$ combine like terms
$7 = 7x$
$\frac{7}{7} = \frac{7x}{7}$ divide by 7
$1 = x$

39. $11 - 18 = 7x$ combine like terms
$-7 = 7x$
$\frac{-7}{7} = \frac{7x}{7}$ divide by 7
$-1 = x$

41. $10p - 11p = 25$ combine like terms
$-1p = 25$
$\frac{-1p}{-1} = \frac{25}{-1}$ divide by -1
$p = -25$

43. $6x - x = 4 - 14$ combine like terms
$5x = -10$
$\frac{5x}{5} = \frac{-10}{5}$ divide by 5
$x = -2$

45. $10 = 7t - 12t$ combine like terms
$10 = -5t$
$\frac{10}{-5} = \frac{-5t}{-5}$ divide by -5
$-2 = t$

47. $5 - 5 = 3x + 2x$ combine like terms
$0 = 5x$
$\frac{0}{5} = \frac{5x}{5}$ divide by 5
$0 = x$

49. $4r - 9r = -20$ combine like terms
$-5r = -20$
$\frac{-5r}{-5} = \frac{-20}{-5}$ divide by -5
$r = 4$

51. $-3x + 4x = -1 + 8$ combine like terms
$x = 7$

53. $3w - 12w = -27$ combine like terms
$-9w = -27$
$\frac{-9w}{-9} = \frac{-27}{-9}$ divide by -9
$w = 3$

55. $-36 = 9u - 10u$ combine like terms
$-36 = -1u$
$\frac{-36}{-1} = \frac{-1u}{-1}$ divide by -1
$36 = u$

57. $23x - 25x = 7 - 9$ combine like terms
$-2x = -2$
$\frac{-2x}{-2} = \frac{-2}{-2}$ divide by -2
$x = 1$

59. Let x represent a number. The product of 4 and a number is $4x$. Thus, seven added to the product of 4 and a number is $4x + 7$.

61. Let x represent a number. Twice a number is $2x$. Thus, twice a number, decreased by 17 is $2x - 17$.

63. Let x represent a number. The sum of a number and 15 is $x + 15$. Thus, the product of -6 and the sum of a number and 15 is $-6(x + 15)$.

65. Let x represent a number. The product of a number and -5 is $-5x$. Thus, the quotient of 45 and the product of a number and -5 is $\frac{45}{-5x}$.

67. $3x + 10 = 3(-5) + 10 = -15 + 10 = -5$

69. $\frac{x-5}{2} = \frac{-5-5}{2} = \frac{-10}{2} = -5$

71. $\frac{3x+4}{x+4} = \frac{3(-5)+4}{-5+4} = \frac{-15+4}{-1} = \frac{-11}{-1} = 11$

73. Answers may vary.

75. $-25x = 900$
$\frac{-25x}{-25} = \frac{900}{-25}$
$x = -36$

77. $d = r \cdot t$
$780 = 65t$
$\frac{780}{65} = \frac{65t}{65}$
$12 = t$
It will take 12 hours.

79.
$$d = r \cdot t$$
$$232 = r \cdot 4$$
$$\frac{232}{4} = \frac{r \cdot 4}{4}$$
$$58 = r$$
The driver should drive at 58 miles per hour.

Integrated Review

1. $y - x = 3 - (-1)$
$= 3 + 1$
$= 4$

2. $\frac{y}{x} = \frac{3}{-1} = -3$

3. $5x + 2y = 5(-1) + 2(3)$
$= -5 + 6$
$= 1$

4. $\frac{y^2 + x}{2x} = \frac{3^2 + (-1)}{2(-1)}$
$= \frac{9-1}{-2}$
$= \frac{8}{-2}$
$= -4$

5. $7x + x = 7x + 1x$
$= (7+1)x$
$= 8x$

6. $6y - 10y = (6-10)y$
$= -4y$

7. $2a + 5a - 9a - 2 = (2 + 5 - 9)a - 2$
$= -2a - 2$

8. $-2(4x + 7) = -2 \cdot 4x + (-2)(7) = -8x - 14$

9. $-(y - 4) + 5(y + 2)$
$= -1(y - 4) + 5 \cdot y + 5 \cdot 2$
$= -1 \cdot y + (-1)(-4) + 5y + 10$
$= -y + 4 + 5y + 10$
$= -y + 5y + 4 + 10$
$= 4y + 14$

10. $A = \text{length} \cdot \text{width}$
$= (4x - 2) \cdot 3$
$= 3(4x - 2)$
$= 3 \cdot 4x - 3 \cdot 2$
$= 12x - 6$
The area is $(12x - 6)$ square meters.

11.
$$x + 7 = 20$$
$$x + 7 - 7 = 20 - 7$$
$$x = 13$$
Check: $x + 7 = 20$
$13 + 7 \stackrel{?}{=} 20$
$20 = 20$ True
The solution is 13.

12.
$$-11 = x - 2$$
$$-11 + 2 = x - 2 + 2$$
$$-9 = x$$
Check: $-11 = x - 2$
$-11 \stackrel{?}{=} -9 - 2$
$-11 = -11$ True
The solution is -9.

13.
$$11x = 55$$
$$\frac{11x}{11} = \frac{55}{11}$$
$$x = 5$$
Check: $11x = 55$
$11(5) \stackrel{?}{=} 55$
$55 = 55$ True
The solution is 5.

14.
$$-7y = 0$$
$$\frac{-7y}{-7} = \frac{0}{-7}$$
$$y = 0$$
Check: $-7y = 0$
$-7(0) \stackrel{?}{=} 0$
$0 = 0$ True
The solution is 0.

15. $12 = 11x - 14x$
$12 = -3x$
$\frac{12}{-3} = \frac{-3x}{-3}$
$-4 = x$
Check: $12 = 11x - 14x$
$12 \stackrel{?}{=} 11(-4) - 14(-4)$
$12 \stackrel{?}{=} -44 - (-56)$
$12 \stackrel{?}{=} -44 + 56$
$12 = 12$ True
The solution is -4.

16. $-3x = -15$
$\frac{-3x}{-3} = \frac{-15}{-3}$
$x = 5$
Check: $-3x = -15$
$-3(5) \stackrel{?}{=} -15$
$-15 = -15$ True
The solution is 5.

17. $x - 12 = -45 + 23$
$x - 12 = -22$
$x - 12 + 12 = -22 + 12$
$x = -1$
Check: $x - 12 = -45 + 23$
$-10 - 12 \stackrel{?}{=} -45 + 23$
$-22 \stackrel{?}{=} -22$ True
The solution is -10.

18. $8y + 7y = -45$
$15y = -45$
$\frac{15y}{15} = \frac{-45}{15}$
$y = -3$
Check: $8y + 7y = -45$
$8(-3) + 7(-3) \stackrel{?}{=} -45$
$-24 + (-21) \stackrel{?}{=} -45$
$-45 = -45$ True
The solution is -3.

19. $6 - (-5) = x + 5$
$6 + 5 = x + 5$
$11 = x + 5$
$11 - 5 = x + 5 - 5$
$6 = x$
Check: $6 - (-5) = x + 5$
$6 - (-5) \stackrel{?}{=} 6 + 5$
$6 + 5 \stackrel{?}{=} 11$
$11 = 11$ True
The solution is 6.

20. $-2m = -1.6$
$\frac{-2m}{-2} = \frac{-16}{-2}$
$m = 8$
Check: $-2m = -16$
$-2(8) \stackrel{?}{=} -16$
$-16 = -16$ True
The solution is 8.

21. $13x = -1 + 12x$
$13x - 12x = -1 + 12x - 12x$
$x = -1$
Check: $13x = -1 + 12x$
$13(-1) \stackrel{?}{=} -1 + 12(-1)$
$-13 \stackrel{?}{=} -1 - 12$
$-13 = -13$ True
The solution is -1.

22. $6(3x - 4) = 19x$
$6 \cdot 3x - 6 \cdot 4 = 19x$
$18x - 24 = 19x$
$18x - 24 - 18x = 19x - 18x$
$-24 = x$
Check: $6(3x - 4) = 19x$
$6(3(-24) - 4) \stackrel{?}{=} 19(-24)$
$6(-72 - 4) \stackrel{?}{=} -456$
$6(-76) \stackrel{?}{=} -456$
$-456 = -456$ True
The solution is -24.

Section 3.4

Practice Problems

1. $$\begin{aligned} 5y+2&=17\\ 5y+2-2&=17-2\\ 5y&=15\\ \frac{5y}{5}&=\frac{15}{5}\\ y&=3 \end{aligned}$$

2. $$\begin{aligned} 45&=-10-y\\ 45+10&=-10-y+10\\ 55&=-1y\\ \frac{55}{-1}&=\frac{-1y}{-1}\\ -55&=y \end{aligned}$$

3. $$\begin{aligned} 7x+12&=3x-4\\ 7x+12-12&=3x-4-12\\ 7x&=3x-16\\ 7x-3x&=3x-16-3x\\ 4x&=-16\\ \frac{4x}{4}&=\frac{-16}{4}\\ x&=-4 \end{aligned}$$

4. $$\begin{aligned} 6(a-5)&=4(a+1)\\ 6a-30&=4a+4\\ 6a-30-4a&=4a+4-4a\\ 2a-30&=4\\ 2a-30+30&=4+30\\ 2a&=34\\ \frac{2a}{2}&=\frac{34}{2}\\ a&=17 \end{aligned}$$

5. $$\begin{aligned} 4(x+3)&=12\\ 4x+12&=12\\ 4x+12-12&=12-12\\ 4x&=0\\ \frac{4x}{4}&=\frac{0}{4}\\ x&=0 \end{aligned}$$

6. **a.** $110-80=30$

   b. $3(-9+11)=6$

   c. $\frac{2(12)}{-6}=-4$

Calculator Explorations

1. $76(12-25) \stackrel{?}{=} -988$
   yes

2. $-47(35)+862 \stackrel{?}{=} -783$
   yes

3. $-170+562 \stackrel{?}{=} 3(-170)+900$
   $392 \stackrel{?}{=} 390$
   no

4. $55(-18+10) \stackrel{?}{=} 75(-18)+910$
   $-440 \stackrel{?}{=} -440$
   yes

5. $29(-21)-1034 \stackrel{?}{=} 61(-21)-362$
   $-1643 \stackrel{?}{=} -1643$
   yes

6. $-38(25)+205 \stackrel{?}{=} 25(25)+120$
   $-745 \stackrel{?}{=} 745$
   no

Exercise Set 3.4

1. $$\begin{aligned} 2x-6&=0\\ 2x-6+6&=0+6\\ 2x&=6\\ \frac{2x}{2}&=\frac{6}{2}\\ x&=3 \end{aligned}$$

3. $$\begin{aligned} 5n+10&=0\\ 5n+10-10&=0-10\\ 5n&=-10\\ \frac{5n}{5}&=\frac{-10}{5}\\ n&=-2 \end{aligned}$$

5. $$\begin{aligned} 6-n&=10\\ 6-n-6&=10-6\\ -n&=4\\ \frac{-n}{-1}&=\frac{4}{-1}\\ n&=-4 \end{aligned}$$

7.
$$\begin{aligned} 10x+15 &= 6x+3 \\ 10x+15-15 &= 6x+3-15 \\ 10x &= 6x-12 \\ 10x-6x &= 6x-12-6x \\ 4x &= -12 \\ \frac{4x}{4} &= \frac{-12}{4} \\ x &= -3 \end{aligned}$$

9.
$$\begin{aligned} 3x-7 &= 4x+5 \\ 3x-7+7 &= 4x+5+7 \\ 3x &= 4x+12 \\ 3x-4x &= 4x+12-4x \\ -x &= 12 \\ \frac{-x}{-1} &= \frac{12}{-1} \\ x &= -12 \end{aligned}$$

11.
$$\begin{aligned} 3(x-1) &= 12 \\ 3x-3 &= 12 \\ 3x-3+3 &= 12+3 \\ 3x &= 15 \\ \frac{3x}{3} &= \frac{15}{3} \\ x &= 5 \end{aligned}$$

13.
$$\begin{aligned} -2(y+4) &= 2 \\ -2y-8 &= 2 \\ -2y-8+8 &= 2+8 \\ -2y &= 10 \\ \frac{-2y}{-2} &= \frac{10}{-2} \\ y &= -5 \end{aligned}$$

15.
$$\begin{aligned} 35 &= 17+3(x-2) \\ 35 &= 17+3x-6 \\ 35 &= 11+3x \\ 35-11 &= 11+3x-11 \\ 24 &= 3x \\ \frac{24}{3} &= \frac{3x}{3} \\ 8 &= x \end{aligned}$$

17.
$$\begin{aligned} 8-t &= 3 \\ 8-t-8 &= 3-8 \\ -t &= -5 \\ \frac{-t}{-1} &= \frac{-5}{-1} \\ t &= 5 \end{aligned}$$

19.
$$\begin{aligned} 0 &= 4x-4 \\ 0+4 &= 4x-4+4 \\ 4 &= 4x \\ \frac{4}{4} &= \frac{4x}{4} \\ 1 &= x \end{aligned}$$

21.
$$\begin{aligned} 2n+8 &= 0 \\ 2n+8-8 &= 0-8 \\ 2n &= -8 \\ \frac{2n}{2} &= \frac{-8}{2} \\ n &= -4 \end{aligned}$$

23.
$$\begin{aligned} 7 &= 4c-1 \\ 7+1 &= 4c-1+1 \\ 8 &= 4c \\ \frac{8}{4} &= \frac{4c}{4} \\ 2 &= c \end{aligned}$$

25.
$$\begin{aligned} 3r+4 &= 19 \\ 3r+4-4 &= 19-4 \\ 3r &= 15 \\ \frac{3r}{3} &= \frac{15}{3} \\ r &= 5 \end{aligned}$$

27.
$$\begin{aligned} 2x-1 &= -7 \\ 2x-1+1 &= -7+1 \\ 2x &= -6 \\ \frac{2x}{2} &= \frac{-6}{2} \\ x &= -3 \end{aligned}$$

29.
$$\begin{aligned} 2 &= 3z-4 \\ 2+4 &= 3z-4+4 \\ 6 &= 3z \\ \frac{6}{3} &= \frac{3z}{3} \\ 2 &= z \end{aligned}$$

31.
$$\begin{aligned} 5x-2&=-12\\ 5x-2+2&=-12+2\\ 5x&=-10\\ \frac{5x}{5}&=\frac{-10}{5}\\ x&=-2 \end{aligned}$$

33.
$$\begin{aligned} -7c+1&=-20\\ -7c+1-1&=-20-1\\ -7c&=-21\\ \frac{-7c}{-7}&=\frac{-21}{-7}\\ c&=3 \end{aligned}$$

35.
$$\begin{aligned} 9&=4x-15\\ 9+15&=4x-15+15\\ 24&=4x\\ \frac{24}{4}&=\frac{4x}{4}\\ 6&=x \end{aligned}$$

37.
$$\begin{aligned} 8m+79&=-1\\ 8m+79-79&=-1-79\\ 8m&=-80\\ \frac{8m}{8}&=\frac{-80}{8}\\ m&=-10 \end{aligned}$$

39.
$$\begin{aligned} 10+4v&=-6\\ 10+4v-10&=-6-10\\ 4v&=-16\\ \frac{4v}{4}&=\frac{-16}{4}\\ v&=-4 \end{aligned}$$

41.
$$\begin{aligned} -5&=-13-8k\\ -5+13&=-13-8k+13\\ 8&=-8k\\ \frac{8}{-8}&=\frac{-8k}{-8}\\ -1&=k \end{aligned}$$

43.
$$\begin{aligned} 4x+3&=2x+11\\ 4x+3-3&=2x+11-3\\ 4x&=2x+8\\ 4x-2x&=2x+8-2x\\ 2x&=8\\ \frac{2x}{2}&=\frac{8}{2}\\ x&=4 \end{aligned}$$

45.
$$\begin{aligned} -2y-10&=5y+18\\ -2y-10+10&=5y+18+10\\ -2y&=5y+28\\ -2y-5y&=5y+28-5y\\ -7y&=28\\ \frac{-7y}{-7}&=\frac{28}{-7}\\ y&=-4 \end{aligned}$$

47.
$$\begin{aligned} -8n+1&=-6n-5\\ -8n+1-1&=-6n-5-1\\ -8n&=-6n-6\\ -8n+6n&=-6n-6+6n\\ -2n&=-6\\ \frac{-2n}{-2}&=\frac{-6}{-2}\\ n&=3 \end{aligned}$$

49.
$$\begin{aligned} 9-3x&=14+2x\\ 9-3x-9&=14+2x-9\\ -3x&=5+2x\\ -3x-2x&=5+2x-2x\\ -5x&=5\\ \frac{-5x}{-5}&=\frac{5}{-5}\\ x&=-1 \end{aligned}$$

51.
$$\begin{aligned} 2(y-3)&=y-6\\ 2y-6&=y-6\\ 2y-6+6&=y-6+6\\ 2y&=y\\ 2y-y&=y-y\\ y&=0 \end{aligned}$$

53.
$$\begin{aligned} 2t-1 &= 3(t+7) \\ 2t-1 &= 3t+21 \\ 2t-1+1 &= 3t+21+1 \\ 2t &= 3t+22 \\ 2t-3t &= 3t+22-3t \\ -1t &= 22 \\ \frac{-t}{-1} &= \frac{22}{-1} \\ t &= -22 \end{aligned}$$

55.
$$\begin{aligned} 3(5c-1)-2 &= 13c+3 \\ 15c-3-2 &= 13c+3 \\ 15c-5 &= 13c+3 \\ 15c-5+5 &= 13c+3+5 \\ 15c &= 13c+8 \\ 15c-13c &= 13c+8-13c \\ 2c &= 8 \\ \frac{2c}{2} &= \frac{8}{2} \\ c &= 4 \end{aligned}$$

57.
$$\begin{aligned} 10+5(z-2) &= 4z+1 \\ 10+5z-10 &= 4z+1 \\ 5z &= 4z+1 \\ 5z-4z &= 4z+1-4z \\ z &= 1 \end{aligned}$$

59.
$$\begin{aligned} 7(6+w) &= 6(2+w) \\ 42+7w &= 12+6w \\ 42+7w-6w &= 12+6w-6w \\ 42+w &= 12 \\ 42+w-42 &= 12-42 \\ w &= -30 \end{aligned}$$

61. The sum of –42 and 16 is –26 translates to $-42 + 16 = -26$.

63. The product of –5 and –29 gives 145 translates to $-5(-29) = 145$.

65. Three times the difference of –14 and 2 amounts to –48 translates to $3(-14-2) = -48$.

67. The quotient of 100 and twice 50 is equal to 1 translates to $\frac{100}{2(50)} = 1$.

69. $x^3 - 2xy$ substitute $x = 3$, $y = -1$
$$\begin{aligned} &= 3^3 - 2(3)(-1) \\ &= 27-(-6) \\ &= 27+6 \\ &= 33 \end{aligned}$$

71. $y^5 - 4x^2$ substitute $x = 3$, $y = -1$
$$\begin{aligned} &= (-1)^5 - 4(3)^2 \\ &= (-1)(-1)(-1)(-1)(-1) - 4(3)(3) \\ &= -1-36 \\ &= -37 \end{aligned}$$

73. $(2x-y)^2$ substitute $x = 3$, $y = -1$
$$\begin{aligned} &= [2(3)-(-1)]^2 \\ &= (6+1)^2 \\ &= (7)^2 \\ &= 49 \end{aligned}$$

75. $4xy + 6y^2$ substitute $x = 3$, $y = -1$
$$\begin{aligned} &= 4(3)(-1) + 6(-1)^2 \\ &= -12 + 6 \cdot 1 \\ &= -12+6 \\ &= -6 \end{aligned}$$

77.
$$\begin{aligned} (-8)^2 + 3x &= 5x + 4^3 \\ 64+3x &= 5x+64 \\ 64+3x-5x &= 5x+64-5x \\ 64-2x &= 64 \\ 64-2x-64 &= 64-64 \\ -2x &= 0 \\ \frac{-2x}{-2} &= \frac{0}{-2} \\ x &= 0 \end{aligned}$$

79.
$$\begin{aligned} x+45^2 &= 54^2 \\ x+2025 &= 2916 \\ x+2025-2025 &= 2916-2025 \\ x &= 891 \end{aligned}$$

81. Answers may vary.

Section 3.5

Practice Problems

1. a. Five times a number is $5x$. Five times a number is 20 is $5x = 20$.

   b. The sum of a number and –5 is $x + (-5)$. The sum of a number and –5 yields 14 is $x + (-5) = 14$.

   c. Ten subtracted from a number is $x - 10$. Ten subtracted from a number amounts to –23 is $x - 10 = -23$.

   d. Five times a number is $5x$. Five times a number added to 7 is equal to –8 is $7 + 5x = -8$.

   e. The sum of a number and 4 is $x + 4$. The quotient of 6 and the sum of a number and 4 gives 1 is $\frac{6}{x+4} = 1$.

2. Let x = the unknown number.

$$\begin{aligned} x+2 &= 3x+6 \\ x+2-2 &= 3x+6-2 \\ x &= 3x+4 \\ x-3x &= 3x+4-3x \\ -2x &= 4 \\ \frac{-2x}{-2} &= \frac{4}{-2} \\ x &= -2 \end{aligned}$$

The unknown number is –2.

3. Let x = the number of delegates the U.S. sent. Then, $x - 19$ = the number of delegates Japan sent.

$$\begin{aligned} x+x-19 &= 121 \\ 2x-19 &= 121 \\ 2x-19+19 &= 121+19 \\ 2x &= 140 \\ \frac{2x}{2} &= \frac{140}{2} \\ x &= 70 \end{aligned}$$

The United States sent 70 delegates.

4. Let x = the amount son receives. Then $2x$ = the amount husband receives.

$$\begin{aligned} x+2x &= 21{,}000 \\ 3x &= 21{,}000 \\ \frac{3x}{3} &= \frac{21{,}000}{3} \\ x &= 7000 \\ 2x &= 2 \cdot 7000 \\ 2x &= 14{,}000 \end{aligned}$$

Her son receives $7000 and her husband receives $14,000.

Exercise Set 3.5

1. A number added to –5 is –7 translates to $-5 + x = -7$.

3. Three times a number yields 27 translates to $3x = 27$.

5. A number subtracted from –20 amounts to 104 translates to $-20 - x = 104$.

7. Twice the sum of a number and –1 is equal to 50 translates to $2[x + (-1)] = 50$.

9. $$\begin{aligned} 3x+9 &= 33 \\ 3x &= 24 \\ x &= 8 \end{aligned}$$

11. $$\begin{aligned} 9x &= 54 \\ x &= 6 \end{aligned}$$

13. $$\begin{aligned} 3+4+x &= 16 \\ 7+x &= 16 \\ x &= 9 \end{aligned}$$

15. $$\begin{aligned} 72 &= 24+8x \\ 48 &= 8x \\ 6 &= x \end{aligned}$$

17. $$\begin{aligned} x-3 &= 45-x \\ 2x &= 48 \\ x &= 24 \end{aligned}$$

19. $$\begin{aligned} 3(x-5) &= 9 \\ 3x-15 &= 9 \\ 3x &= 24 \\ x &= 8 \end{aligned}$$

21. $8 - x = \frac{15}{5}$
$8 - x = 3$
$-x = -5$
$x = 5$

23. $3x + 13 = 3 + 5x$
$10 = 2x$
$5 = x$

25. $5x - 40 = x + 8$
$4x = 48$
$x = 12$

27. $x - 3 = \frac{10}{5}$
$x - 3 = 2$
$x = 5$

29. Let x = the number of electoral votes for Bob Dole. Then
$x + 220$ = the number of electoral votes for Bill Clinton.
$x + x + 220 = 538$
$2x + 220 = 538$
$2x = 318$
$x = 159$
Dole received 159 votes and Clinton received $159 + 220 = 379$ votes.

31. Let x = the number of coins that Stuart has. Then $2x$ = the number of coins that Mark has.
$x + 2x = 120$
$3x = 120$
$x = 40$
Mark has $2 \cdot 40 = 80$ coins.

33. Let x = the speed of the Dodge truck. Then $2x$ = the speed of the Toyota Camry.
$x + 2x = 105$
$3x = 105$
$x = 35$
The truck's speed is 35 mph and the car's speed is $2 \cdot 35 = 70$ mph.

35. Let x = the amount received for the accessories. Then
$5x$ = the amount received for the bike.
$x + 5x = 270$
$6x = 270$
$x = 45$
He received $5 \cdot 45 = \$225$ for the bike.

37. Let x = the capacity of Neyland Stadium. Then $x + 4647$ = the capacity of Michigan Stadium.
$x + x + 4647 = 210,355$
$2x + 4647 = 210,355$
$2x = 205,708$
$x = 102,854$
Neyland Stadium has a capacity of 102,854 and Michigan Stadium has a capacity of $102,854 + 4647 = 107,501$.

39. Let x = the native American population of Washington state. Then
$3x$ = the native American population of California.
$x + 3x = 412$
$4x = 412$
$x = 103$
The native American population of Washington state is 103 thousand. The native American population of California is $3 \cdot 103 = 309$ thousand.

41. Let x = the points scored by the Tennessee Lady Volunteers. Then
$x + 19$ = the points scored by the Connecticut Huskies.
$x + x + 19 = 123$
$2x + 19 = 123$
$2x = 104$
$x = 52$
The Connecticut Huskies scored $52 + 19 = 71$ points.

43. To round 586 to the nearest ten, observe that the digit in the ones place is 6. Since this digit is at least 5, we need to add 1 to the digit in the tens place. The number 586 rounded to the nearest ten is 590.

45. To round 1026 to the nearest hundred, observe that the digit in the tens place is 2. Since this digit is less than 5, we do not add 1 to the digit in the hundreds place. The number 1026 rounded to the nearest hundred is 1000.

47. To round 2986 to the nearest thousand observe that the digit in the hundreds place is 9. Since this digit is at least 5, we add 1 to the digit in the thousands place. The number 2986 rounded to the nearest thousand is 3000.

49. Answers may vary.

51.
$$P = A + C$$
$$165,000 = 156,750 + C$$
$$8250 = C$$
The agent's commission will be $8250.

53.
$$P = C + M$$
$$29 = 12 + M$$
$$29 - 12 = 12 + M - 12$$
$$17 = M$$
The markup on the blouse is $17.

55. Answers may vary.

Chapter 3 Review

1. $3y + 7y - 15 = 10y - 15$

2. $2y - 10 - 8y = -6y - 10$

3. $8a + a - 7 - 15a = -6a - 7$

4. $y + 3 - 9y - 1 = -8y + 2$

5. $2(x + 5) = 2x + 10$

6. $-3(y + 8) = -3y - 24$

7. $7x + 3(x - 4) + x = 7x + 3x - 12 + x$
$= 11x - 12$

8. $-(3m + 2) - m - 10$
$= -1 \cdot 3m + (-1)(2) - m - 10$
$= -3m - 2 - m - 10$
$= -4m - 12$

9. $3(5a - 2) - 20a + 10 = 15a - 6 - 20a + 10$
$= -5a + 4$

10. $6y + 3 + 2(3y - 6) = 6y + 3 + 6y - 12$
$= 12y - 9$

11. $P = 3 + 2x + 3 + 2x$
$= 4x + 6$
The perimeter is $(4x + 6)$ yards.

12. $P = 5y + 5y + 5y + 5y = 20y$
The perimeter is $20y$ meters.

13. $A = 3(2x - 1) = 6x - 3$
The area is $(6x - 3)$ square yards.

14. $A = 10(x - 2) + 7(5x + 4)$
$= 10x - 20 + 35x + 28$
$= 45x + 8$
The area is $(45x + 8)$ square centimeters.

15. $85(7068x - 108) + 42x$
$= 600,780x - 9180 + 42x$
$= 600,822x - 9180$

16. $-4268y + 120(63y - 32)$
$= -4268y + 7560y - 3840$
$= 3292y - 3840$

17. $5(2 - 4) \stackrel{?}{=} -10$
$5(-2) \stackrel{?}{=} -10$
$-10 = -10$ True
Yes, it is a solution.

18. $6(0) + 2 \stackrel{?}{=} 23 + 4(0)$
$0 + 2 \stackrel{?}{=} 23 + 0$
$2 = 23$ False
No, it is not a solution.

19.
$$z - 5 = -7$$
$$z - 5 + 5 = -7 + 5$$
$$z = -2$$

20.
$$3x + 10 = 4x$$
$$3x + 10 - 3x = 4x - 3x$$
$$10 = x$$

21. $$\begin{aligned} n+18 &= 10-(-2) \\ n+18 &= 10+2 \\ n+18 &= 12 \\ n+18-18 &= 12-18 \\ n &= -6 \end{aligned}$$

22. $$\begin{aligned} c-5 &= -13+7 \\ c-5 &= -6 \\ c-5+5 &= -6+5 \\ c &= -1 \end{aligned}$$

23. $$\begin{aligned} 7x+5-6x &= -20 \\ x+5 &= -20 \\ x+5-5 &= -20-5 \\ x &= -25 \end{aligned}$$

24. $$\begin{aligned} 17x &= 2(8x-4) \\ 17x &= 2\cdot 8x-2\cdot 4 \\ 17x &= 16x-8 \\ 17x-16x &= 16x-8-16x \\ x &= -8 \end{aligned}$$

25. $$\begin{aligned} -3y &= -21 \\ \frac{-3y}{-3} &= \frac{-21}{-3} \\ y &= 7 \end{aligned}$$

26. $$\begin{aligned} -8x &= 72 \\ \frac{-8x}{-8} &= \frac{72}{-8} \\ x &= -9 \end{aligned}$$

27. $$\begin{aligned} -5n &= -5 \\ \frac{-5n}{-5} &= \frac{-5}{-5} \\ n &= 1 \end{aligned}$$

28. $$\begin{aligned} -3a &= 15 \\ \frac{-3a}{-3} &= \frac{15}{-3} \\ a &= -5 \end{aligned}$$

29. $$\begin{aligned} 0 &= 12x \\ \frac{0}{12} &= \frac{12x}{12} \\ 0 &= x \end{aligned}$$

30. $$\begin{aligned} -14 &= 7x \\ \frac{-14}{7} &= \frac{7x}{7} \\ -2 &= x \end{aligned}$$

31. $$\begin{aligned} -5t+32+4t &= 32 \\ -t+32 &= 32 \\ -t+32-32 &= 32-32 \\ -t &= 0 \\ t &= 0 \end{aligned}$$

32. $$\begin{aligned} 3z+72-2z &= -56 \\ z+72 &= -56 \\ z+72-72 &= -56-72 \\ z &= -128 \end{aligned}$$

33. $2x+11$

34. $-5x-50$

35. $\frac{70}{x+6}$

36. $2(x-13)$

37. $$\begin{aligned} 3x-4 &= 11 \\ 3x-4+4 &= 11+4 \\ 3x &= 15 \\ \frac{3x}{3} &= \frac{15}{3} \\ x &= 5 \end{aligned}$$

38. $$\begin{aligned} 6y+1 &= 73 \\ 6y+1-1 &= 73-1 \\ 6y &= 72 \\ \frac{6y}{6} &= \frac{72}{6} \\ y &= 12 \end{aligned}$$

39. $$\begin{aligned} 14-y &= -3 \\ 14-y-14 &= -3-14 \\ -y &= -17 \\ \frac{-y}{-1} &= \frac{-17}{-1} \\ y &= 17 \end{aligned}$$

40. $$\begin{aligned} 7-z &= 0 \\ 7-z+z &= 0+z \\ 7 &= z \end{aligned}$$

41. $4z - z = -6$
$3z = -6$
$\frac{3z}{3} = \frac{-6}{3}$
$z = -2$

42. $t - 9t = -64$
$-8t = -64$
$\frac{-8t}{-8} = \frac{-64}{-8}$
$t = 8$

43. $5(n - 3) = 7 + 3n$
$5n - 15 = 7 + 3n$
$5n - 15 + 15 = 7 + 3n + 15$
$5n = 22 + 3n$
$5n - 3n = 22 + 3n - 3n$
$2n = 22$
$\frac{2n}{2} = \frac{22}{2}$
$n = 11$

44. $7(2 + x) = 4x - 1$
$14 + 7x = 4x - 1$
$14 + 7x - 14 = 4x - 1 - 14$
$7x = 4x - 15$
$7x - 4x = 4x - 15 - 4x$
$3x = -15$
$\frac{3x}{3} = \frac{-15}{3}$
$x = -5$

45. $2x + 7 = 6x - 1$
$2x + 7 - 7 = 6x - 1 - 7$
$2x = 6x - 8$
$2x - 6x = 6x - 8 - 6x$
$-4x = -8$
$\frac{-4x}{-4} = \frac{-8}{-4}$
$x = 2$

46. $5x - 18 = -4x$
$5x - 18 + 18 = -4x + 18$
$5x = -4x + 18$
$5x + 4x = -4x + 18 + 4x$
$9x = 18$
$\frac{9x}{9} = \frac{18}{9}$
$x = 2$

47. $20 - (-8) = 28$

48. $5(2 + (-6)) = -20$

49. $\frac{-75}{5 + 20} = -3$

50. $-2 - 19 = -21$

51. $2x - 8 = 40$

52. $\frac{x}{2} - 12 = 10$

53. $x - 3 = x \div 4$ or $x - 3 = \frac{x}{4}$

54. $6x = x + 2$

55. $40 - 5x = 3x$
$40 - 5x + 5x = 3x + 5x$
$40 = 8x$
$\frac{40}{8} = \frac{8x}{8}$
$5 = x$

56. $3x = 2(x - 8)$
$3x = 2x - 16$
$3x - 2x = 2x - 16 - 2x$
$x = -16$

57. Let I = votes recieved by the Independent candidate.
$I + 272$ = votes received by the Democratic candidate.

$$18{,}500 - 14{,}000 = 4500 \text{ votes}$$
$$I + (I + 272) = 4500$$
$$2I + 272 = 4500$$
$$2I + 272 - 272 = 4500 - 272$$
$$2I = 4228$$
$$\frac{2I}{2} = \frac{4228}{2}$$
$$I = 2114$$
$$I + 272 = 2114 + 272 = 2386$$

The Democratic candidate received 2386 votes.

58. Let x = number of cassette tapes
Let $2x$ = number of CDs

$$x + 2x = 126$$
$$3x = 126$$
$$\frac{3x}{3} = \frac{126}{3}$$
$$x = 42$$
$$2x = 2 \cdot 42 = 84$$

Rajiv has 84 CDs.

Chapter 3 Test

1. $7x - 5 - 12x + 10 = -5x + 5$

2. $-2(3y + 7) = -6y - 14$

3. $-(3z + 2) - 5z - 18 = -1 \cdot 3z - 1 \cdot 2 - 5z - 18$
$= -3z - 2 - 5z - 18$
$= -8z - 20$

4. $P = a + b + c$
$= 5x + 5 + 5x + 5 + 5x + 5$
$= 15x + 15$
The perimeter is $(15x + 15)$ inches.

5. $A = 3(3x - 1)$
$= 9x - 3$
The area is $(9x - 3)$ square meters.

6.
$$9x + x = -60$$
$$10x = -60$$
$$\frac{10x}{10} = \frac{-60}{10}$$
$$x = -6$$

7.
$$12 = y - 3y$$
$$12 = -2y$$
$$\frac{12}{-2} = \frac{-2y}{-2}$$
$$-6 = y$$

8.
$$2n - 3 = 17$$
$$2n - 3 + 3 = 17 + 3$$
$$2n = 20$$
$$\frac{2n}{2} = \frac{20}{2}$$
$$n = 10$$

9.
$$5 + 4z = 37$$
$$5 + 4z - 5 = 37 - 5$$
$$4z = 32$$
$$\frac{4z}{4} = \frac{32}{4}$$
$$z = 8$$

10.
$$5x + 12 - 4x - 14 = 22$$
$$x - 2 = 22$$
$$x - 2 + 2 = 22 + 2$$
$$x = 24$$

11.
$$2 - c + 4c = 5$$
$$2 + 3c = 5$$
$$2 + 3c - 2 = 5 - 2$$
$$3c = 3$$
$$\frac{3c}{3} = \frac{3}{3}$$
$$c = 1$$

12.
$$3x - 5 = -11$$
$$3x - 5 + 5 = -11 + 5$$
$$3x = -6$$
$$\frac{3x}{3} = \frac{-6}{3}$$
$$x = -2$$

13.
$$-4x + 7 = 15$$
$$-4x + 7 - 7 = 15 - 7$$
$$-4x = 8$$
$$\frac{-4x}{-4} = \frac{8}{-4}$$
$$x = -2$$

14.
$$\begin{aligned} 2(x-6) &= 0 \\ 2x-12 &= 0 \\ 2x-12+12 &= 0+12 \\ 2x &= 12 \\ \frac{2x}{2} &= \frac{12}{2} \\ x &= 6 \end{aligned}$$

15.
$$\begin{aligned} 3(4+2y) &= 12 \\ 12+6y &= 12 \\ 12+6y-12 &= 12-12 \\ 6y &= 0 \\ \frac{6y}{6} &= \frac{0}{6} \\ y &= 0 \end{aligned}$$

16.
$$\begin{aligned} 5x-2 &= x-10 \\ 5x-2+2 &= x-10+2 \\ 5x &= x-8 \\ 5x-x &= x-8-x \\ 4x &= -8 \\ \frac{4x}{4} &= \frac{-8}{4} \\ x &= -2 \end{aligned}$$

17.
$$\begin{aligned} 10y-1 &= 7y+20 \\ 10y-1+1 &= 7y+20+1 \\ 10y &= 7y+21 \\ 10y-7y &= 7y+21-7y \\ 3y &= 21 \\ \frac{3y}{3} &= \frac{21}{3} \\ y &= 7 \end{aligned}$$

18.
$$\begin{aligned} 6+2(3n-1) &= 28 \\ 6+6n-2 &= 28 \\ 4+6n &= 28 \\ 4+6n-4 &= 28-4 \\ 6n &= 24 \\ \frac{6n}{6} &= \frac{24}{6} \\ n &= 4 \end{aligned}$$

19.
$$\begin{aligned} 4(5x+3) &= 2(7x+6) \\ 20x+12 &= 14x+12 \\ 20x+12-12 &= 14x+12-12 \\ 20x &= 14x \\ 20x-14x &= 14x-14x \\ 6x &= 0 \\ \frac{6x}{6} &= \frac{0}{6} \\ x &= 0 \end{aligned}$$

20.
$$\begin{aligned} A &= \frac{1}{2}\cdot h\cdot(B+b) \\ &= \frac{1}{2}\cdot 60(130+70) \\ &= \frac{1}{2}\cdot 60(200) \\ &= 30(200) \\ &= 6000 \end{aligned}$$

The area is 6000 square feet.

21. **a.** $17x$

b. $20-2x$

22.
$$\begin{aligned} 3x-5x &= 4 \\ -2x &= 4 \\ \frac{-2x}{-2} &= \frac{4}{-2} \\ x &= -2 \end{aligned}$$

23. Let x = the number of free throws Maria made.
Then $2x$ = the number of free throws Paula made.
$$\begin{aligned} 2x+x &= 12 \\ 3x &= 12 \\ \frac{3x}{3} &= \frac{12}{3} \\ x &= 4 \end{aligned}$$

Maria made 4 free throws, so Paula made 8 free throws.

24. Let x = the number of women entered in the race.
Then $x + 112$ = the number of men entered in the space.

$$\begin{aligned} x+(x+112) &= 600 \\ 2x+112 &= 600 \\ 2x+112-112 &= 600-112 \\ 2x &= 488 \\ x &= 244 \end{aligned}$$

244 women entered the race.

Chapter 4

Pretest

1. The figure has 8 equal parts. Three parts are shaded. The fraction that is shaded is $\frac{3}{8}$.

2. $\frac{5}{6}=\frac{5}{6}\cdot 1=\frac{5}{6}\cdot\frac{3}{3}=\frac{5\cdot 3}{6\cdot 3}=\frac{15}{18}$

3. $\frac{0}{-4}=0\div -4=0$

4. $$\begin{aligned}140&=2\cdot 70\\&=2\cdot 2\cdot 35\\&=2\cdot 2\cdot 5\cdot 7 \text{ or } 2^2\cdot 5\cdot 7\end{aligned}$$

5. $\frac{30}{54}=\frac{6\cdot 5}{6\cdot 9}=\frac{5}{9}$

6. $$\frac{\text{Number of milligrams}}{\text{Number of milligrams in a gram}}\begin{matrix}\rightarrow\\\rightarrow\end{matrix}\frac{450}{1000}$$
$$=\frac{5\cdot 9\cdot 10}{5\cdot 2\cdot 10\cdot 10}=\frac{9}{20}$$

7. $\frac{3}{4}\cdot\frac{24}{15}=\frac{3\cdot 24}{4\cdot 15}=\frac{3\cdot 6\cdot 4}{4\cdot 3\cdot 5}=\frac{6}{5}$

8. $$\begin{aligned}\frac{5x}{7}\div 10x^2&=\frac{5x}{7}\cdot\frac{1}{10x^2}\\&=\frac{5x\cdot 1}{7\cdot 10x^2}\\&=\frac{5x}{7\cdot 5x\cdot 2x}\\&=\frac{1}{14x}\end{aligned}$$

9. $\frac{8}{11}-\frac{3}{11}=\frac{8-3}{11}=\frac{5}{11}$

10. $-\frac{3}{10}+\frac{2}{10}=\frac{-3+2}{10}=\frac{-1}{10}=-\frac{1}{10}$

11. $$\begin{aligned}\left(-\frac{2}{3}\right)^3&=\left(-\frac{2}{3}\right)\left(-\frac{2}{3}\right)\left(-\frac{2}{3}\right)\\&=-\frac{2\cdot 2\cdot 2}{3\cdot 3\cdot 3}\\&=-\frac{8}{27}\end{aligned}$$

12. $$\begin{aligned}x\div y&=\frac{2}{9}\div -\frac{2}{3}\\&=\frac{2}{9}\cdot -\frac{3}{2}\\&=-\frac{2\cdot 3}{9\cdot 2}\\&=-\frac{2\cdot 3}{3\cdot 3\cdot 2}\\&=-\frac{1}{3}\end{aligned}$$

13. $$\begin{aligned}\frac{5}{9}+\frac{1}{12}&=\frac{5}{9}\cdot 1+\frac{1}{12}\cdot 1\\&=\frac{5}{9}\cdot\frac{4}{4}+\frac{1}{12}\cdot\frac{3}{3}\\&=\frac{5\cdot 4}{9\cdot 4}+\frac{1\cdot 3}{12\cdot 3}\\&=\frac{20}{36}+\frac{3}{36}\\&=\frac{20+3}{36}\\&=\frac{23}{36}\end{aligned}$$

14.
$$\begin{aligned} x-y &= -\frac{3}{14}-\left(-\frac{2}{7}\right) \\ &= -\frac{3}{14}+\frac{2}{7} \\ &= -\frac{3}{14}+\frac{2}{7}\cdot 1 \\ &= -\frac{3}{14}+\frac{2}{7}\cdot\frac{2}{2} \\ &= -\frac{3}{14}+\frac{2\cdot 2}{7\cdot 2} \\ &= -\frac{3}{14}+\frac{4}{14} \\ &= \frac{-3+4}{14} \\ &= \frac{1}{14} \end{aligned}$$

15.
$$\begin{aligned} \frac{\frac{x}{3}}{\frac{7}{9}} &= \frac{x}{3}\div\frac{7}{9} \\ &= \frac{x}{3}\cdot\frac{9}{7} \\ &= \frac{x\cdot 9}{3\cdot 7} \\ &= \frac{x\cdot 3\cdot 3}{3\cdot 7} \\ &= \frac{3x}{7} \end{aligned}$$

16.
$$\begin{aligned} \left(\frac{2}{5}\right)^2-2 &= \left(\frac{2}{5}\right)\left(\frac{2}{5}\right)-2 \\ &= \frac{2\cdot 2}{5\cdot 5}-2\cdot 1 \\ &= \frac{4}{25}-2\cdot\frac{25}{25} \\ &= \frac{4}{25}-\frac{50}{25} \\ &= \frac{4-50}{25} \\ &= \frac{-46}{25} \\ &= -\frac{46}{25} \end{aligned}$$

17.
$$\begin{aligned} \frac{x}{4}+3 &= \frac{1}{8} \\ 8\left(\frac{x}{4}+3\right) &= 8\cdot\frac{1}{8} \\ 8\cdot\frac{x}{4}+8\cdot 3 &= 1 \\ 2x+24 &= 1 \\ 2x+24-24 &= 1-24 \\ 2x &= -23 \\ \frac{2x}{2} &= \frac{-23}{2} \\ x &= -\frac{23}{2} \end{aligned}$$

18. $2\frac{3}{5}=\frac{5\cdot 2+3}{5}=\frac{10+3}{5}=\frac{13}{5}$

19.
$$\begin{aligned} 3\frac{1}{5}\cdot 2\frac{3}{4} &= \frac{5\cdot 3+1}{5}\cdot\frac{4\cdot 2+3}{4} \\ &= \frac{15+1}{5}\cdot\frac{8+3}{4} \\ &= \frac{16}{5}\cdot\frac{11}{4} \\ &= \frac{16\cdot 11}{5\cdot 4} \\ &= \frac{4\cdot 4\cdot 11}{5\cdot 4} \\ &= \frac{44}{5} \end{aligned}$$

20.
$$\begin{array}{rcrcr} 5\frac{2}{3} & = & 5\frac{2\cdot 2}{3\cdot 2} & = & 5\frac{4}{6} \\ +\ 4\frac{1}{6} & = & +\ 4\frac{1}{6} & = & +\ 4\frac{1}{6} \\ \hline & & & & 9\frac{5}{6} \end{array}$$

Section 4.1

Practice Problems

1. $\frac{9}{2}$ ← numerator (9), ← denominator (2)

2. $\frac{10y}{17}$ ← numerator ($10y$), ← denominator (17)

3. 3 out of 8 equal parts are shaded: $\frac{3}{8}$.

4. 1 out of 6 equal parts is shaded: $\frac{1}{6}$.

5. Each part is $\frac{1}{3}$, and there are 8 parts shaded, or $\frac{8}{3}$.

6. Each part is $\frac{1}{4}$, and there are 5 parts shaded, or $\frac{5}{4}$.

7. 7 out of 9 planets: $\frac{7}{9}$.

8. a.

b.

c.

9. a.

b.

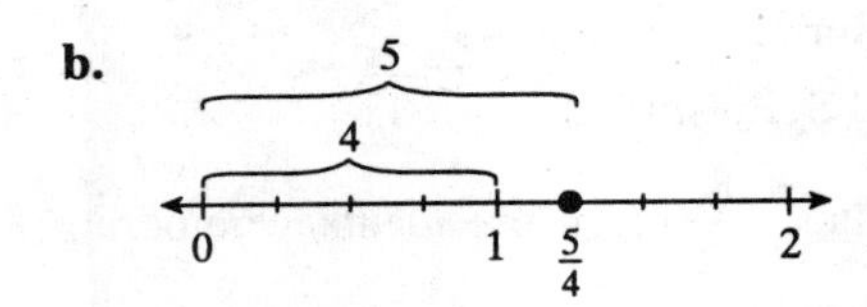

c.

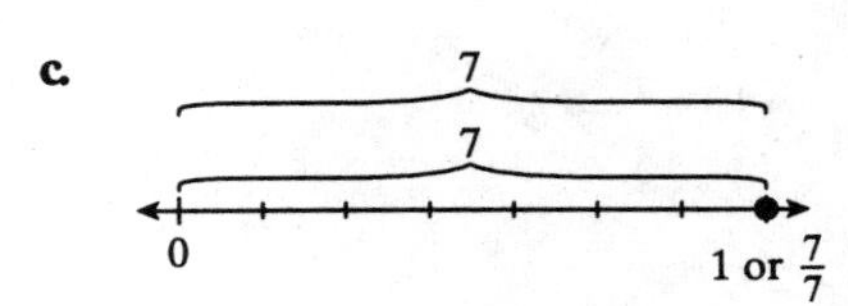

10. Since $4 \cdot 5 = 20$, we multiply the numerator and denominator of $\frac{1}{4}$ by 5.

$$\frac{1}{4} = \frac{1 \cdot 5}{4 \cdot 5} = \frac{5}{20}$$

Then $\frac{1}{4}$ is equivalent to $\frac{5}{20}$.

11. Since $7 \cdot 6 = 42$, we multiply the numerator and denominator of $\frac{3x}{7}$ by 6.

$$\frac{3x}{7} = \frac{3x \cdot 6}{7 \cdot 6} = \frac{18x}{42}$$

Then $\frac{3x}{7}$ is equivalent to $\frac{18x}{42}$.

12. Recall that $4 = \frac{4}{1}$ and $1 \cdot 6 = 6$.

$$\frac{4}{1} = \frac{4 \cdot 6}{1 \cdot 6} = \frac{24}{6}$$

13. $4x \cdot 5 = 20x$. Multiply the numerator and denominator by 5.

$$\frac{9}{4x} = \frac{9 \cdot 5}{4x \cdot 5} = \frac{45}{20x}$$

14. $\frac{9}{9} = 9 \div 9 = 1$

15. $\frac{-6}{-6} = -6 \div -6 = 1$

16. $\frac{-1}{-1} = -1 \div -1 = 1$

17. $\frac{4}{1} = 4 \div 1 = 4$

18. $\frac{-30}{1} = -30 \div 1 = -30$

19. $\frac{13}{1} = 13 \div 1 = 13$

20. $\frac{0}{14} = 0$

21. $\frac{0}{-9} = 0$

22. $\frac{6}{0}$ is undefined.

Mental Math

1. $\frac{1}{2}$ $\frac{\leftarrow \text{ numerator}}{\leftarrow \text{ denominator}}$

2. $\frac{1}{4}$ $\frac{\leftarrow \text{ numerator}}{\leftarrow \text{ denomiantor}}$

3. $\frac{10}{3}$ $\frac{\leftarrow \text{ numerator}}{\leftarrow \text{ denominator}}$

4. $\frac{53}{21}$ $\frac{\leftarrow \text{ numerator}}{\leftarrow \text{ denominator}}$

5. $\frac{3z}{7}$ $\frac{\leftarrow \text{ numerator}}{\leftarrow \text{ denominator}}$

6. $\frac{11x}{15}$ $\frac{\leftarrow \text{ numerator}}{\leftarrow \text{ denominator}}$

Exercise Set 4.1

1. 1 out of 3 equal parts is shaded: $\frac{1}{3}$.

3. 4 out of 7 equal parts are shaded: $\frac{4}{7}$

5. 7 out of 12 equal parts are shaded: $\frac{7}{12}$.

7. 3 out of 7 equal parts are shaded: $\frac{3}{7}$.

9. 4 out of 9 equal parts are shaded: $\frac{4}{9}$.

11. 5 out of 8 pieces are gone: $\frac{5}{8}$.

13. $\frac{11}{4}$

15. $\frac{11}{3}$

17. $\frac{3}{2}$

19. $\frac{4}{3}$

21. $\frac{17}{6}$

23. $\frac{14}{9}$

25. $\frac{\text{Freshmen} \rightarrow 42}{\text{Students} \rightarrow 131}$

Thus, $\frac{42}{131}$ of the students are freshmen.

27. $131 - 42 = 89$ non-freshmen

$\frac{\text{Non-freshmen} \rightarrow 89}{\text{Students} \rightarrow 131}$

Thus, $\frac{89}{131}$ of the students are not freshmen.

29. $\frac{\text{Injury-related visits} \rightarrow 4}{\text{Total vists} \rightarrow 10}$

Thus, $\frac{4}{10}$ of the visits are injury-related.

31. $\frac{\text{Born in Virginia} \rightarrow 8}{\text{U.S. Presidents} \rightarrow 42}$

Thus, $\frac{8}{42}$ of the presidents were born in Virginia.

33. Number of inches $\rightarrow$ 5
Inches in a foot $\rightarrow$ 12

Thus, 5 inches represents $\frac{5}{12}$ of a foot.

35. Number of hours $\rightarrow$ 11
Hours in a day $\rightarrow$ 24

Thus, 11 hours represents $\frac{11}{24}$ of a day.

37. Number of girls $\rightarrow$ 7
Number of children $\rightarrow$ 11

Thus, the girls are $\frac{7}{11}$ of the children.

39. a. Number of legal states $\rightarrow$ 40
Number of states $\rightarrow$ 50

Thus, consumer fireworks are legal in $\frac{40}{50}$ of the states.

b. $50 - 40 = 10$
Consumer fireworks are illegal in 10 states.

c. Number of illegal states $\rightarrow$ 10
Number of states $\rightarrow$ 50

Thus, consumer fireworks are illegal in $\frac{10}{50}$ of the states.

41. To graph $\frac{1}{4}$ on a number line, divide the distance from 0 to 1 into 4 equal parts. Then start at 0 and count over 1 part.

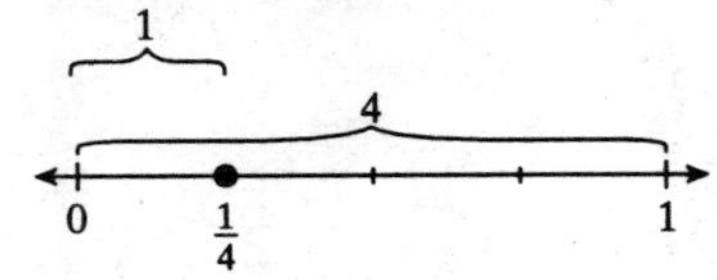

43. To graph $\frac{4}{7}$ on a number line, divide the distance from 0 to 1 into 7 equal parts. Then start at 0 and count over 4 parts.

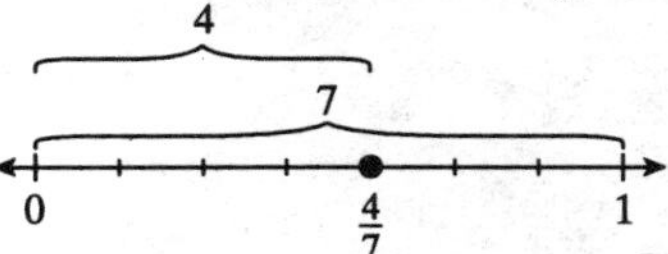

45. To graph $\frac{8}{5}$ on a number line, divide the distance from 0 to 1 into 5 equal parts and divide the distance from 1 to 2 into 5 equal parts. Then start at 0 and count over 8 parts.

47. To graph $\frac{7}{3}$ on a number line, divide the distances from 0 to 1, 1 to 2, and 2 to 3 into 3 equal parts. Then start at 0 and count over 7 parts.

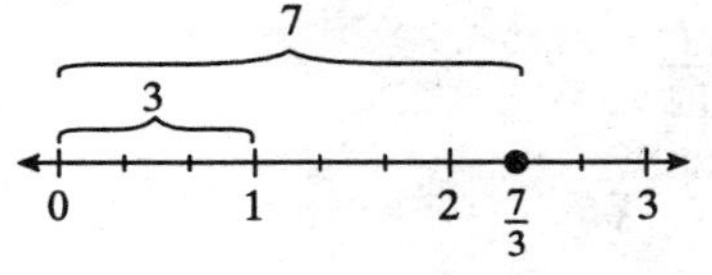

49. To graph $\frac{3}{8}$ on a number line, divide the distance from 0 to 1 into 8 equal parts. Then start at 0 and count over 3 parts.

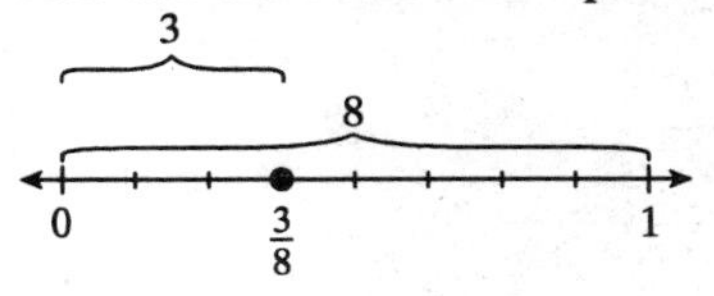

51. $\frac{4}{7} = \frac{?}{35}$

$\frac{4 \cdot 5}{7 \cdot 5} = \frac{20}{35}$

53. $\frac{2}{3} = \frac{?}{21}$

$\frac{2 \cdot 7}{3 \cdot 7} = \frac{14}{21}$

55. $\frac{2y}{5}=\frac{?}{25}$

$\frac{2y\cdot 5}{5\cdot 5}=\frac{10y}{25}$

57. $\frac{1}{2}=\frac{?}{30}$

$\frac{1\cdot 15}{2\cdot 15}=\frac{15}{30}$

59. $\frac{10}{7x}=\frac{?}{21x}$

$\frac{10\cdot 3}{7x\cdot 3}=\frac{30}{21x}$

61. $2=\frac{?}{5}$

$\frac{2}{1}=\frac{?}{5}$

$\frac{2\cdot 5}{1\cdot 5}=\frac{10}{5}$

63. $\frac{3}{4}=\frac{?}{12}$

$\frac{3\cdot 3}{4\cdot 3}=\frac{9}{12}$

65. $\frac{2y}{3}=\frac{?}{12}$

$\frac{2y\cdot 4}{3\cdot 4}=\frac{8y}{12}$

67. $\frac{1}{2}=\frac{?}{12}$

$\frac{1\cdot 6}{2\cdot 6}=\frac{6}{12}$

69. $\frac{4}{3}=\frac{?}{36x}$

$\frac{4\cdot 12x}{3\cdot 12x}=\frac{48x}{36x}$

71. $\frac{5}{9}=\frac{?}{36x}$

$\frac{5\cdot 4x}{9\cdot 4x}=\frac{20x}{36x}$

73. $1=\frac{?}{36x}$

$\frac{1}{1}=\frac{?}{36x}$

$\frac{1\cdot 36x}{1\cdot 36x}=\frac{36x}{36x}$

75. Denmark: $\frac{7}{20}=\frac{7\cdot 5}{20\cdot 5}=\frac{35}{100}$

Finland: $\frac{57}{100}$

Hong Kong: $\frac{43}{100}$

Israel: $\frac{37}{100}$

Italy: $\frac{9}{25}=\frac{9\cdot 4}{25\cdot 4}=\frac{36}{100}$

Japan: $\frac{31}{100}$

Norway: $\frac{12}{25}=\frac{12\cdot 4}{25\cdot 4}=\frac{48}{100}$

Singapore: $\frac{8}{25}=\frac{8\cdot 4}{25\cdot 4}=\frac{32}{100}$

South Korea: $\frac{3}{10}=\frac{3\cdot 10}{10\cdot 10}=\frac{30}{100}$

Sweden: $\frac{23}{50}=\frac{23\cdot 2}{50\cdot 2}=\frac{46}{100}$

United States: $\frac{13}{50}=\frac{13\cdot 2}{50\cdot 2}=\frac{26}{100}$

77. The smallest numerator of all the fractions with a denominator of 100 is 26, which is the United States.

79. $\frac{12}{12}=12\div 12=1$

81. $\frac{-5}{1}=-5\div 1=-5$

83. $\frac{0}{-2}=0\div -2=0$

85. $\frac{-8}{-8}=-8\div(-8)=1$

87. $\frac{-9}{0}$ is undefined.

89. $\frac{3}{1} = 3 \div 1 = 3$

91. $3^2 = 3 \cdot 3 = 9$

93. $5^3 = 5 \cdot 5 \cdot 5 = 125$

95. $7^2 = 7 \cdot 7 = 49$

97. $2^3 \cdot 3 = 2 \cdot 2 \cdot 2 \cdot 3 = 24$

99.
$$\begin{array}{r} 232 \\ 9\overline{)2088} \\ \underline{18} \\ 28 \\ \underline{27} \\ 18 \\ \underline{18} \\ 0 \end{array}$$

$\frac{2}{9} = \frac{2 \cdot 232}{2 \cdot 232} = \frac{464}{2088}$

101. Answers may vary.

103. Total affiliates = 1300 + 285 = 1585
$\frac{1300}{1585}$ of Habitat for Humanity's affiliates are in the U.S.

105. Total number of licensees
= 87 + 56 + 21 + 8
= 172
$\frac{56}{172}$ of the licensees are colleges or universities.

Section 4.2

Practice Problems

1. 28 → 2, 14; 14 → 2, 7

$28 = 2 \cdot 2 \cdot 7$ or $2^2 \cdot 7$

2. 60 → $6 \cdot 10$ → $2 \cdot 3 \cdot 2 \cdot 5$

$60 = 2 \cdot 2 \cdot 3 \cdot 5$ or $2^2 \cdot 3 \cdot 5$

3. 297 → $9 \cdot 33$ → $3 \cdot 3 \cdot 3 \cdot 11$

$297 = 3 \cdot 3 \cdot 3 \cdot 11$ or $3^3 \cdot 11$

4. $\frac{30}{45} = \frac{2 \cdot 3 \cdot 5}{3 \cdot 3 \cdot 5} = \frac{2}{3}$

5. $\frac{39}{51} = \frac{3 \cdot 13}{3 \cdot 17} = \frac{13}{17}$

6. $\frac{45}{105y} = \frac{3 \cdot 3 \cdot 5}{3 \cdot 5 \cdot 7 \cdot y} = \frac{3}{7y}$

7. $\frac{9a}{50a} = \frac{3 \cdot 3 \cdot a}{2 \cdot 5 \cdot 5 \cdot a} = \frac{9}{50}$

8. $\frac{38}{4} = \frac{2 \cdot 19}{2 \cdot 2} = \frac{19}{2}$

9. $\frac{7a^3}{56a^2} = \frac{7 \cdot a \cdot a \cdot a}{2 \cdot 2 \cdot 2 \cdot 7 \cdot a \cdot a} = \frac{a}{8}$

10. $\frac{12}{80} = \frac{2 \cdot 2 \cdot 3}{2 \cdot 2 \cdot 2 \cdot 2 \cdot 5} = \frac{3}{2 \cdot 2 \cdot 5} = \frac{3}{20}$

Calculator Explorations

1. $\frac{128}{224} = \frac{4}{7}$

2. $\frac{231}{396} = \frac{7}{12}$

3. $\frac{340}{459} = \frac{20}{27}$

4. $\frac{999}{1350} = \frac{37}{50}$

5. $\frac{810}{432} = \frac{15}{8}$

6. $\frac{315}{225} = \frac{7}{5}$

7. $\frac{243}{54} = \frac{9}{2}$

8. $\frac{689}{455} = \frac{53}{35}$

Mental Math

1. Yes, we know that 2430 is divisible by 2 because its ones digit is 0.
   Yes, we know that 2430 is divisible by 3 because the sum of its digits, $2 + 4 + 3 + 0 = 9$, is divisible by 3.
   Yes, we know that 2430 is divisible by 5 because its ones digit is 0.

2. $15 = 3 \cdot 5$

3. $10 = 2 \cdot 5$

4. $6 = 2 \cdot 3$

5. $21 = 3 \cdot 7$

6. $4 = 2 \cdot 2$ or 2^2

7. $9 = 3 \cdot 3 = 3^2$

8. $14 = 2 \cdot 7$

Exercise Set 4.2

1. $20 = 2 \cdot 10$
   $2 \cdot 2 \cdot 5 = 2^2 \cdot 5$

3. $48 = 6 \cdot 8$
   $2 \cdot 3 \cdot 2 \cdot 4$
   $2 \cdot 3 \cdot 2 \cdot 2 \cdot 2 = 2^4 \cdot 3$

5. $64 = 8 \cdot 8$
   $2 \cdot 4 \cdot \quad 2 \cdot 4$
   $2 \cdot 2 \cdot 2 \cdot 2 \cdot 2 \cdot 2 = 2^6$

7. $240 = 2 \cdot 120$
   $2 \cdot 2 \cdot 60$
   $2 \cdot 2 \cdot 2 \cdot 30$
   $2 \cdot 2 \cdot 2 \cdot 2 \cdot 15$
   $2 \cdot 2 \cdot 2 \cdot 2 \cdot 3 \cdot 5 = 2^4 \cdot 3 \cdot 5$

9. $\frac{3}{12} = \frac{3}{3 \cdot 4} = \frac{1}{4}$

11. $\frac{7x}{35} = \frac{7 \cdot x}{7 \cdot 5} = \frac{x}{5}$

13. $\frac{14}{16} = \frac{2 \cdot 7}{2 \cdot 8} = \frac{7}{8}$

15. $\frac{24a}{30a} = \frac{2 \cdot 2 \cdot 2 \cdot 3 \cdot a}{2 \cdot 3 \cdot 5 \cdot a} = \frac{4}{5}$

17. $\frac{35}{42} = \frac{7 \cdot 5}{7 \cdot 2 \cdot 3} = \frac{5}{6}$

19. $\frac{30x^2}{36x} = \frac{2 \cdot 3 \cdot 5 \cdot x \cdot x}{2 \cdot 2 \cdot 3 \cdot 3 \cdot x} = \frac{5x}{6}$

21. $\frac{16}{24} = \frac{2 \cdot 2 \cdot 2 \cdot 2}{2 \cdot 2 \cdot 2 \cdot 3} = \frac{2}{3}$

23. $\frac{45xz}{60z} = \frac{3 \cdot 3 \cdot 5 \cdot x \cdot z}{2 \cdot 2 \cdot 3 \cdot 5 \cdot z} = \frac{3x}{4}$

25. $\frac{39ab}{26a^2} = \frac{3 \cdot 13 \cdot a \cdot b}{2 \cdot 13 \cdot a \cdot a} = \frac{3b}{2a}$

27. $\frac{63}{72} = \frac{3 \cdot 3 \cdot 7}{2 \cdot 2 \cdot 2 \cdot 3 \cdot 3} = \frac{7}{8}$

29. $\frac{21}{49} = \frac{3 \cdot 7}{7 \cdot 7} = \frac{3}{7}$

31. $\frac{24y}{40} = \frac{2 \cdot 2 \cdot 2 \cdot 3 \cdot y}{2 \cdot 2 \cdot 2 \cdot 5} = \frac{3y}{5}$

33. $\frac{36z}{63z} = \frac{3 \cdot 3 \cdot 2 \cdot 2 \cdot z}{3 \cdot 3 \cdot 7 \cdot z} = \frac{4}{7}$

35. $\frac{72x^3y^2}{90xy} = \frac{2 \cdot 2 \cdot 2 \cdot 3 \cdot 3 \cdot x \cdot x \cdot x \cdot y \cdot y}{2 \cdot 3 \cdot 3 \cdot 5 \cdot x \cdot y}$
$= \frac{4x^2y}{5}$

37. $\frac{12}{15} = \frac{2 \cdot 2 \cdot 3}{3 \cdot 5} = \frac{4}{5}$

39. $\frac{25x^2}{40x} = \frac{5 \cdot 5 \cdot x \cdot x}{2 \cdot 2 \cdot 2 \cdot 5 \cdot x} = \frac{5x}{8}$

41. $\frac{27xy}{90y} = \frac{3 \cdot 3 \cdot 3 \cdot x \cdot y}{2 \cdot 3 \cdot 3 \cdot 5 \cdot y} = \frac{3x}{10}$

43. $\frac{36a^3bc^2}{24ab^4c^2} = \frac{2 \cdot 2 \cdot 3 \cdot 3 \cdot a \cdot a \cdot a \cdot b \cdot c \cdot c}{2 \cdot 2 \cdot 2 \cdot 3 \cdot a \cdot b \cdot b \cdot b \cdot b \cdot c \cdot c}$
$= \frac{3a^2}{2b^3}$

45. $\frac{40xy}{64xyz} = \frac{2 \cdot 2 \cdot 2 \cdot 5 \cdot x \cdot y}{2 \cdot 2 \cdot 2 \cdot 2 \cdot 2 \cdot 2 \cdot x \cdot y \cdot z}$
$= \frac{5}{8z}$

47. $\frac{6 \text{ hours}}{8 \text{ hours}} = \frac{2 \cdot 3}{2 \cdot 2 \cdot 2} = \frac{3}{2 \cdot 2} = \frac{3}{4}$
6 hours represents $\frac{3}{4}$ of a work shift.

49. $\frac{2640 \text{ feet}}{5280 \text{ feet}} = \frac{2 \cdot 2 \cdot 2 \cdot 2 \cdot 3 \cdot 5 \cdot 11}{2 \cdot 2 \cdot 2 \cdot 2 \cdot 2 \cdot 3 \cdot 5 \cdot 11} = \frac{1}{2}$
2640 feet represents $\frac{1}{2}$ mile

51. $\frac{257}{408} = \frac{257}{2 \cdot 2 \cdot 2 \cdot 3 \cdot 17} = \frac{257}{408}$
$\frac{257}{408}$ of individuals who have flown in space were Americans.

53. $16{,}000 - 8800 = 7200$
Number of male students $\rightarrow$ $\frac{7200}{16{,}000}$
Number of students $\rightarrow$
$\frac{7200}{16{,}000} = \frac{9 \cdot 800}{20 \cdot 800}$
$= \frac{9}{20}$
$\frac{9}{20}$ of the students are male.

55. **a.** $\frac{15}{50} = \frac{3 \cdot 5}{2 \cdot 5 \cdot 5} = \frac{3}{2 \cdot 5} = \frac{3}{10}$
$\frac{3}{10}$ of states have at least one Ritz-Carlton.

b. $50 - 15 = 35$
35 states have no Ritz-Carlton.

c. $\frac{35}{50} = \frac{5 \cdot 7}{2 \cdot 5 \cdot 5} = \frac{7}{2 \cdot 5} = \frac{7}{10}$
$\frac{7}{10}$ of states have no Ritz-Carlton.

57. $\frac{10}{24} = \frac{2 \cdot 5}{2 \cdot 2 \cdot 2 \cdot 3} = \frac{5}{12}$
$\frac{5}{12}$ of the width is concrete.

59. Education is $\frac{1}{10}$.

61. Answers will vary.

63. $\frac{x^3}{9} = \frac{(-3)^3}{9} = \frac{(-3)(-3)(-3)}{9} = \frac{-27}{9} = -3$

65. $2y = 2(-7) = -14$

67. $3z - y = 3(2) - 6 = 6 - 6 = 0$

69. $a^2 + 2b + 3 = 4^2 + 2(5) + 3$
$= 16 + 10 + 3$
$= 29$

71. d

0 $\frac{1}{2}$ 1

0 $\frac{1}{3}$ 1

0 $\frac{1}{4}$ 1

0 $\frac{1}{5}$ 1

73. False, $\frac{14}{42} = \frac{2 \cdot 7}{2 \cdot 3 \cdot 7} = \frac{1}{3}$

75. True

77. $\frac{372}{620} = \frac{2 \cdot 2 \cdot 3 \cdot 31}{2 \cdot 2 \cdot 5 \cdot 31} = \frac{3}{5}$

79. $\frac{36}{100} = \frac{2 \cdot 2 \cdot 3 \cdot 3}{2 \cdot 2 \cdot 5 \cdot 5} = \frac{3 \cdot 3}{5 \cdot 5} = \frac{9}{25}$
$\frac{9}{25}$ of the donors have A Rh-positive blood type.

81. $3 + 1 = 4$
$\frac{4}{100} = \frac{2 \cdot 2}{2 \cdot 2 \cdot 5 \cdot 5} = \frac{1}{5 \cdot 5} = \frac{1}{25}$
$\frac{1}{25}$ of the donors have AB blood type (either Rh-positive or Rh-negative).

83. $7 + 6 + 1 + 1 = 15$
$\frac{15}{100} = \frac{3 \cdot 5}{2 \cdot 2 \cdot 5 \cdot 5} = \frac{3}{20}$
$\frac{3}{20}$ of the donors have the negative Rh factor.

Section 4.3

Practice Problems

1. $\frac{3}{8} \cdot \frac{5}{7} = \frac{3 \cdot 5}{8 \cdot 7} = \frac{15}{56}$

2. $\frac{1}{3} \cdot \frac{1}{6} = \frac{1 \cdot 1}{3 \cdot 6} = \frac{1}{18}$

3. $\frac{6}{11} \cdot \frac{5}{8} = \frac{6 \cdot 5}{11 \cdot 8} = \frac{2 \cdot 3 \cdot 5}{11 \cdot 2 \cdot 4} = \frac{3 \cdot 5}{11 \cdot 4} = \frac{15}{44}$

4. $\frac{4}{15} \cdot \frac{3}{8} = \frac{4 \cdot 3}{15 \cdot 8} = \frac{4 \cdot 3}{3 \cdot 5 \cdot 4 \cdot 2} = \frac{1 \cdot 1}{5 \cdot 2} = \frac{1}{10}$

5. $\frac{1}{22} \cdot \left(-\frac{11}{28}\right) = -\frac{1 \cdot 11}{22 \cdot 28} = -\frac{1 \cdot 11}{2 \cdot 11 \cdot 28} = -\frac{1}{56}$

6. $\frac{9}{5} \cdot \frac{20}{12} = \frac{3 \cdot 3 \cdot 4 \cdot 5}{5 \cdot 3 \cdot 4} = \frac{3}{1} = 3$

7. $\frac{2}{3} \cdot \frac{3y}{2} = \frac{2 \cdot 3 \cdot y}{3 \cdot 2} = \frac{y}{1} = y$

8. $\frac{a^3}{b^2} \cdot \frac{b}{a^2} = \frac{a^3 \cdot b}{b^2 \cdot a^2} = \frac{a \cdot a \cdot a \cdot b}{b \cdot b \cdot a \cdot a} = \frac{a}{b}$

9. a. $\left(\frac{3}{4}\right)^3 = \frac{3}{4} \cdot \frac{3}{4} \cdot \frac{3}{4} = \frac{3 \cdot 3 \cdot 3}{4 \cdot 4 \cdot 4} = \frac{27}{64}$

b. $\left(-\frac{4}{5}\right)^2 = \left(-\frac{4}{5}\right)\left(-\frac{4}{5}\right) = \frac{4 \cdot 4}{5 \cdot 5} = \frac{16}{25}$

10. $\frac{3}{2} \div \frac{14}{5} = \frac{3}{2} \cdot \frac{5}{14} = \frac{3 \cdot 5}{2 \cdot 14} = \frac{15}{28}$

11. $\frac{4}{9} \div \frac{1}{2} = \frac{4}{9} \cdot \frac{2}{1} = \frac{4 \cdot 2}{9 \cdot 1} = \frac{8}{9}$

12. $\frac{10}{4}\div\frac{2}{9}=\frac{10}{4}\cdot\frac{9}{2}=\frac{2\cdot 5\cdot 9}{4\cdot 2}=\frac{45}{4}$

13. $\frac{3y}{4}\div 5y^3=\frac{3y}{4}\cdot\frac{1}{5y^3}=\frac{3\cdot y\cdot 1}{4\cdot 5\cdot y\cdot y\cdot y}=\frac{3}{20y^2}$

14.
$$\begin{aligned}\left(-\frac{2}{3}\cdot\frac{9}{14}\right)\div\frac{7}{15}&=\left(-\frac{2\cdot 3\cdot 3}{3\cdot 2\cdot 7}\right)\div\frac{7}{15}\\&=\left(-\frac{3}{7}\right)\div\frac{7}{15}\\&=-\frac{3}{7}\cdot\frac{15}{7}\\&=-\frac{45}{49}\end{aligned}$$

15. a. $xy=\left(-\frac{3}{4}\right)\cdot\frac{9}{2}=-\frac{3\cdot 9}{4\cdot 2}=-\frac{27}{8}$

b. $x\div y=-\frac{3}{4}\div\frac{9}{2}=-\frac{3}{4}\cdot\frac{2}{9}=-\frac{3\cdot 2}{2\cdot 2\cdot 3\cdot 3}$
$=-\frac{1}{6}$

16.
$$\begin{aligned}2x&=-\frac{9}{4}\\2\cdot-\frac{9}{8}&\stackrel{?}{=}-\frac{9}{4}\\-\frac{2\cdot 9}{2\cdot 4}&\stackrel{?}{=}-\frac{9}{4}\\-\frac{9}{4}&\stackrel{?}{=}-\frac{9}{4}\ \text{True.}\end{aligned}$$

The statement is true, so $-\frac{9}{8}$ is a solution.

17.
$$\begin{aligned}\frac{1}{3}\cdot 20{,}439&=\frac{1}{3}\cdot\frac{20{,}439}{1}\\&=\frac{1\cdot 20{,}439}{3\cdot 1}\\&=\frac{1\cdot 3\cdot 6813}{3\cdot 1}\\&=\frac{1\cdot 6813}{1}\\&=\frac{6813}{1}\\&=6813\end{aligned}$$

6813 species are at risk.

Mental Math

1. $\frac{1}{3}\cdot\frac{2}{5}=\frac{2}{15}$

2. $\frac{2}{3}\cdot\frac{4}{7}=\frac{8}{21}$

3. $\frac{6}{5}\cdot\frac{1}{7}=\frac{6}{35}$

4. $\frac{7}{3}\cdot\frac{2}{3}=\frac{14}{9}$

5. $\frac{3}{1}\cdot\frac{3}{8}-=\frac{9}{8}$

6. $\frac{2}{1}\cdot\frac{7}{11}=\frac{14}{11}$

Exercise Set 4.3

1. $\frac{7}{8}\cdot\frac{2}{3}=\frac{7\cdot 2}{8\cdot 3}=\frac{7\cdot 2}{2\cdot 2\cdot 2\cdot 3}=\frac{7}{2\cdot 2\cdot 3}=\frac{7}{12}$

3. $-\frac{2}{7}\cdot\frac{5}{8}=-\frac{2\cdot 5}{7\cdot 8}=-\frac{2\cdot 5}{7\cdot 2\cdot 2\cdot 2}=-\frac{5}{28}$

5. $-\frac{1}{2}\cdot-\frac{2}{15}=\frac{1\cdot 2}{2\cdot 15}=\frac{1}{15}$

7.
$$\begin{aligned}\frac{18x}{20}\cdot\frac{36}{99}&=\frac{18\cdot x\cdot 36}{20\cdot 99}\\&=\frac{9\cdot 2\cdot x\cdot 4\cdot 9}{4\cdot 5\cdot 9\cdot 11}\\&=\frac{2\cdot x\cdot 9}{5\cdot 11}=\frac{18x}{55}\end{aligned}$$

9. $3a^2\cdot\frac{1}{4}=\frac{3\cdot a^2\cdot 1}{4}=\frac{3a^2}{4}$

11. $\frac{x^3}{y^3}\cdot\frac{y^2}{x}=\frac{x^2\cdot x\cdot y^2}{y^2\cdot y\cdot x}=\frac{x^2}{y}$

13. $\left(\frac{1}{5}\right)^3=\frac{1}{5}\cdot\frac{1}{5}\cdot\frac{1}{5}=\frac{1\cdot 1\cdot 1}{5\cdot 5\cdot 5}=\frac{1}{125}$

15. $\left(-\frac{2}{3}\right)^2 = \left(-\frac{2}{3}\right)\cdot\left(-\frac{2}{3}\right) = \frac{2\cdot 2}{3\cdot 3} = \frac{4}{9}$

17. $\left(-\frac{2}{3}\right)^3\cdot\frac{1}{2} = \left(-\frac{2}{3}\right)\cdot\left(-\frac{2}{3}\right)\cdot\left(-\frac{2}{3}\right)\cdot\left(\frac{1}{2}\right)$
$= -\frac{2\cdot 2\cdot 2\cdot 1}{3\cdot 3\cdot 3\cdot 2}$
$= -\frac{4}{27}$

19. $\frac{2}{3} \div \frac{5}{6} = \frac{2}{3}\cdot\frac{6}{5} = \frac{2\cdot 2\cdot 3}{3\cdot 5} = \frac{4}{5}$

21. $-\frac{6}{15} \div \frac{12}{5} = -\frac{6}{15}\cdot\frac{5}{12}$
$= -\frac{6\cdot 5}{3\cdot 5\cdot 2\cdot 6}$
$= -\frac{1}{6}$

23. $\frac{8}{9} \div \frac{x}{2} = \frac{8}{9}\cdot\frac{2}{x} = \frac{2\cdot 2\cdot 2\cdot 2}{3\cdot 3\cdot x} = \frac{16}{9x}$

25. $\frac{11y}{20} \div \frac{3}{11} = \frac{11y}{20}\cdot\frac{11}{3} = \frac{11\cdot y\cdot 11}{2\cdot 2\cdot 5\cdot 3} = \frac{121y}{60}$

27. $-\frac{2}{3} \div 4 = -\frac{2}{3}\cdot\frac{1}{4} = -\frac{2\cdot 1}{3\cdot 2\cdot 2} = -\frac{1}{6}$

29. $\frac{1}{5x} \div \frac{5}{x^2} = \frac{1}{5x}\cdot\frac{x^2}{5} = \frac{x\cdot x}{5\cdot x\cdot 5} = \frac{x}{25}$

31. $\frac{2}{3}\cdot\frac{5}{9} = \frac{2\cdot 5}{3\cdot 3\cdot 3} = \frac{10}{27}$

33. $\frac{3x}{7} \div \frac{5}{6x} = \frac{3x}{7}\cdot\frac{6x}{5} = \frac{3\cdot x\cdot 2\cdot 3\cdot x}{7\cdot 5} = \frac{18x^2}{35}$

35. $-\frac{5}{28}\cdot\frac{35}{25} = -\frac{5\cdot 5\cdot 7}{2\cdot 2\cdot 7\cdot 5\cdot 5} = -\frac{1}{4}$

37. $-\frac{3}{5} \div -\frac{4}{5} = \left(-\frac{3}{5}\right)\cdot\left(-\frac{5}{4}\right) = \frac{3\cdot 5}{5\cdot 2\cdot 2} = \frac{3}{4}$

39. $\left(-\frac{3}{4}\right)^2 = \left(-\frac{3}{4}\right)\cdot\left(-\frac{3}{4}\right) = \frac{3\cdot 3}{2\cdot 2\cdot 2\cdot 2} = \frac{9}{16}$

41. $\frac{x^2}{y}\cdot\frac{y^3}{x} = \frac{x\cdot x\cdot y\cdot y^2}{y\cdot x} = \frac{x\cdot y^2}{1} = xy^2$

43. $7 \div \frac{2}{11} = \frac{7}{1}\cdot\frac{11}{2} = \frac{7\cdot 11}{1\cdot 2} = \frac{77}{2}$

45. $-3x \div \frac{x^2}{12} = -\frac{3x}{1}\cdot\frac{12}{x^2} = -\frac{3\cdot x\cdot 12}{1\cdot x\cdot x} = -\frac{36}{x}$

47. $\left(\frac{2}{7} \div \frac{7}{2}\right)\cdot\frac{3}{4} = \frac{2}{7}\cdot\frac{2}{7}\cdot\frac{3}{4} = \frac{2\cdot 2\cdot 3}{7\cdot 7\cdot 2\cdot 2} = \frac{3}{49}$

49. $-\frac{19}{63y}\cdot 9y^2 = -\frac{19}{63y}\cdot\frac{9y^2}{1}$
$= \frac{-19\cdot 9\cdot y\cdot y}{7\cdot 9\cdot y\cdot 1}$
$= -\frac{19y}{7}$

51. $-\frac{2}{3}\cdot -\frac{6}{11} = \frac{2\cdot 2\cdot 3}{3\cdot 11} = \frac{4}{11}$

53. $\frac{4}{8} \div \frac{3}{16} = \frac{4}{8}\cdot\frac{16}{3} = \frac{4\cdot 8\cdot 2}{8\cdot 3} = \frac{8}{3}$

55. $\frac{21x^2}{10y} \div \frac{14x}{25y} = \frac{21x^2}{10y}\cdot\frac{25y}{14x}$
$= \frac{3\cdot 7\cdot x\cdot x\cdot 5\cdot 5\cdot y}{2\cdot 5\cdot y\cdot 2\cdot 7\cdot x}$
$= \frac{3\cdot x\cdot 5}{2\cdot 2}$
$= \frac{15x}{4}$

57. $\left(1 \div \frac{3}{4}\right)\cdot\frac{2}{3} = \frac{1}{1}\cdot\frac{4}{3}\cdot\frac{2}{3} = \frac{1\cdot 2\cdot 2\cdot 2}{1\cdot 3\cdot 3} = \frac{8}{9}$

59. $\frac{a^3}{2} \div 30a^3 = \frac{a^3}{2}\cdot\frac{1}{30a^3}$
$= \frac{a^3}{2\cdot 30\cdot a^3}$
$= \frac{1}{60}$

61. $\frac{ab^2}{c} \cdot \frac{c}{ab} = \frac{a \cdot b \cdot b \cdot c}{c \cdot a \cdot b} = \frac{b}{1} = b$

63.
$$\begin{aligned}\left(\frac{1}{2} \cdot \frac{2}{3}\right) \div \frac{5}{6} &= \left(\frac{1 \cdot 2}{2 \cdot 3}\right) \cdot \frac{6}{5}\\ &= \frac{1}{3} \cdot \frac{6}{5}\\ &= \frac{1 \cdot 2 \cdot 3}{3 \cdot 5}\\ &= \frac{2}{5}\end{aligned}$$

65.
$$\begin{aligned}-\frac{4}{7} \div \left(\frac{4}{5} \cdot \frac{3}{7}\right) &= -\frac{4}{7} \div \left(\frac{2 \cdot 2 \cdot 3}{5 \cdot 7}\right)\\ &= -\frac{4}{7} \cdot \frac{5 \cdot 7}{2 \cdot 2 \cdot 3}\\ &= -\frac{4 \cdot 5 \cdot 7}{7 \cdot 4 \cdot 3}\\ &= -\frac{5}{3}\end{aligned}$$

67. **a.** $xy = \frac{2}{5} \cdot \frac{5}{6} = \frac{2 \cdot 5}{5 \cdot 2 \cdot 3} = \frac{1}{3}$

b. $x \div y = \frac{2}{5} \div \frac{5}{6} = \frac{2}{5} \cdot \frac{6}{5} = \frac{2 \cdot 2 \cdot 3}{5 \cdot 5} = \frac{12}{25}$

69. $xy = -\frac{4}{5} \cdot \frac{9}{11} = -\frac{2 \cdot 2 \cdot 3 \cdot 3}{5 \cdot 11} = -\frac{36}{55}$

$$\begin{aligned}x \div y &= -\frac{4}{5} \div \frac{9}{11}\\ &= -\frac{4}{5} \cdot \frac{11}{9}\\ &= -\frac{2 \cdot 2 \cdot 11}{5 \cdot 3 \cdot 3}\\ &= -\frac{44}{45}\end{aligned}$$

71.
$$\begin{aligned}3x &= -\frac{5}{6}\\ 3\left(-\frac{5}{18}\right) &= -\frac{5}{6} \quad \text{replace } x \text{ with } -\frac{5}{18}\\ -\frac{3 \cdot 5}{2 \cdot 3 \cdot 3} &= -\frac{5}{6}\\ -\frac{5}{2 \cdot 3} &= -\frac{5}{6}\\ -\frac{5}{6} &= -\frac{5}{6} \quad \text{True}\end{aligned}$$

Yes, it is a solution.

73.
$$\begin{aligned}-\frac{1}{2}z &= \frac{1}{10}\\ -\frac{1}{2} \cdot \frac{2}{5} &= \frac{1}{10} \quad \text{replace } z \text{ with } \frac{2}{5}\\ -\frac{1 \cdot 2}{2 \cdot 5} &= \frac{1}{10}\\ -\frac{1}{5} &= \frac{1}{10} \quad \text{False}\end{aligned}$$

No, it is not a solution.

75. $\frac{5}{6}(36) = \frac{5 \cdot 6 \cdot 6}{6 \cdot 1} = \frac{30}{1}$

30 gallons are normally in the vat.

77. $8\left(\frac{3}{16}\right) = \frac{24}{16} = \frac{3}{2}$

The screw is $\frac{3}{2}$ inches deep.

79. $\frac{7}{10} \cdot 31{,}050 = \frac{7 \cdot 10 \cdot 3{,}105}{10} = 21{,}735$

21,735 tornadoes occurred during that time.

81. Let x = size of Jorge's wrist

(wrist size)

= (fraction of waist) · (waist size)

$$\begin{aligned}x &= \frac{1}{4} \cdot \frac{34}{1}\\ &= \frac{1 \cdot 2 \cdot 17}{2 \cdot 2 \cdot 1}\\ &= \frac{17}{2}\end{aligned}$$

Jorge's wrist is about $\frac{17}{2}$ inches.

83. $9 \cdot \frac{1}{3} = \frac{9}{1} \cdot \frac{1}{3} = \frac{3 \cdot 3}{3} = 3$
3 feet of the 9-foot post should be buried.

85. $\frac{5}{14} \cdot \frac{1}{5} = \frac{5 \cdot 1}{14 \cdot 5} = \frac{1}{14}$
The area is $\frac{1}{14}$ square foot.

87. $90 = 2 \cdot 3 \cdot 3 \cdot 5$ or $2 \cdot 3^2 \cdot 5$

89. $65 = 5 \cdot 13$

91. $126 = 2 \cdot 3 \cdot 3 \cdot 7$ or $2 \cdot 3^2 \cdot 7$

93. $10,300,000 \cdot \frac{7}{10} = 7,210,000$ households

95. Answers may vary.

97. $5144 \div \frac{1}{3} = 5144 \cdot 3 = 15,432$
There are 15,432 flowering plant species that are native to the United States.

Section 4.4

Practice Problems

1. $\frac{5}{9} + \frac{2}{9} = \frac{5+2}{9} = \frac{7}{9}$

2. $\frac{5}{8} + \frac{1}{8} = \frac{5+1}{8} = \frac{6}{8} = \frac{3 \cdot 2}{4 \cdot 2} = \frac{3}{4}$

3. $\frac{10}{11} + \frac{1}{11} + \frac{7}{11} = \frac{10+1+7}{11} = \frac{18}{11}$

4. $\frac{7}{12} - \frac{2}{12} = \frac{7-2}{12} = \frac{5}{12}$

5. $\frac{9}{10} - \frac{1}{10} = \frac{9-1}{10} = \frac{8}{10} = \frac{4 \cdot 2}{5 \cdot 2} = \frac{4}{5}$

6. $-\frac{8}{5} + \frac{4}{5} = \frac{-8+4}{5} = \frac{-4}{5}$ or $-\frac{4}{5}$

7. $\frac{2}{5} - \frac{3y}{5} = \frac{2-3y}{5}$

8. $\frac{4}{11} - \frac{6}{11} - \frac{3}{11} = \frac{4-6-3}{11} = \frac{-5}{11}$ or $-\frac{5}{11}$.

9. $x + y = -\frac{10}{12} + \frac{5}{12} = \frac{-10+5}{12} = \frac{-5}{12}$ or $-\frac{5}{12}$

10. Solve: $\frac{7}{10} = x - \frac{1}{10}$

$$\frac{7}{10} + \frac{1}{10} = x - \frac{1}{10} + \frac{1}{10}$$
$$\frac{8}{10} = x$$
$$\frac{4}{5} = x$$

11. $\frac{3}{4} + \frac{1}{4} = \frac{3+1}{4} = \frac{4}{4} = 1$
She practiced 1 hour.

12. 9 laps on Monday
$= \frac{1}{8} + \frac{1}{8} + \frac{1}{8} + \frac{1}{8} + \frac{1}{8} + \frac{1}{8} + \frac{1}{8} + \frac{1}{8} + \frac{1}{8}$ miles
3 laps on Wednesday $= \frac{1}{8} + \frac{1}{8} + \frac{1}{8} = \frac{3}{8}$ miles
$\frac{9}{8} - \frac{3}{8} = \frac{9-3}{8} = \frac{6}{8} = \frac{2 \cdot 3}{2 \cdot 4} = \frac{3}{4}$ miles
He went $\frac{3}{4}$ mile farther on Monday.

13. Since 15 is not divisible by 12, we check multiples of 15.
$2 \cdot 15 = 30$
30 is not divisible by 12.
$3 \cdot 15 = 45$
45 is not divisible by 12.
$4 \cdot 15 = 60$
60 is divisible by 12.
60 is the LCD.

14. $14 = \textcircled{2 \cdot 7}$

$35 = \textcircled{5} \cdot 7$

LCD $= 2 \cdot 5 \cdot 7 = 70$

15. $4 = (2 \cdot 2)$

$15 = (3) \cdot (5)$

$10 = 2 \cdot 5$

$\text{LCD} = 2 \cdot 2 \cdot 3 \cdot 5 = 60$

16. $y = (y)$

$11 = (11)$

$\text{LCD} = 11 \cdot y = 11y$

Mental Math

1. Since $\frac{7}{8}$ and $\frac{7}{10}$ have different denominators, they are unlike fractions.

2. Since $\frac{2}{3}$ and $\frac{2}{9}$ have different denominators, they are unlike fractions.

3. Since $\frac{9}{10}$ and $\frac{1}{10}$ have the same denominator, they are like fractions.

4. Since $\frac{8}{11}$ and $\frac{2}{11}$ have the same denominator, they are like fractions.

5. Since $\frac{2}{31}$, $\frac{30}{31}$, and $\frac{19}{31}$ have the same denominator, they are like fractions.

6. Since $\frac{3}{10}$, $\frac{3}{11}$, and $\frac{3}{13}$ have different denominators, they are unlike fractions.

7. Since $\frac{5}{12}$, $\frac{7}{12}$, and $\frac{12}{11}$ have different denominators, they are unlike fractions.

8. Since $\frac{1}{5}$, $\frac{2}{5}$, and $\frac{4}{5}$ have the same denominator, they are like fractions.

9. $\frac{3}{7} + \frac{2}{7} = \frac{3+2}{7} = \frac{5}{7}$

10. $\frac{5}{9} + \frac{2}{9} = \frac{5+2}{9}$
$= \frac{7}{9}$

11. $\frac{10}{11} - \frac{4}{11} = \frac{10-4}{11} = \frac{6}{11}$

12. $\frac{9}{13} - \frac{5}{13} = \frac{9-5}{13} = \frac{4}{13}$

13. $\frac{5}{11} + \frac{2}{11} = \frac{5+2}{11} = \frac{7}{11}$

14. $\frac{4}{7} + \frac{2}{7} = \frac{4+2}{7} = \frac{6}{7}$

15. $\frac{9}{15} - \frac{1}{15} = \frac{9-1}{15} = \frac{8}{15}$

16. $\frac{3}{15} - \frac{1}{15} = \frac{3-1}{15} = \frac{2}{15}$

Exercise Set 4.4

1. $-\frac{1}{2} + \frac{1}{2} = \frac{-1+1}{2} = \frac{0}{2} = 0$

3. $\frac{2}{9x} + \frac{4}{9x} = \frac{2+4}{9x} = \frac{6}{9x} = \frac{2 \cdot 3}{3 \cdot 3 \cdot x} = \frac{2}{3x}$

5. $-\frac{4}{13} + \frac{2}{13} + \frac{1}{13} = \frac{-4+2+1}{13} = \frac{-1}{13} = -\frac{1}{13}$

7. $\frac{7}{18} + \frac{3}{18} + \frac{2}{18} = \frac{7+3+2}{18} = \frac{12}{18} = \frac{2 \cdot 6}{3 \cdot 6} = \frac{2}{3}$

9. $\frac{1}{y} - \frac{4}{y} = \frac{1-4}{y} = \frac{-3}{y} = -\frac{3}{y}$

11. $\frac{7a}{4} - \frac{3}{4} = \frac{7a-3}{4}$

13. $\frac{1}{8}-\frac{7}{8}=\frac{1-7}{8}=\frac{-6}{8}=-\frac{3\cdot 2}{4\cdot 2}=-\frac{3}{4}$

15. $\frac{20}{21}-\frac{10}{21}-\frac{17}{21}=\frac{20-10-17}{21}$
$=\frac{-7}{21}$
$=-\frac{7}{3\cdot 7}$
$=-\frac{1}{3}$

17. $\frac{9x}{15}+\frac{1x}{15}=\frac{9x+1x}{15}=\frac{10x}{15}=\frac{5\cdot 2x}{5\cdot 3}=\frac{2x}{3}$

19. $\frac{7x}{16}-\frac{15x}{16}=\frac{7x-15x}{16}$
$=\frac{-8x}{16}$
$=-\frac{8\cdot x}{2\cdot 8}$
$=-\frac{x}{2}$

21. $\frac{15}{16z}-\frac{3}{16z}=\frac{15-3}{16z}=\frac{12}{16z}=\frac{4\cdot 3}{4\cdot 4\cdot z}=\frac{3}{4z}$

23. $\frac{3}{10}-\frac{6}{10}=\frac{3-6}{10}=\frac{-3}{10}=-\frac{3}{10}$

25. $\frac{15}{17}+\frac{5}{17}+\frac{14}{17}=\frac{15+5+14}{17}$
$=\frac{34}{17}$
$=\frac{17\cdot 2}{17}$
$=\frac{2}{1}$
$=2$

27. $\frac{9}{12}-\frac{7}{12}-\frac{10}{12}=\frac{9-7-10}{12}$
$=\frac{-8}{12}$
$=-\frac{2\cdot 4}{3\cdot 4}$
$=-\frac{2}{3}$

29. $\frac{x}{4}+\frac{3x}{4}-\frac{2x}{4}+\frac{x}{4}=\frac{x+3x-2x+x}{4}=\frac{3x}{4}$

31. $x+y=\frac{3}{4}+\frac{2}{4}=\frac{5}{4}$

33. $x-y=-\frac{1}{5}-\frac{3}{5}=\frac{-1-3}{5}=\frac{-4}{5}=-\frac{4}{5}$

35. $x-y+z=\frac{3}{12}-\frac{5}{12}+\left(-\frac{7}{12}\right)$
$=\frac{3-5-7}{12}$
$=\frac{-9}{12}$
$=-\frac{3\cdot 3}{3\cdot 4}$
$=-\frac{3}{4}$

37. $x+\frac{1}{3}=-\frac{1}{3}$
$x+\frac{1}{3}-\frac{1}{3}=-\frac{1}{3}-\frac{1}{3}$
$x=\frac{-1-1}{3}=\frac{-2}{3}=-\frac{2}{3}$

Check:
$x+\frac{1}{3}=-\frac{1}{3}$
$-\frac{2}{3}+\frac{1}{3}=-\frac{1}{3}$
$\frac{-2+1}{3}=-\frac{1}{3}$
$\frac{-1}{3}=-\frac{1}{3}$
$-\frac{1}{3}=-\frac{1}{3}$ True

39.
$$y-\frac{3}{13}=-\frac{2}{13}$$
$$y-\frac{3}{13}+\frac{3}{13}=-\frac{2}{13}+\frac{3}{13}$$
$$y=\frac{-2+3}{13}=\frac{1}{13}$$
Check:
$$y-\frac{3}{13}=-\frac{2}{13}$$
$$\frac{1}{13}-\frac{3}{13}=-\frac{2}{13}$$
$$\frac{1-3}{13}=-\frac{2}{13}$$
$$\frac{-2}{13}=-\frac{2}{13}$$
$$-\frac{2}{13}=-\frac{2}{13} \quad \text{True}$$

41.
$$3x-\frac{1}{5}-2x=\frac{1}{5}+\frac{2}{5}$$
$$(3x-2x)-\frac{1}{5}=\frac{1+2}{5}$$
$$x-\frac{1}{5}=\frac{3}{5}$$
$$x-\frac{1}{5}+\frac{1}{5}=\frac{3}{5}+\frac{1}{5}$$
$$x=\frac{4}{5}$$
Check:
$$3x-\frac{1}{5}-2x=\frac{1}{5}+\frac{2}{5}$$
$$3\cdot\frac{4}{5}-\frac{1}{5}-2\cdot\frac{4}{5}=\frac{1+2}{5}$$
$$\frac{12}{5}-\frac{1}{5}-\frac{8}{5}=\frac{3}{5}$$
$$\frac{12-1-8}{5}=\frac{3}{5}$$
$$\frac{11-8}{5}=\frac{3}{5}$$
$$\frac{3}{5}=\frac{3}{5} \quad \text{True}$$

43. The perimeter is the distance around. Add the lengths of the sides.
$$\frac{4}{20}+\frac{7}{20}+\frac{9}{20}=\frac{4+7+9}{20}=\frac{20}{20}=1$$
The perimeter is 1 inch.

45. The perimeter is the distance around. A rectangle has 2 sets of equal sides. Add the lengths of the sides.
$$\frac{5}{12}+\frac{7}{12}+\frac{5}{12}+\frac{7}{12}=\frac{5+7+5+7}{12}$$
$$=\frac{24}{12}$$
$$=\frac{2\cdot 12}{1\cdot 12}$$
$$=2$$
The perimeter is 2 meters.

47. Find the total workout time. Add the times in the morning and in the evening.
$$\frac{7}{8}+\frac{5}{8}=\frac{7+5}{8}=\frac{12}{8}=\frac{3\cdot 4}{2\cdot 4}=\frac{3}{2}$$
He worked out $\frac{3}{2}$ hours.

49. The fraction of employees enrolled in each plan in order from smallest to largest are $\frac{3}{20}$, $\frac{4}{20}$, $\frac{6}{20}$, and $\frac{7}{20}$.
Thus, the health plans in order from the smallest fraction of employees to the largest fraction of employees are Traditional fee-for-service, Point-of-Service, Health Maintenance Organization, and Preferred Provider Organization.

51.
$$\frac{6}{20}+\frac{4}{20}+\frac{3}{20}=\frac{6+4+3}{20}=\frac{13}{20}$$
$\frac{13}{20}$ of the employees are not covered by Preferred Provider Organization.

53. Subtract $\frac{16}{50}$ from $\frac{39}{50}$.
$$\frac{39}{50}-\frac{16}{50}=\frac{39-16}{50}=\frac{23}{50}$$
$\frac{23}{50}$ of the states had maximum speed limits that were less than 70 mph.

55. $3=3$
$4=2\cdot 2$
$\text{LCD}=3\cdot 2\cdot 2=12$

57. $9 = 3 \cdot 3$
$15 = 3 \cdot 5$
$\text{LCD} = 3 \cdot 3 \cdot 5 = 45$

59. $12 = 2 \cdot 2 \cdot 3$
$18 = 2 \cdot 3 \cdot 3$
$\text{LCD} = 2 \cdot 2 \cdot 3 \cdot 3 = 36$

61. $24 = 2 \cdot 2 \cdot 2 \cdot 3$
$x = x$
$\text{LCD} = 2 \cdot 2 \cdot 2 \cdot 3 \cdot \text{x} = 24x$

63. $25 = 5 \cdot 5$
$15 = 3 \cdot 5$
$6 = 2 \cdot 3$
$\text{LCD} = 2 \cdot 3 \cdot 5 \cdot 5 = 150$

65. $18 = 2 \cdot 3 \cdot 3$
$21 = 3 \cdot 7$
$\text{LCD} = 2 \cdot 3 \cdot 3 \cdot 7 = 126$

67. $15 = 3 \cdot 5$
$25 = 5 \cdot 5$
$\text{LCD} = 3 \cdot 5 \cdot 5 = 75$

69. $8 = 2 \cdot 2 \cdot 2$
$24 = 2 \cdot 2 \cdot 2 \cdot 3$
$\text{LCD} = 2 \cdot 2 \cdot 2 \cdot 3 = 24$

71. $25 = 5 \cdot 5$
$10 = 2 \cdot 5$
$\text{LCD} = 5 \cdot 5 \cdot 2 = 50$

73. $a = a$
$12 = 2 \cdot 2 \cdot 3$
$\text{LCD} = 2 \cdot 2 \cdot 3 \cdot \text{a} = 12a$

75. $3 = 3$
$21 = 3 \cdot 7$
$56 = 2 \cdot 2 \cdot 2 \cdot 7$
$\text{LCD} = 2 \cdot 2 \cdot 2 \cdot 3 \cdot 7 = 168$

77. $11 = 11$
$33 = 3 \cdot 11$
$121 = 11 \cdot 11$
$\text{LCD} = 3 \cdot 11 \cdot 11 = 363$

79. $\frac{4}{5} \cdot \frac{3}{7} = \frac{4 \cdot 3}{5 \cdot 7} = \frac{12}{35}$

81. $-2 + 10 = 8$

83. $\frac{2}{5} \div \frac{1}{2} = \frac{2}{5} \cdot \frac{2}{1} = \frac{2 \cdot 2}{5 \cdot 1} = \frac{4}{5}$

85. $-12 - 16 = -12 + (-16) = -28$

87. $\frac{4}{11} + \frac{5}{11} - \frac{3}{11} + \frac{2}{11} = \frac{4+5-3+2}{11} = \frac{8}{11}$

89. Subtract the sum of $\frac{38}{50}$ and $\frac{7}{50}$ from the total of $\frac{50}{50}$.

$$\begin{aligned} \frac{50}{50} - \left(\frac{38}{50} + \frac{7}{50}\right) &= \frac{50}{50} - \frac{45}{50} \\ &= \frac{50-45}{45} \\ &= \frac{5}{50} \\ &= \frac{1 \cdot 5}{10 \cdot 5} \\ &= \frac{1}{10} \end{aligned}$$

$\frac{1}{10}$ of American men over age 65 are either single or divorced.

91. Answers may vary.

Section 4.5

Practice Problems

1. $\frac{4}{7} + \frac{3}{14}$:

Step 1. The LCD for the denominators 7 and 14 is 14.

Step 2. $\frac{4}{7} = \frac{4 \cdot 2}{7 \cdot 2} = \frac{8}{14}$; $\frac{3}{14}$ has a denominator of 14.

Step 3. $\frac{4}{7} + \frac{3}{14} = \frac{8}{14} + \frac{3}{14} = \frac{11}{14}$

Step 4. $\frac{11}{14}$ is in simplest form.

2. $\frac{3}{7}-\frac{9}{10}$:

Step 1. The LCD for denominators 7 and 10 is 70.

Step 2.
$$\frac{3}{7}=\frac{3\cdot 10}{7\cdot 10}=\frac{30}{70} \text{ and } \frac{9}{10}=\frac{9\cdot 7}{10\cdot 7}=\frac{63}{70}$$

Step 3.
$$\frac{3}{7}-\frac{9}{10}=\frac{30}{70}-\frac{63}{70}=\frac{30-63}{70}=\frac{-33}{70}$$
$$\text{or } -\frac{33}{70}$$

Step 4. $-\frac{33}{70}$ is in simplest form.

3. $-\frac{1}{5}+\frac{3}{20}$:

The LCD denominators 5 and 20 is 20.
$$-\frac{1}{5}+\frac{3}{20}=\frac{-1\cdot 4}{5\cdot 4}+\frac{3}{20}$$
$$=\frac{-4}{20}+\frac{3}{20}$$
$$=\frac{-1}{20} \text{ or } -\frac{1}{20}$$

4. $5+\frac{3y}{4}$:

Recall that $5=\frac{5}{1}$. The LCD for denominators 1 and 4 is 4.
$$\frac{5}{1}+\frac{3y}{4}=\frac{5\cdot 4}{1\cdot 4}+\frac{3y}{4}=\frac{20}{4}+\frac{3y}{4}=\frac{20+3y}{4}$$

5. $\frac{5}{8}-\frac{1}{3}-\frac{1}{12}$:

The LCD of 8, 3 and 12 is 24.
$$\frac{5}{8}-\frac{1}{3}-\frac{1}{12}=\frac{5\cdot 3}{8\cdot 3}-\frac{1\cdot 8}{3\cdot 8}-\frac{1\cdot 2}{12\cdot 2}$$
$$=\frac{15}{24}-\frac{8}{24}-\frac{2}{24}$$
$$=\frac{5}{24}$$

6. $x+y=\frac{5}{11}+\frac{4}{9}$

The LCD of 11 and 9 is 99.
$$\frac{5}{11}+\frac{4}{9}=\frac{5\cdot 9}{11\cdot 9}+\frac{4\cdot 11}{9\cdot 11}=\frac{45}{99}+\frac{44}{99}=\frac{89}{99}$$

7.
$$y-\frac{2}{3}=\frac{5}{12}$$
$$y-\frac{2}{3}+\frac{2}{3}=\frac{5}{12}+\frac{2}{3}$$
$$y=\frac{5}{12}+\frac{2\cdot 4}{3\cdot 4}$$
$$y=\frac{5}{12}+\frac{8}{12}$$
$$y=\frac{13}{12}$$

8. The phrase "total amount" tells us to add.

Add the three amounts: $\frac{3}{5}+\frac{2}{10}+\frac{2}{15}$.

The LCD for the denominators 5, 10, and 15 is 30.
$$\frac{3}{5}=\frac{3\cdot 6}{5\cdot 6}=\frac{18}{30},\ \frac{2}{10}=\frac{2\cdot 3}{10\cdot 3}=\frac{6}{30},$$
$$\frac{2}{15}=\frac{2\cdot 2}{15\cdot 2}=\frac{4}{30}$$
$$\frac{3}{5}+\frac{2}{10}+\frac{2}{15}=\frac{18}{30}+\frac{6}{30}+\frac{4}{30}=\frac{28}{30}$$
$$\frac{28}{30}=\frac{14\cdot 2}{15\cdot 2}=\frac{14}{15}$$

The homeowner needs $\frac{14}{15}$ of a cubic yard.

Yes, she bought enough because 1 or $\frac{15}{15}$ is greater than $\frac{14}{15}$.

9. The phrase "find the difference" tells us to subtract. Subtract $\frac{2}{3}$ from $\frac{4}{5}$: $\frac{4}{5}-\frac{2}{3}$.
The LCD for the denominators 5 and 3 is 15.
$\frac{4}{5}=\frac{4\cdot 3}{5\cdot 3}=\frac{12}{15}$, $\frac{2\cdot 5}{3\cdot 5}=\frac{10}{15}$
$\frac{4}{5}-\frac{2}{3}=\frac{12}{15}-\frac{10}{15}=\frac{2}{15}$
$\frac{2}{15}$ is in simplest form.
The difference in the lengths is $\frac{2}{15}$ of a foot.

Calculator Explorations

1. $\frac{1}{16}+\frac{2}{5}=\frac{37}{80}$
2. $\frac{3}{20}+\frac{2}{25}=\frac{23}{100}$
3. $\frac{4}{9}+\frac{7}{8}=\frac{95}{72}$
4. $\frac{9}{11}+\frac{5}{12}=\frac{163}{132}$
5. $\frac{10}{17}+\frac{12}{19}=\frac{394}{323}$
6. $\frac{14}{31}+\frac{15}{21}=\frac{253}{217}$

Mental Math

1. The LCD of 2 and 3 is $2 \cdot 3 = 6$.
2. The LCD of 2 and 4 is 4.
3. The LCD of 6 and 12 is 12.
4. The LCD of 5 and 10 is 10.
5. The LCD of 7 and 8 is $7\cdot 2^3 = 56$.
6. The LCD of 24 and 3 is 24.
7. The LCD of 12 and 4 is 12.
8. The LCD of 3 and 11 is $3 \cdot 11 = 33$.

Exercise Set 4.5

1. The LCD of 3 and 6 is 6.
$\frac{2}{3}+\frac{1}{6}=\frac{2\cdot 2}{3\cdot 2}+\frac{1}{6}=\frac{4}{6}+\frac{1}{6}=\frac{5}{6}$

3. The LCD of 2 and 3 is 6.
$\frac{1}{2}-\frac{1}{3}=\frac{1\cdot 3}{2\cdot 3}-\frac{1\cdot 2}{3\cdot 2}=\frac{3}{6}-\frac{2}{6}=\frac{1}{6}$

5. The LCD of 11 and 33 is 33.
$$-\frac{2}{11}+\frac{2}{33}=-\frac{2\cdot 3}{11\cdot 3}+\frac{2}{33}$$
$$=-\frac{6}{33}+\frac{2}{33}$$
$$=-\frac{4}{33}$$

7. The LCD of 14 and 7 is 14.
$\frac{3x}{14}-\frac{3}{7}=\frac{3x}{14}-\frac{3\cdot 2}{7\cdot 2}=\frac{3x}{14}-\frac{6}{14}=\frac{3x-6}{14}$

9. The LCD of 35 and 7 is 35.
$$\frac{11}{35}+\frac{2}{7}=\frac{11}{35}+\frac{2\cdot 5}{7\cdot 5}$$
$$=\frac{11}{35}+\frac{10}{35}$$
$$=\frac{21}{35}$$
$$=\frac{3\cdot 7}{5\cdot 7}$$
$$=\frac{3}{5}$$

11. The LCD of 1 and 12 is 12.
$$\frac{2y}{1}-\frac{5}{12}=\frac{2y\cdot 12}{1\cdot 12}-\frac{5}{12}$$
$$=\frac{24y}{12}-\frac{5}{12}$$
$$=\frac{24y-5}{12}$$

13. The LCD of 12 and 9 is 36.
$\frac{5}{12}-\frac{1}{9}=\frac{5\cdot 3}{12\cdot 3}-\frac{1\cdot 4}{9\cdot 4}=\frac{15}{36}-\frac{4}{36}=\frac{11}{36}$

15. The LCD of 7 and 1 is 7.
$\frac{5}{7}+1=\frac{5}{7}+\frac{1}{1}=\frac{5}{7}+\frac{1\cdot 7}{1\cdot 7}=\frac{5}{7}+\frac{7}{7}=\frac{12}{7}$

17. The LCD of 11 and 9 is 99.

$$\frac{5a}{11}+\frac{4a}{9}=\frac{5a\cdot 9}{11\cdot 9}+\frac{4a\cdot 11}{9\cdot 11}=\frac{45a}{99}+\frac{44a}{99}=\frac{89a}{99}$$

19. The LCD of 3 and 6 is 6.

$$\frac{2y}{3}-\frac{1}{6}=\frac{2y\cdot 2}{3\cdot 2}-\frac{1}{6}=\frac{4y}{6}-\frac{1}{6}=\frac{4y-1}{6}$$

21. The LCD of 2 and x is $2x$.

$$\frac{1}{2}+\frac{3}{x}=\frac{1\cdot x}{2\cdot x}+\frac{3\cdot 2}{x\cdot 2}=\frac{x}{2x}+\frac{6}{2x}=\frac{x+6}{2x}$$

23. The LCD of 11 and 33 is 33.

$$-\frac{2}{11}-\frac{2}{33}=-\frac{2\cdot 3}{11\cdot 3}-\frac{2}{33}=-\frac{6}{33}-\frac{2}{33}=-\frac{8}{33}$$

25. The LCD of 14 and 7 is 14.

$$\frac{9}{14}-\frac{3}{7}=\frac{9}{14}-\frac{3\cdot 2}{7\cdot 2}=\frac{9}{14}-\frac{6}{14}=\frac{3}{14}$$

27. The LCD of 35 and 7 is 35.

$$\frac{11y}{35}-\frac{2}{7}=\frac{11y}{35}-\frac{2\cdot 5}{7\cdot 5}=\frac{11y}{35}-\frac{10}{35}=\frac{11y-10}{35}$$

29. The LCD of 9 and 12 is 36.

$$\frac{1}{9}-\frac{5}{12}=\frac{1\cdot 4}{9\cdot 4}-\frac{5\cdot 3}{12\cdot 3}=\frac{4}{36}-\frac{15}{36}=-\frac{11}{36}$$

31. The LCD of 15 and 12 is 60.

$$\frac{7}{15}-\frac{5}{12}=\frac{7\cdot 4}{15\cdot 4}-\frac{5\cdot 5}{12\cdot 5}=\frac{28}{60}-\frac{25}{60}=\frac{3}{60}=\frac{3}{3\cdot 20}=\frac{1}{20}$$

33. The LCD of 7 and 8 is 56.

$$\frac{5}{7}-\frac{1}{8}=\frac{5\cdot 8}{7\cdot 8}-\frac{1\cdot 7}{8\cdot 7}=\frac{40}{56}-\frac{7}{56}=\frac{33}{56}$$

35. The LCD of 8 and 16 is 16.

$$\frac{7}{8}+\frac{3}{16}=\frac{7\cdot 2}{8\cdot 2}+\frac{3}{16}=\frac{14}{16}+\frac{3}{16}=\frac{17}{16}$$

37. $\frac{5}{9}+\frac{3}{9}=\frac{8}{9}$

39. The LCD of 11 and 3 is 33.

$$\frac{5}{11}+\frac{y}{3}=\frac{5\cdot 3}{11\cdot 3}+\frac{y\cdot 11}{3\cdot 11}=\frac{15}{33}+\frac{11y}{33}=\frac{15+11y}{33}$$

41. The LCD of 6 and 7 is 42.

$$-\frac{5}{6}-\frac{3}{7}=-\frac{5\cdot 7}{6\cdot 7}-\frac{3\cdot 6}{7\cdot 6}=-\frac{35}{42}-\frac{18}{42}=-\frac{53}{42}$$

43. The LCD of 9 and 6 is 18.

$$\frac{7}{9}-\frac{1}{6}=\frac{7\cdot 2}{9\cdot 2}-\frac{1\cdot 3}{6\cdot 3}=\frac{14}{18}-\frac{3}{18}=\frac{11}{18}$$

45. The LCD of 3 and 13 is 39.

$$\frac{2a}{3}+\frac{6a}{13}=\frac{2a\cdot 13}{3\cdot 13}+\frac{6a\cdot 3}{13\cdot 3}=\frac{26a}{39}+\frac{18a}{39}=\frac{44a}{39}$$

47. The LCD of 30 and 12 is 60.

$$\frac{7}{30}-\frac{5}{12}=\frac{7\cdot 2}{30\cdot 2}-\frac{5\cdot 5}{12\cdot 5}=\frac{14}{60}-\frac{25}{60}=-\frac{11}{60}$$

49. The LCD of 9 and y is $9y$.

$$\frac{5}{9}+\frac{1}{y}=\frac{5\cdot y}{9\cdot y}+\frac{1\cdot 9}{y\cdot 9}=\frac{5y}{9y}+\frac{9}{9y}=\frac{5y+9}{9y}$$

51. The LCD of 5 and 9 is 45.

$$\frac{4}{5}+\frac{4}{9}=\frac{4\cdot 9}{5\cdot 9}+\frac{4\cdot 5}{9\cdot 5}=\frac{36}{45}+\frac{20}{45}=\frac{56}{45}$$

53. The LCD of $9x$ and 8 is $72x$.

$$\frac{5}{9x}+\frac{1}{8}=\frac{5\cdot 8}{9x\cdot 8}+\frac{1\cdot 9x}{8\cdot 9x}=\frac{40}{72x}+\frac{9x}{72x}=\frac{40+9x}{72x}$$

55. $\frac{9}{16}+\frac{5}{16}+\frac{2}{16}=\frac{9+5+2}{16}=\frac{16}{16}=1$

57. The LCD of 5, 3, and 10 is 30.

$$-\frac{2}{5}+\frac{1}{3}-\frac{3}{10}=-\frac{2\cdot 6}{5\cdot 6}+\frac{1\cdot 10}{3\cdot 10}-\frac{3\cdot 3}{10\cdot 3}=-\frac{12}{30}+\frac{10}{30}-\frac{9}{30}=-\frac{11}{30}$$

59. The LCD of 2, 4, and 16 is 16.

$$\frac{x}{2}+\frac{x}{4}+\frac{2x}{16}=\frac{x\cdot 8}{2\cdot 8}+\frac{x\cdot 4}{4\cdot 4}+\frac{2x}{16}=\frac{8x}{16}+\frac{4x}{16}+\frac{2x}{16}=\frac{14x}{16}=\frac{2\cdot 7\cdot x}{2\cdot 4}=\frac{7x}{8}$$

61. The LCD of 2, 4, and 16 is 16.

$$-\frac{1}{2}-\frac{1}{4}-\frac{1}{16}=-\frac{1\cdot 8}{2\cdot 8}-\frac{1\cdot 4}{4\cdot 4}-\frac{1}{16}=-\frac{8}{16}-\frac{4}{16}-\frac{1}{16}=-\frac{13}{16}$$

63. The LCD of 5, 4, and 2 is 20.

$$\frac{6}{5}-\frac{3}{4}+\frac{1}{2}=\frac{6\cdot 4}{5\cdot 4}-\frac{3\cdot 5}{4\cdot 5}+\frac{1\cdot 10}{2\cdot 10}=\frac{24}{20}-\frac{15}{20}+\frac{10}{20}=\frac{19}{20}$$

65. The LCD of 12, 24, and 6 is 24.

$$-\frac{9}{12}+\frac{17}{24}-\frac{1}{6}=-\frac{9\cdot 2}{12\cdot 2}+\frac{17}{24}-\frac{1\cdot 4}{6\cdot 4}=-\frac{18}{24}+\frac{17}{24}-\frac{4}{24}=-\frac{5}{24}$$

67. The LCD of 16, 18, and 24 is 144.

$$\frac{3}{16}+\frac{7x}{18}+\frac{5}{24}=\frac{3\cdot 9}{16\cdot 9}+\frac{7x\cdot 8}{18\cdot 8}+\frac{5\cdot 6}{24\cdot 6}=\frac{27}{144}+\frac{56x}{144}+\frac{30}{144}=\frac{57+56x}{144}$$

69. The LCD of 8, 7, and 14 is 56.

$$\begin{aligned}\frac{3x}{8}+\frac{2x}{7}-\frac{5}{14}&=\frac{3x\cdot 7}{8\cdot 7}+\frac{2x\cdot 8}{7\cdot 8}-\frac{5\cdot 4}{14\cdot 4}\\&=\frac{21x}{56}+\frac{16x}{56}-\frac{20}{56}\\&=\frac{37x-20}{56}\end{aligned}$$

71.
$$\begin{aligned}x+y&=\frac{1}{3}+\frac{3}{4}\\&=\frac{1\cdot 4}{3\cdot 4}+\frac{3\cdot 3}{4\cdot 3}\\&=\frac{4}{12}+\frac{9}{12}\\&=\frac{13}{12}\end{aligned}$$

73. $xy=\frac{1}{3}\cdot\frac{3}{4}=\frac{1\cdot 3}{3\cdot 4}=\frac{1}{4}$

75.
$$\begin{aligned}2y+x&=2\cdot\frac{3}{4}+\frac{1}{3}\\&=\frac{2}{1}\cdot\frac{3}{4}+\frac{1}{3}\\&=\frac{2\cdot 3}{1\cdot 2\cdot 2}+\frac{1}{3}\\&=\frac{3}{2}+\frac{1}{3}\\&=\frac{3\cdot 3}{2\cdot 3}+\frac{1\cdot 2}{3\cdot 2}\\&=\frac{9}{6}+\frac{2}{6}\\&=\frac{11}{6}\end{aligned}$$

77.
$$\begin{aligned}x-\frac{1}{12}&=\frac{5}{6}\\x-\frac{1}{12}+\frac{1}{12}&=\frac{5}{6}+\frac{1}{12}\\x&=\frac{5\cdot 2}{6\cdot 2}+\frac{1}{12}\\x&=\frac{10}{12}+\frac{1}{12}\\&=\frac{11}{12}\end{aligned}$$

Check:
$$\begin{aligned}x-\frac{1}{12}&=\frac{5}{6}\\\frac{11}{12}-\frac{1}{12}&=\frac{5}{6}\\\frac{10}{12}&=\frac{5}{6}\\\frac{5\cdot 2}{6\cdot 2}&=\frac{5}{6}\\\frac{5}{6}&=\frac{5}{6}\ \text{True}\end{aligned}$$

79.
$$\begin{aligned}\frac{2}{5}+y&=-\frac{3}{10}\\\frac{2}{5}+y-\frac{2}{5}&=-\frac{3}{10}-\frac{2}{5}\\y&=-\frac{3}{10}-\frac{2\cdot 2}{5\cdot 2}\\y&=-\frac{3}{10}-\frac{4}{10}\\&=-\frac{7}{10}\end{aligned}$$

Check:
$$\begin{aligned}\frac{2}{5}+y&=-\frac{3}{10}\\\frac{2}{5}+\left(-\frac{7}{10}\right)&=-\frac{3}{10}\\\frac{2\cdot 2}{5\cdot 2}-\frac{7}{10}&=-\frac{3}{10}\\\frac{4}{10}-\frac{7}{10}&=-\frac{3}{10}\\-\frac{3}{10}&=-\frac{3}{10}\ \text{True}\end{aligned}$$

81.
$$7z+\frac{1}{16}-6z=\frac{3}{4}$$
$$(7z-6z)+\frac{1}{16}=\frac{3}{4}$$
$$z+\frac{1}{16}=\frac{3}{4}$$
$$z+\frac{1}{16}-\frac{1}{16}=\frac{3}{4}-\frac{1}{16}$$
$$z=\frac{3\cdot 4}{4\cdot 4}-\frac{1}{16}$$
$$z=\frac{12}{16}-\frac{1}{16}$$
$$=\frac{11}{16}$$

Check:
$$7z+\frac{1}{16}-6z=\frac{3}{4}$$
$$7\cdot\frac{11}{16}+\frac{1}{16}-6\cdot\frac{11}{16}=\frac{3}{4}$$
$$\frac{77}{16}+\frac{1}{16}-\frac{66}{16}=\frac{3}{4}$$
$$\frac{12}{16}=\frac{3}{4}$$
$$\frac{3\cdot 4}{4\cdot 4}=\frac{3}{4}$$
$$\frac{3}{4}=\frac{3}{4} \text{ True}$$

83.
$$-\frac{2}{9}=x-\frac{5}{6}$$
$$-\frac{2}{9}+\frac{5}{6}=x-\frac{5}{6}+\frac{5}{6}$$
$$-\frac{2\cdot 2}{9\cdot 2}+\frac{5\cdot 3}{6\cdot 3}=x$$
$$-\frac{4}{18}+\frac{15}{18}=x$$
$$\frac{11}{18}=x$$

Check:
$$-\frac{2}{9}=x-\frac{5}{6}$$
$$-\frac{2}{9}=\frac{11}{18}-\frac{5}{6}$$
$$-\frac{2}{9}=\frac{11}{18}-\frac{5\cdot 3}{6\cdot 3}$$
$$-\frac{2}{9}=\frac{11}{18}-\frac{15}{18}$$
$$-\frac{2}{9}=-\frac{4}{18}$$
$$-\frac{2}{9}=-\frac{2\cdot 2}{2\cdot 9}$$
$$-\frac{2}{9}=-\frac{2}{9} \text{ True}$$

85. Add the lengths of the 4 sides. A parallelogram has 2 sets of equal sides.
$$\frac{1}{3}+\frac{4}{5}+\frac{1}{3}+\frac{4}{5}=\frac{1\cdot 5}{3\cdot 5}+\frac{4\cdot 3}{5\cdot 3}+\frac{1\cdot 5}{3\cdot 5}+\frac{4\cdot 3}{5\cdot 3}$$
$$=\frac{5}{15}+\frac{12}{15}+\frac{5}{15}+\frac{12}{15}$$
$$=\frac{34}{15}$$

The perimeter is $\frac{34}{15}$ centimeters.

87. Add the lengths of the 4 sides.

$$\frac{1}{4}+\frac{1}{5}+\frac{1}{2}+\frac{3}{4}=\frac{1\cdot 5}{4\cdot 5}+\frac{1\cdot 4}{5\cdot 4}+\frac{1\cdot 10}{2\cdot 10}+\frac{3\cdot 5}{4\cdot 5}$$
$$=\frac{5}{20}+\frac{4}{20}+\frac{10}{20}+\frac{15}{20}$$
$$=\frac{34}{20}$$
$$=\frac{2\cdot 17}{2\cdot 10}$$
$$=\frac{17}{10}$$

The perimeter is $\frac{17}{10}$ meters.

89. Subtract $\frac{5}{264}$ from $\frac{1}{4}$.

$$\frac{1}{4}-\frac{5}{264}=\frac{1\cdot 66}{4\cdot 66}-\frac{5}{264}=\frac{66-5}{264}=\frac{61}{264}$$

A killer bee will chase a person $\frac{61}{264}$ mile farther.

91. Subtract $\frac{4}{25}$ from $\frac{13}{20}$.

$$\frac{13}{20}-\frac{4}{25}=\frac{13\cdot 5}{20\cdot 5}-\frac{4\cdot 4}{25\cdot 4}$$
$$=\frac{65}{100}-\frac{16}{100}$$
$$=\frac{49}{100}$$

$\frac{49}{100}$ of students name math or science as their favorite subject.

93. Add the fractions for 1 or 2 times per week and 3 times per week.

$$\frac{23}{50}+\frac{31}{100}=\frac{23\cdot 2}{50\cdot 2}+\frac{31}{100}=\frac{46}{100}+\frac{81}{100}=\frac{77}{100}$$

$\frac{77}{100}$ of Americans eat pasta 1, 2, or 3 times per week.

95. Less than 50 miles per week includes 10 to 49 miles and less than 10 miles. Add these amounts.

$$\frac{23}{100}+\frac{3}{50}=\frac{23}{100}+\frac{3\cdot 2}{50\cdot 2}=\frac{23}{100}+\frac{6}{100}=\frac{29}{100}$$

$\frac{29}{100}$ of adults drive less than 50 miles in an average week.

97. $\left(\frac{5}{6}\right)^2=\frac{5}{6}\cdot\frac{5}{6}=\frac{5\cdot 5}{6\cdot 6}=\frac{25}{36}$

99. $\left(-\frac{5}{6}\right)^2=-\frac{5}{6}\cdot-\frac{5}{6}=\frac{5\cdot 5}{6\cdot 6}=\frac{25}{36}$

101. 57,236 rounded to the nearest hundred is 57,200.

103. 327 rounded to the nearest ten is 330.

105.
$$\frac{30}{55}+\frac{1000}{1760}=\frac{30\cdot 32}{55\cdot 32}+\frac{1000}{1760}$$
$$=\frac{960}{1760}+\frac{1000}{1760}$$
$$=\frac{1960}{1760}$$
$$=\frac{49\cdot 40}{44\cdot 40}$$
$$=\frac{49}{44}$$

107. Answers may vary.

109. Find the sum of the fractions for Asia and Europe.

$$\frac{58}{193}+\frac{38}{579}=\frac{58\cdot 3}{193\cdot 3}+\frac{38}{579}$$
$$=\frac{174}{579}+\frac{38}{579}$$
$$=\frac{212}{579}$$

$\frac{212}{579}$ of the world's land area is accounted for by Asia and Europe.

111. Multiply 57,900,000 by the sum of the fractions for Asia and Europe $\left(\frac{212}{579}\text{ from Exercise 57}\right)$.

$$\begin{aligned}57{,}900{,}000\cdot\frac{212}{579} &= \frac{57{,}900{,}000}{1}\cdot\frac{212}{579}\\ &= \frac{579\cdot 100{,}000\cdot 212}{1\cdot 579}\\ &= \frac{21{,}200{,}000}{1}\\ &= 21{,}200{,}000\end{aligned}$$

The combined land area of the European and Asian continents is 21,200,000 square miles.

Integrated Review

1. The figure has 4 equal parts. One part is shaded. The fraction shaded is $\frac{1}{4}$.

2. The figure has 6 equal parts. Three parts are shaded. The fraction shaded is $\frac{3}{6}$ or $\frac{1}{2}$.

3. Each figure has 4 equal parts. Seven parts are shaded. The fraction shaded is $\frac{7}{4}$.

4. Number of people getting less than 8 hours of sleep → 73
   Total number of people surveyed → 85
   $\frac{73}{85}$

5. $6 = 2\cdot 3$

6. $70 = 2\cdot 35 = 2\cdot 5\cdot 7$

7. $$\begin{aligned}252 &= 2\cdot 126\\ &= 2\cdot 2\cdot 63\\ &= 2\cdot 2\cdot 7\cdot 9\\ &= 2\cdot 2\cdot 7\cdot 3\cdot 3\\ &= 2^2\cdot 3^2\cdot 7\end{aligned}$$

8. $\frac{2}{14} = \frac{2}{2\cdot 7} = \frac{1}{7}$

9. $\frac{20}{24} = \frac{4\cdot 5}{4\cdot 6} = \frac{5}{6}$

10. $\frac{18}{38} = \frac{2\cdot 9}{2\cdot 19} = \frac{9}{19}$

11. $\frac{42}{110} = \frac{2\cdot 21}{2\cdot 55} = \frac{21}{55}$

12. $\frac{32}{64} = \frac{32}{2 \cdot 32} = \frac{1}{2}$

13. $\frac{72}{80} = \frac{8 \cdot 9}{8 \cdot 10} = \frac{9}{10}$

14. Number of states that are not adjacent to any other state → 2
Number of states → 50

$\frac{2}{50} = \frac{2}{2 \cdot 25} = \frac{1}{25}$

15. $\frac{1}{5} + \frac{3}{5} = \frac{1+3}{5} = \frac{4}{5}$

16. $\frac{1}{5} - \frac{3}{5} = \frac{1-3}{5} = \frac{-2}{5} = -\frac{2}{5}$

17. $\frac{1}{5} \cdot \frac{3}{5} = \frac{1 \cdot 3}{5 \cdot 5} = \frac{3}{25}$

18. $\frac{1}{5} \div \frac{3}{5} = \frac{1}{5} \cdot \frac{5}{3} = \frac{1 \cdot 5}{5 \cdot 3} = \frac{1}{3}$

19. $\frac{2}{3} \div \frac{5}{6} = \frac{2}{3} \cdot \frac{6}{5} = \frac{2 \cdot 6}{3 \cdot 5} = \frac{2 \cdot 2 \cdot 3}{3 \cdot 5} = \frac{4}{5}$

20. $\frac{2}{3} \cdot \frac{5}{6} = \frac{2 \cdot 5}{3 \cdot 6} = \frac{2 \cdot 5}{3 \cdot 2 \cdot 3} = \frac{5}{9}$

21. $\frac{2}{3} - \frac{5}{6} = \frac{2 \cdot 2}{3 \cdot 2} - \frac{5}{6} = \frac{4}{6} - \frac{5}{6} = \frac{4-5}{6} = \frac{-1}{6} = -\frac{1}{6}$

22.
$$\begin{aligned} \frac{2}{3} + \frac{5}{6} &= \frac{2 \cdot 2}{3 \cdot 2} + \frac{5}{6} \\ &= \frac{4}{6} + \frac{5}{6} \\ &= \frac{4+5}{6} \\ &= \frac{9}{6} \\ &= \frac{3 \cdot 3}{3 \cdot 2} \\ &= \frac{3}{2} \end{aligned}$$

23. $-\frac{1}{7} \cdot -\frac{7}{18} = \frac{1 \cdot 7}{7 \cdot 18} = \frac{1}{18}$

24. $-\frac{4}{9}\cdot -\frac{3}{7}=\frac{4\cdot 3}{9\cdot 7}=\frac{4\cdot 3}{3\cdot 3\cdot 7}=\frac{4}{21}$

25. $-\frac{7}{8}\div 6=-\frac{7}{8}\div\frac{6}{1}=-\frac{7}{8}\cdot\frac{1}{6}=-\frac{7\cdot 1}{8\cdot 6}=-\frac{7}{48}$

26. $-\frac{9}{10}\div 5=-\frac{9}{10}\div\frac{5}{1}$
$=-\frac{9}{10}\cdot\frac{1}{5}$
$=-\frac{9\cdot 1}{10\cdot 5}$
$=-\frac{9}{50}$

27. $\frac{7}{8}+\frac{1}{20}=\frac{7\cdot 5}{8\cdot 5}+\frac{1\cdot 2}{20\cdot 2}=\frac{35}{40}+\frac{2}{40}=\frac{37}{40}$

28. $\frac{5}{12}-\frac{1}{9}=\frac{5\cdot 3}{12\cdot 3}-\frac{1\cdot 4}{9\cdot 4}=\frac{15}{36}-\frac{4}{36}=\frac{11}{36}$

29. $\frac{9}{11}-\frac{2}{3}=\frac{9\cdot 3}{11\cdot 3}-\frac{2\cdot 11}{3\cdot 11}=\frac{27}{33}-\frac{22}{33}=\frac{5}{33}$

30. $\frac{2}{9}+\frac{1}{18}=\frac{2\cdot 2}{9\cdot 2}+\frac{1}{18}=\frac{4}{18}+\frac{1}{18}=\frac{5}{18}$

31. $\frac{2}{9}+\frac{1}{18}+\frac{1}{3}=\frac{2\cdot 2}{9\cdot 2}+\frac{1}{18}+\frac{1\cdot 6}{3\cdot 6}$
$=\frac{4}{18}+\frac{1}{18}+\frac{6}{18}$
$=\frac{11}{18}$

32. $\frac{3}{10}+\frac{1}{5}+\frac{6}{25}=\frac{3\cdot 5}{10\cdot 5}+\frac{1\cdot 10}{5\cdot 10}+\frac{6\cdot 2}{25\cdot 2}$
$=\frac{15}{50}+\frac{10}{50}+\frac{12}{50}$
$=\frac{37}{50}$

Section 4.6

Practice Problems

1. $\frac{\frac{7y}{10}}{\frac{1}{5}}=\frac{7y}{10}\div\frac{1}{5}=\frac{7y}{10}\cdot\frac{5}{1}=\frac{7\cdot y\cdot 5}{2\cdot 5}=\frac{7y}{2}$

2. $\frac{\frac{1}{2}+\frac{1}{6}}{\frac{3}{4}-\frac{2}{3}}=\frac{\frac{1\cdot 3}{2\cdot 3}+\frac{1}{6}}{\frac{3\cdot 3}{4\cdot 3}-\frac{2\cdot 4}{3\cdot 4}}$
$=\frac{\frac{3}{6}+\frac{1}{6}}{\frac{9}{12}+\frac{8}{12}}$
$=\frac{\frac{4}{6}}{\frac{1}{12}}$
$=\frac{4}{6}\div\frac{1}{12}$
$=\frac{4}{6}\cdot\frac{12}{1}$
$=\frac{4\cdot 6\cdot 2}{6\cdot 1}$
$=\frac{8}{1}$ or 8

3. $\frac{\frac{1}{2}+\frac{1}{6}}{\frac{3}{4}-\frac{2}{3}}=\frac{12\left(\frac{1}{2}+\frac{1}{6}\right)}{12\left(\frac{3}{4}-\frac{2}{3}\right)}$
$=\frac{\left(12\cdot\frac{1}{2}\right)+\left(12\cdot\frac{1}{6}\right)}{\left(12\cdot\frac{3}{4}\right)-\left(12\cdot\frac{2}{3}\right)}$
$=\frac{6+2}{9-8}$
$=\frac{8}{1}$ or 8

4. $\frac{\frac{3}{4}}{\frac{x}{5}-1}=\frac{20\left(\frac{3}{4}\right)}{20\left(\frac{x}{5}-1\right)}$
$=\frac{20\cdot\frac{3}{4}}{\left(20\cdot\frac{x}{5}\right)-(20\cdot 1)}$
$=\frac{15}{4x-20}$

5. $\left(2-\frac{2}{3}\right)^3$:

First evaluate $2-\frac{2}{3}$.

$2-\frac{2}{3}=\frac{2}{1}-\frac{2}{3}=\frac{2\cdot 3}{1\cdot 3}-\frac{2}{3}=\frac{6}{3}-\frac{2}{3}=\frac{4}{3}$

Next, evaluate $\left(\frac{4}{3}\right)^3$.

$\left(\frac{4}{3}\right)^3=\frac{4}{3}\cdot\frac{4}{3}\cdot\frac{4}{3}=\frac{4\cdot 4\cdot 4}{3\cdot 3\cdot 3}=\frac{64}{27}$

6. $\left(-\frac{1}{2}+\frac{1}{5}\right)\left(\frac{7}{8}+\frac{1}{8}\right)=\left(\frac{-1\cdot 5}{2\cdot 5}+\frac{1\cdot 2}{5\cdot 2}\right)\left(\frac{8}{8}\right)$
$=\left(\frac{-5}{10}+\frac{2}{10}\right)(1)$
$=\left(-\frac{3}{10}\right)(1)$
$=-\frac{3}{10}$

7. $-\frac{3}{5}-xy=-\frac{3}{5}-\left(\frac{3}{10}\cdot\frac{2}{3}\right)$
$=-\frac{3}{5}-\left(\frac{3\cdot 2}{2\cdot 5\cdot 3}\right)$
$=-\frac{3}{5}-\frac{1}{5}$
$=\frac{-3-1}{5}$
$=\frac{-4}{5}$
$=-\frac{4}{5}$

Exercise Set 4.6

1. $\frac{\frac{1}{8}}{\frac{3}{4}}=\frac{1}{8}\div\frac{3}{4}=\frac{1}{8}\cdot\frac{4}{3}=\frac{1\cdot 4}{2\cdot 4\cdot 3}=\frac{1}{6}$

3. $\frac{\frac{9}{10}}{\frac{21}{10}}=\frac{9}{10}\div\frac{21}{10}=\frac{9}{10}\cdot\frac{10}{21}=\frac{3\cdot 3\cdot 10}{10\cdot 3\cdot 7}=\frac{3}{7}$

5. $\frac{\frac{2x}{27}}{\frac{4}{9}}=\frac{2x}{27}\div\frac{4}{9}=\frac{2x}{27}\cdot\frac{9}{4}=\frac{2\cdot x\cdot 3\cdot 3}{3\cdot 3\cdot 3\cdot 2\cdot 2}=\frac{x}{6}$

7. $\frac{\frac{3}{4}+\frac{2}{5}}{\frac{1}{2}+\frac{3}{5}}=\frac{20\left(\frac{3}{4}+\frac{2}{5}\right)}{20\left(\frac{1}{2}+\frac{3}{5}\right)}=\frac{15+8}{10+12}=\frac{23}{22}$

9. $\frac{\frac{3x}{4}}{5-\frac{1}{8}}=\frac{8\cdot\left(\frac{3x}{4}\right)}{8\cdot\left(\frac{5}{1}-\frac{1}{8}\right)}$
$=\frac{6x}{8\cdot\left(\frac{5}{1}\right)-8\left(\frac{1}{8}\right)}$
$=\frac{6x}{40-1}$
$=\frac{6x}{39}$
$=\frac{3\cdot 2\cdot x}{3\cdot 13}$
$=\frac{2x}{13}$

11. $2^2-\left(\frac{1}{3}\right)^2=4-\frac{1}{9}$
$=\frac{4\cdot 9}{1\cdot 9}-\frac{1}{9}$
$=\frac{36}{9}-\frac{1}{9}$
$=\frac{35}{9}$

13. $\left(\frac{2}{9}+\frac{4}{9}\right)\left(\frac{1}{3}-\frac{9}{10}\right)=\left(\frac{6}{9}\right)\left(\frac{1\cdot 10}{3\cdot 10}-\frac{9\cdot 3}{10\cdot 3}\right)$
$=\left(\frac{6}{9}\right)\left(\frac{10}{30}-\frac{27}{30}\right)$
$=\frac{6}{9}\cdot-\frac{17}{30}$
$=-\frac{6\cdot 17}{9\cdot 5\cdot 6}$
$=-\frac{17}{45}$

15. $\left(\frac{7}{8}-\frac{1}{2}\right)\div\frac{3}{11}=\left(\frac{7}{8}-\frac{1\cdot 4}{2\cdot 4}\right)\div\frac{3}{11}$
$=\left(\frac{7}{8}-\frac{4}{8}\right)\cdot\frac{11}{3}$
$=\frac{3}{8}\cdot\frac{11}{3}$
$=\frac{3\cdot 11}{8\cdot 3}$
$=\frac{11}{8}$

17. $5y-z=5\left(\frac{2}{5}\right)-\left(\frac{5}{6}\right)$
$=2-\frac{5}{6}$
$=\frac{2\cdot 6}{6}-\frac{5}{6}$
$=\frac{12}{6}-\frac{5}{6}$
$=\frac{7}{6}$

19. $\frac{x}{z}=\frac{-\frac{1}{3}}{\frac{5}{6}}$
$=-\frac{1}{3}\div\frac{5}{6}$
$=-\frac{1}{3}\cdot\frac{6}{5}$
$=-\frac{1\cdot 2\cdot 3}{3\cdot 5}$
$=-\frac{2}{5}$

21. $x^2-yz=\left(-\frac{1}{3}\right)^2-\left(\frac{2}{5}\right)\left(\frac{5}{6}\right)$
$=\frac{1}{9}-\left(\frac{2}{5}\right)\left(\frac{5}{6}\right)$
$=\frac{1}{9}-\frac{2\cdot 5}{5\cdot 2\cdot 3}$
$=\frac{1}{9}-\frac{1}{3}$
$=\frac{1}{9}-\frac{3}{9}$
$=-\frac{2}{9}$

23. $\frac{\frac{5}{24}}{\frac{1}{12}}=\frac{5}{24}\div\frac{1}{12}=\frac{5}{24}\cdot\frac{12}{1}=\frac{5\cdot 12}{2\cdot 12}=\frac{5}{2}$

25. $\left(\frac{3}{2}\right)^3+\left(\frac{1}{2}\right)^3=\frac{27}{8}+\frac{1}{8}=\frac{28}{8}=\frac{7\cdot 4}{2\cdot 4}=\frac{7}{2}$

27. $\left(-\frac{1}{3}\right)^2+\frac{1}{3}=\frac{1}{9}+\frac{1}{3}=\frac{1}{9}+\frac{3}{9}=\frac{4}{9}$

29. $\frac{2+\frac{1}{6}}{1-\frac{4}{3}}=\frac{6\left(2+\frac{1}{6}\right)}{6\left(1-\frac{4}{3}\right)}=\frac{12+1}{6-8}=\frac{13}{-2}=-\frac{13}{2}$

31. $\left(1-\frac{2}{5}\right)^2=\left(\frac{5}{5}-\frac{2}{5}\right)^2=\left(\frac{3}{5}\right)^2=\frac{9}{25}$

33. $\left(\frac{3}{4}-1\right)\left(\frac{1}{8}+\frac{1}{2}\right)=\left(\frac{3}{4}-\frac{4}{4}\right)\left(\frac{1}{8}+\frac{4}{8}\right)$
$=\left(-\frac{1}{4}\right)\left(\frac{5}{8}\right)$
$=-\frac{1\cdot 5}{4\cdot 8}$
$=-\frac{5}{32}$

35. $\left(-\frac{2}{9}-\frac{7}{9}\right)^4=\left(-\frac{9}{9}\right)^4=(-1)^4=1$

37. $\dfrac{\left(\frac{1}{2}-\frac{3}{8}\right)}{\left(\frac{3}{4}+\frac{1}{2}\right)}=\dfrac{8\left(\frac{1}{2}-\frac{3}{8}\right)}{8\left(\frac{3}{4}+\frac{1}{2}\right)}$
$=\dfrac{8\cdot\frac{1}{2}-8\cdot\frac{3}{8}}{8\cdot\frac{3}{4}+8\cdot\frac{1}{2}}$
$=\dfrac{4-3}{6+4}$
$=\dfrac{1}{10}$

39. $\left(\dfrac{3}{4}\div\dfrac{6}{5}\right)-\left(\dfrac{3}{4}\cdot\dfrac{6}{5}\right)=\left(\dfrac{3}{4}\cdot\dfrac{5}{6}\right)-\left(\dfrac{3}{2\cdot 2}\cdot\dfrac{2\cdot 3}{5}\right)$
$=\dfrac{3\cdot 5}{4\cdot 3\cdot 2}-\dfrac{3\cdot 2\cdot 3}{2\cdot 2\cdot 5}$
$=\dfrac{5}{8}-\dfrac{9}{10}$
$=\dfrac{5\cdot 5}{8\cdot 5}-\dfrac{9\cdot 4}{10\cdot 4}$
$=\dfrac{25}{40}-\dfrac{36}{40}$
$=-\dfrac{11}{40}$

41. $\dfrac{\frac{x}{3}+2}{5+\frac{1}{3}}=\dfrac{3\left(\frac{x}{3}+2\right)}{3\left(5+\frac{1}{3}\right)}$
$=\dfrac{3\cdot\frac{x}{3}+3\cdot 2}{3\cdot 5+3\cdot\frac{1}{3}}$
$=\dfrac{x+6}{15+1}$
$=\dfrac{x+6}{16}$

43. $2^3=2\cdot 2\cdot 2=8$

45. $5^2=5\cdot 5=25$

47. $\dfrac{1}{3}(3x)=\dfrac{1}{3}\cdot\dfrac{3x}{1}=\dfrac{1\cdot 3\cdot x}{3\cdot 1}=x$

49. $\dfrac{2}{3}\left(\dfrac{3}{2}a\right)=\dfrac{2}{3}\cdot\dfrac{3}{2}\cdot\dfrac{a}{1}=\dfrac{2\cdot 3\cdot a}{3\cdot 2\cdot 1}=\dfrac{a}{1}=a$

51. $\dfrac{2+x}{y}=\dfrac{2+\frac{3}{4}}{-\frac{4}{7}}$
$=\dfrac{28\left(2+\frac{3}{4}\right)}{28\left(-\frac{4}{7}\right)}$
$=\dfrac{28\cdot 2+28\cdot\frac{3}{4}}{-16}$
$=\dfrac{56+21}{-16}$
$=-\dfrac{77}{16}$

53. $x^2+7y=\left(\dfrac{3}{4}\right)^2+7\left(-\dfrac{4}{7}\right)$
$=\dfrac{9}{16}+\dfrac{7}{1}\cdot-\dfrac{4}{7}$
$=\dfrac{9}{16}+\left(-\dfrac{7\cdot 4}{1\cdot 7}\right)$
$=\dfrac{9}{16}-\dfrac{4}{1}$
$=\dfrac{9}{16}-\dfrac{4\cdot 16}{1\cdot 16}$
$=\dfrac{9}{16}-\dfrac{64}{16}$
$=-\dfrac{55}{16}$

55. $\dfrac{\frac{1}{2}+\frac{3}{4}}{2}=\dfrac{4\left(\frac{1}{2}+\frac{3}{4}\right)}{4\cdot 2}=\dfrac{4\cdot\frac{1}{2}+4\cdot\frac{3}{4}}{8}=\dfrac{2+3}{8}=\dfrac{5}{8}$

57. $\dfrac{\frac{1}{4}+\frac{2}{14}}{2}=\dfrac{28\left(\frac{1}{4}+\frac{2}{14}\right)}{28\cdot 2}$
$=\dfrac{28\cdot\left(\frac{1}{4}\right)+28\cdot\left(\frac{2}{14}\right)}{56}$
$=\dfrac{7+4}{56}$
$=\dfrac{11}{56}$

59. The average is halfway between *a* and *b*.

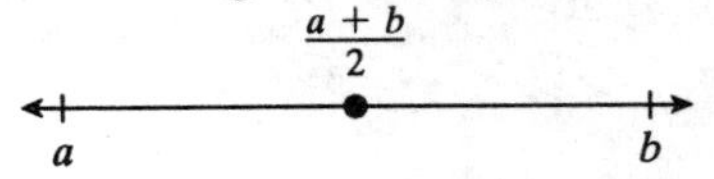

61. False, the average of two numbers is between the two numbers.

63. False, consider $-\frac{2}{3}+\frac{1}{3}=-\frac{1}{3}$.

65. True, consider $\frac{9}{4}-\frac{5}{4}=\frac{4}{4}=1$.

Section 4.7

Practice Problems

1.
$$\begin{aligned} \frac{1}{5}y &= 2 \\ 5\cdot\frac{1}{2}y &= 5\cdot 2 \\ 1\cdot y &= 10 \text{ or } y = 10 \end{aligned}$$

2.
$$\begin{aligned} \frac{5}{7}b &= 25 \\ \frac{7}{5}\cdot\frac{5}{7}b &= \frac{7}{5}\cdot 25 \\ 1b &= \frac{7\cdot 25}{5} \\ b &= 35 \end{aligned}$$

3.
$$\begin{aligned} -\frac{7}{10}x &= \frac{2}{5} \\ -\frac{10}{7}\cdot-\frac{7}{10}x &= -\frac{10}{7}\cdot\frac{2}{5} \\ 1x &= -\frac{10\cdot 2}{7\cdot 5} \\ x &= -\frac{4}{7} \end{aligned}$$

4.
$$\begin{aligned} 5x &= -\frac{3}{4} \\ \frac{1}{5}\cdot 5x &= \frac{1}{5}\cdot-\frac{3}{4} \\ 1x &= -\frac{1\cdot 3}{5\cdot 4} \\ x &= -\frac{3}{20} \end{aligned}$$

5.
$$\begin{aligned} \frac{y}{8}+\frac{3}{4} &= 2 \\ 8\left(\frac{y}{8}+\frac{3}{4}\right) &= 8(2) \\ 8\left(\frac{y}{8}\right)+8\left(\frac{3}{4}\right) &= 8(2) \\ y+6 &= 16 \\ y+6+(-6) &= 16+(-6) \\ y &= 10 \end{aligned}$$

6.
$$\begin{aligned} \frac{x}{5}-x &= \frac{1}{5} \\ 5\left(\frac{x}{5}-x\right) &= 5\cdot\frac{1}{5} \\ 5\left(\frac{x}{5}\right)-5(x) &= 5\cdot\frac{1}{5} \\ x-5x &= 1 \\ -4x &= 1 \\ \frac{-4x}{-4} &= \frac{1}{-4} \\ x &= -\frac{1}{4} \end{aligned}$$

7.
$$\begin{aligned} \frac{y}{2} &= \frac{y}{5}+\frac{3}{2} \\ 10\left(\frac{y}{2}\right) &= 10\left(\frac{y}{5}+\frac{3}{2}\right) \\ 10\left(\frac{y}{2}\right) &= 10\left(\frac{y}{5}\right)+10\left(\frac{3}{2}\right) \\ 5y &= 2y+15 \\ 5y-2y &= 2y+15-2y \\ 3y &= 15 \\ \frac{3y}{3} &= \frac{15}{3} \\ y &= 5 \end{aligned}$$

8.
$$\begin{aligned} \frac{9}{10}-\frac{y}{3} &= \frac{9\cdot 3}{10\cdot 3}-\frac{y\cdot 10}{3\cdot 10} \\ &= \frac{27}{30}-\frac{10y}{30} \\ &= \frac{27-10y}{30} \end{aligned}$$

Exercise Set 4.7

1. $7x = 2$
$$\frac{7x}{7} = \frac{2}{7}$$
$$x = \frac{2}{7}$$

3. $\frac{1}{4}x = 3$
$$4\frac{1}{4}x = 4 \cdot 3$$
$$x = 12$$

5. $\frac{2}{9}y = -6$
$$\frac{9}{2} \cdot \frac{2}{9}y = \frac{9}{2} \cdot -6$$
$$y = -27$$

7. $-\frac{4}{9}z = -\frac{3}{2}$
$$-\frac{9}{4} \cdot -\frac{4}{9}z = -\frac{9}{4} \cdot -\frac{3}{2}$$
$$z = \frac{27}{8}$$

9. $7a = \frac{1}{3}$
$$\frac{1}{7} \cdot 7a = \frac{1}{7} \cdot \frac{1}{3}$$
$$a = \frac{1}{21}$$

11. $-3x = -\frac{6}{11}$
$$-\frac{1}{3} \cdot -3x = -\frac{1}{3} \cdot -\frac{6}{11}$$
$$x = \frac{1 \cdot 6}{3 \cdot 11}$$
$$x = \frac{2 \cdot 3}{3 \cdot 11}$$
$$x = \frac{2}{11}$$

13. Multiply both sides of the equation by 3.
$$\frac{x}{3} + 2 = \frac{7}{3}$$
$$3\left(\frac{x}{3} + 2\right) = 3 \cdot \frac{7}{3}$$
$$x + 6 = 7$$
$$x + 6 - 6 = 7 - 6$$
$$x = 1$$

15. Multiply both sides of the equation by the LCD of 5 and 10, 10.
$$\frac{x}{5} - \frac{5}{10} = 1$$
$$10\left(\frac{x}{5} - \frac{5}{10}\right) = 10 \cdot 1$$
$$2x - 5 = 10$$
$$2x - 5 + 5 = 10 + 5$$
$$2x = 15$$
$$\frac{2x}{2} = \frac{15}{2}$$
$$x = \frac{15}{2}$$

17. Multiply both sides of the equation by the LCD of 2, 5, and 10:10.
$$\frac{1}{2} - \frac{3}{5} = \frac{x}{10}$$
$$10\left(\frac{1}{2} - \frac{3}{5}\right) = 10\left(\frac{x}{10}\right)$$
$$5 - 6 = x$$
$$-1 = x$$

19. Multiply both sides of the equation by the LCD of 3 and 5, 15.
$$\frac{x}{3} = \frac{x}{5} - 2$$
$$15\left(\frac{x}{3}\right) = 15\left(\frac{x}{5} - 2\right)$$
$$5x = 3x - 30$$
$$5x - 3x = 3x - 30 - 3x$$
$$2x = -30$$
$$\frac{2x}{2} = -\frac{30}{2}$$
$$x = -15$$

21. $\frac{x}{7}-\frac{4}{3}=\frac{x\cdot 3}{7\cdot 3}-\frac{4\cdot 7}{3\cdot 7}=\frac{3x}{21}-\frac{28}{21}$
$=\frac{3x-28}{21}$

23. $\frac{y}{2}+5=\frac{y}{2}+\frac{5\cdot 2}{1\cdot 2}=\frac{y}{2}+\frac{10}{2}=\frac{y+10}{2}$

25. $\frac{3x}{10}+\frac{x}{6}=\frac{3x\cdot 3}{10\cdot 3}+\frac{x\cdot 5}{6\cdot 5}$
$=\frac{9x}{30}+\frac{5x}{30}$
$=\frac{14x}{30}$
$=\frac{2\cdot 7x}{2\cdot 15x}$
$=\frac{7x}{15}$

27.
$$\begin{aligned}\frac{3}{8}x&=\frac{1}{2}\\ \frac{8}{3}\cdot\frac{3}{8}x&=\frac{8}{3}\cdot\frac{1}{2}\\ x&=\frac{2\cdot 4}{3\cdot 2}\\ x&=\frac{4}{3}\end{aligned}$$

29.
$$\begin{aligned}\frac{2}{3}-\frac{x}{5}&=\frac{4}{15}\\ 15\left(\frac{2}{3}-\frac{x}{5}\right)&=15\left(\frac{4}{15}\right)\\ 10-3x&=4\\ 10-10-3x&=4-10\\ -3x&=-6\\ \frac{-3x}{-3}&=\frac{-6}{-3}\\ x&=2\end{aligned}$$

31.
$$\begin{aligned}\frac{9}{14}z&=\frac{27}{20}\\ \left(\frac{14}{9}\right)\left(\frac{9}{14}\right)z&=\left(\frac{14}{9}\right)\left(\frac{27}{20}\right)\\ z&=\frac{2\cdot 7\cdot 9\cdot 3}{9\cdot 2\cdot 10}\\ &=\frac{7\cdot 3}{10}\\ &=\frac{21}{10}\end{aligned}$$

33.
$$\begin{aligned}-3m-5m&=\frac{4}{7}\\ -8m&=\frac{4}{7}\\ \left(-\frac{1}{8}\right)(-8m)&=\left(-\frac{1}{8}\right)\left(\frac{4}{7}\right)\\ m&=-\frac{1\cdot 4}{2\cdot 4\cdot 7}\\ m&=-\frac{1}{14}\end{aligned}$$

35.
$$\begin{aligned}\frac{x}{4}+1&=\frac{1}{4}\\ 4\left(\frac{x}{4}+1\right)&=4\cdot\frac{1}{4}\\ x+4&=1\\ x+4-4&=1-4\\ x&=-3\end{aligned}$$

37.
$$\begin{aligned}\frac{1}{5}y&=10\\ 5\left(\frac{1}{5}y\right)&=5\cdot 10\\ y&=50\end{aligned}$$

39. $\frac{5}{9}-\frac{2}{3}=\frac{5}{9}-\frac{2\cdot 3}{3\cdot 3}=\frac{5}{9}-\frac{6}{9}=-\frac{1}{9}$

41.
$$\begin{aligned}-\frac{3}{4}x&=\frac{9}{2}\\ -\frac{4}{3}\cdot-\frac{3}{4}x&=-\frac{4}{3}\cdot\frac{9}{2}\\ x&=-\frac{2\cdot 2\cdot 3\cdot 3}{3\cdot 2}\\ x&=-6\end{aligned}$$

43.
$$\frac{x}{2} - x = -2$$
$$2\left(\frac{x}{2} - x\right) = 2(-2)$$
$$x - 2x = -4$$
$$-x = -4$$
$$\frac{-x}{-1} = \frac{-4}{-1}$$
$$x = 4$$

45.
$$-\frac{5}{8}y = \frac{3}{16} - \frac{9}{16}$$
$$-\frac{5}{8}y = -\frac{6}{16}$$
$$-\frac{5}{8}y = -\frac{2 \cdot 3}{2 \cdot 8}$$
$$-\frac{5}{8}y = -\frac{3}{8}$$
$$\left(-\frac{8}{5}\right)\left(-\frac{5}{8}y\right) = \left(-\frac{8}{5}\right)\left(-\frac{3}{8}\right)$$
$$y = \frac{8 \cdot 3}{5 \cdot 8}$$
$$y = \frac{3}{5}$$

47.
$$17x - 25x = \frac{1}{3}$$
$$-8x = \frac{1}{3}$$
$$\left(-\frac{1}{8}\right)(-8x) = \left(-\frac{1}{8}\right)\left(\frac{1}{3}\right)$$
$$x = -\frac{1}{24}$$

49.
$$\frac{7}{6}x = \frac{1}{4} - \frac{2}{3}$$
$$12\left(\frac{7}{6}x\right) = 12\left(\frac{1}{4} - \frac{2}{3}\right)$$
$$14x = 3 - 8$$
$$14x = -5$$
$$\frac{14x}{14} = \frac{-5}{14}$$
$$x = -\frac{5}{14}$$

51.
$$\frac{b}{4} = \frac{b}{12} + \frac{2}{3}$$
$$12\left(\frac{b}{4}\right) = 12\left(\frac{b}{12} + \frac{2}{3}\right)$$
$$3b = b + 8$$
$$3b - b = b + 8 - b$$
$$2b = 8$$
$$\frac{2b}{2} = \frac{8}{2}$$
$$b = 4$$

53.
$$\frac{x}{3} + 2 = \frac{x}{2} + 8$$
$$6\left(\frac{x}{3} + 2\right) = 6\left(\frac{x}{2} + 8\right)$$
$$2x + 12 = 3x + 48$$
$$2x + 12 - 2x = 3x + 48 - 2x$$
$$12 = x + 48$$
$$12 - 48 = x + 48 - 48$$
$$-36 = x$$

55.
$$3 + \frac{1}{2} = \frac{3}{1} + \frac{1}{2}$$
$$= \frac{3 \cdot 2}{1 \cdot 2} + \frac{1}{2}$$
$$= \frac{6}{2} + \frac{1}{2}$$
$$= \frac{6+1}{2}$$
$$= \frac{7}{2}$$

57.
$$5 + \frac{9}{10} = \frac{5}{1} + \frac{9}{10}$$
$$= \frac{5 \cdot 10}{1 \cdot 10} + \frac{9}{10}$$
$$= \frac{50}{10} + \frac{9}{10}$$
$$= \frac{50+9}{10}$$
$$= \frac{59}{10}$$

59. $9-\frac{5}{6}=\frac{9}{1}-\frac{5}{6}$
$=\frac{9\cdot 6}{1\cdot 6}-\frac{5}{6}$
$=\frac{54}{6}-\frac{5}{6}$
$=\frac{54-5}{6}$
$=\frac{49}{6}$

61. Answers will vary.

63.
$$\frac{19}{53}=\frac{353x}{1431}+\frac{23}{27}$$
$$1431\left(\frac{19}{53}\right)=1431\left(\frac{353x}{1431}+\frac{23}{27}\right)$$
$$513=353x+1219$$
$$513-1219=353x+1219-1219$$
$$-706=353x$$
$$\frac{-706}{353}=\frac{353x}{353}$$
$$-2=x$$

Section 4.8

Practice Problems

1. Each part is $\frac{1}{3}$, and there are 8 parts shaded, or 2 wholes and 2 more parts: $\frac{8}{3}$ or $2\frac{2}{3}$.

2. Each part is $\frac{1}{4}$, and there are 5 parts shaded, or 1 whole and 1 more part: $\frac{5}{4}$ or $1\frac{1}{4}$.

3. a. $2\frac{5}{7}=\frac{7\cdot 2+5}{7}=\frac{19}{7}$

b. $5\frac{1}{3}=\frac{3\cdot 5+1}{3}=\frac{16}{3}$

c. $9\frac{3}{10}=\frac{10\cdot 9+3}{10}=\frac{93}{10}$

d. $1\frac{1}{5}=\frac{5\cdot 1+1}{5}=\frac{6}{5}$

4. a. $5\overline{)8}$ quotient 1; -5; remainder 3

$\frac{8}{5}=1\frac{3}{5}$

b. $6\overline{)17}$ quotient 2; -12; remainder 5

$\frac{17}{6}=2\frac{5}{6}$

c. $4\overline{)48}$ quotient 12; -4; 08; -8; 0

$\frac{48}{4}=12$

d. $4\overline{)35}$ quotient 8; -32; remainder 3

$\frac{35}{4}=8\frac{3}{4}$

e. $7\overline{)51}$ quotient 7; -49; remainder 2

$\frac{51}{7}=7\frac{2}{7}$

f. $20\overline{)\,21}$ quotient 1

-20

1

$$\frac{21}{20} = 1\frac{1}{20}$$

5. $2\frac{1}{2}\cdot\frac{8}{15} = \frac{5}{2}\cdot\frac{8}{15}$
$= \frac{5\cdot 8}{2\cdot 15}$
$= \frac{5\cdot 2\cdot 4}{2\cdot 5\cdot 3}$
$= \frac{4}{3}$ or $1\frac{1}{3}$

6. $3\frac{1}{5}\cdot 2\frac{3}{4} = \frac{16}{5}\cdot\frac{11}{4}$
$= \frac{16\cdot 11}{5\cdot 4}$
$= \frac{4\cdot 4\cdot 11}{5\cdot 4}$
$= \frac{4\cdot 11}{5}$
$= \frac{44}{5}$ or $8\frac{4}{5}$

7. $\frac{2}{3}\cdot 18 = \frac{2}{3}\cdot\frac{18}{1}$
$= \frac{2\cdot 18}{3\cdot 1}$
$= \frac{2\cdot 3\cdot 6}{3\cdot 1}$
$= \frac{2\cdot 6}{1\cdot 1}$
$= \frac{12}{1}$ or 12

8. $\frac{4}{9} \div 5 = \frac{4}{9} \div \frac{5}{1} = \frac{4}{9}\cdot\frac{1}{5} = \frac{4\cdot 1}{9\cdot 5} = \frac{4}{45}$

9. $\frac{8}{15} \div 3\frac{4}{5} = \frac{8}{15} \div \frac{19}{5}$
$= \frac{8}{15}\cdot\frac{5}{19}$
$= \frac{8\cdot 5}{15\cdot 19}$
$= \frac{8\cdot 5}{5\cdot 3\cdot 19}$
$= \frac{8}{3\cdot 19}$
$= \frac{8}{57}$

10. $3\frac{2}{5} \div 2\frac{2}{15} = \frac{17}{5} \div \frac{32}{15}$
$= \frac{17}{5}\cdot\frac{15}{32}$
$= \frac{17\cdot 15}{5\cdot 32}$
$= \frac{17\cdot 5\cdot 3}{5\cdot 32}$
$= \frac{17\cdot 3}{32}$
$= \frac{51}{32}$ or $1\frac{19}{32}$

11.

$$\begin{array}{rcr} 4\frac{2}{5} & = & 4\frac{4}{10} \\ +\ 5\frac{3}{10} & = & +\ 5\frac{3}{10} \\ \hline & & 9\frac{7}{10} \end{array}$$

12.

$$\begin{array}{rcr} 2\frac{5}{14} & = & 2\frac{5}{14} \\ +\ 5\frac{6}{7} & = & +\ 5\frac{12}{14} \\ \hline & & 7\frac{17}{14} = 7 + 1\frac{3}{14} = 8\frac{3}{14} \end{array}$$

13.
$$\begin{array}{rcr} 10 & = & 10 \\ 2\frac{1}{7} & = & 2\frac{5}{35} \\ +\ 3\frac{1}{5} & = & +\ 3\frac{7}{35} \\ \hline & & 15\frac{12}{35} \end{array}$$

14.
$$\begin{array}{rcr} 29\frac{7}{8} & = & 29\frac{14}{16} \\ -\ 13\frac{3}{16} & = & -\ 13\frac{3}{16} \\ \hline & & 16\frac{11}{16} \end{array}$$

15.
$$\begin{array}{rcrcr} 9\frac{7}{15} & = & 9\frac{7}{15} & = & 8\frac{22}{15} \\ -\ 5\frac{4}{5} & = & -\ 5\frac{12}{15} & = & -\ 5\frac{12}{15} \\ \hline & & & & 3\frac{10}{15} = 3\frac{2}{3} \end{array}$$

16.
$$\begin{array}{rcr} 25 & = & 24\frac{9}{9} \\ -10\frac{2}{9} & = & -10\frac{2}{9} \\ \hline & & 14\frac{7}{9} \end{array}$$

17. The phrase "how much larger" tells us to subtract. Subtract $19\frac{5}{12}$ feet from $23\frac{1}{4}$ feet.

$$\begin{array}{rcrcr} 23\frac{1}{4} & = & 23\frac{3}{12} & = & 22\frac{15}{12} \\ -19\frac{5}{12} & = & -19\frac{5}{12} & = & -19\frac{5}{12} \\ \hline & & & & 3\frac{10}{12} = 3\frac{5}{6} \end{array}$$

The girth of the largest known American Beech tree is $3\frac{5}{6}$ feet larger than the girth of the largest known Sugar Maple tree.

18.
$$\begin{aligned} 30 \div 2\frac{1}{7} &= \frac{30}{1} \div \frac{15}{7} \\ &= \frac{30}{1} \cdot \frac{7}{15} \\ &= \frac{30 \cdot 7}{1 \cdot 15} \\ &= \frac{15 \cdot 2 \cdot 7}{1 \cdot 15} \\ &= \frac{2 \cdot 7}{1} \\ &= \frac{14}{1} \\ &= 14 \end{aligned}$$

14 dresses can be made.

Calculator Explorations

1. $25\frac{5}{11} = \frac{280}{11}$

2. $67\frac{14}{15} = \frac{1019}{15}$

3. $107\frac{31}{35} = \frac{3776}{35}$

4. $186\frac{17}{21} = \frac{3923}{21}$

5. $\frac{365}{14} = 26\frac{1}{14}$

6. $\frac{290}{13} = 22\frac{4}{13}$

7. $\frac{2769}{30} = 92\frac{3}{10}$

8. $\frac{3941}{17} = 231\frac{14}{17}$

Exercise Set 4.8

1. Each part is $\frac{1}{4}$, and there are 11 parts shaded, or 2 wholes and 3 more parts.

   a. $\frac{11}{4}$

   b. $2\frac{3}{4}$

3. Each part is $\frac{1}{3}$, and there are 11 parts shaded, or 3 wholes and 2 more parts.

   a. $\frac{11}{3}$

   b. $3\frac{2}{3}$

5. Each part is $\frac{1}{2}$, and there are 3 parts shaded or 1 whole part and 1 more part.

   a. $\frac{3}{2}$

   b. $1\frac{1}{2}$

7. Each part is $\frac{1}{3}$, and there are 4 parts shaded, or 1 whole and 1 more part.

   a. $\frac{4}{3}$

   b. $1\frac{1}{3}$

9. $2\frac{1}{3} = \frac{2 \cdot 3 + 1}{3} = \frac{7}{3}$

11. $3\frac{3}{8} = \frac{3 \cdot 8 + 3}{8} = \frac{27}{8}$

13. $11\frac{6}{7} = \frac{11 \cdot 7 + 6}{7} = \frac{83}{7}$

15. $$\begin{array}{r} 1 \\ 7\overline{)13} \\ \underline{-7} \\ 6 \end{array}$$

    $\frac{13}{7} = 1\frac{6}{7}$

17. $$\begin{array}{r} 3 \\ 15\overline{)\ 47} \\ \underline{-45} \\ 2 \end{array}$$

    $\frac{47}{15} = 3\frac{2}{15}$

19. $$\begin{array}{r} 4 \\ 8\overline{)\ 37} \\ \underline{-32} \\ 5 \end{array}$$

    $\frac{37}{8} = 4\frac{5}{8}$

21. $2\frac{2}{3} \cdot \frac{1}{7} = \frac{8}{3} \cdot \frac{1}{7} = \frac{8 \cdot 1}{3 \cdot 7} = \frac{8}{21}$

23. $$\begin{aligned} 8 \div 1\frac{5}{7} &= 8 \div \frac{12}{7} \\ &= \frac{8}{1} \cdot \frac{7}{12} \\ &= \frac{2 \cdot 4 \cdot 7}{4 \cdot 3} \\ &= \frac{14}{3} \\ &= 4\frac{2}{3} \end{aligned}$$

25. $3\frac{2}{3} \cdot 1\frac{1}{2} = \frac{11}{3} \cdot \frac{3}{2} = \frac{11 \cdot 3}{3 \cdot 2} = \frac{11}{2} = 5\frac{1}{2}$

27. $2\frac{2}{3} \div \frac{1}{7} = \frac{8}{3} \div \frac{1}{7} = \frac{8}{3} \cdot \frac{7}{1} = \frac{8 \cdot 7}{3 \cdot 1} = \frac{56}{3} = 18\frac{2}{3}$

29. $4\frac{7}{10} + 2\frac{1}{10} = 6\frac{8}{10} = 6\frac{4}{5}$

31.
$$\begin{array}{rcl} 15\frac{4}{7} & = & 15\frac{8}{14} \\ +\ 9\frac{11}{14} & = & 9\frac{11}{14} \\ \hline & & 24\frac{19}{14} = 25\frac{5}{14} \end{array}$$

33.
$$\begin{array}{rcl} 3\frac{5}{8} & = & 3\frac{15}{24} \\ 2\frac{1}{6} & = & 2\frac{4}{24} \\ +\ 7\frac{3}{4} & = & 7\frac{18}{24} \\ \hline & & 12\frac{37}{24} = 13\frac{13}{24} \end{array}$$

35.
$$\begin{array}{r} 4\frac{7}{10} \\ -\ 2\frac{1}{10} \\ \hline 2\frac{6}{10} = 2\frac{3}{5} \end{array}$$

37.
$$\begin{array}{rcl} 10\frac{13}{14} & = & 10\frac{13}{14} \\ -\ 3\frac{4}{7} & = & -\ 3\frac{8}{14} \\ \hline & & 7\frac{5}{14} \end{array}$$

39.
$$\begin{array}{rcrcr} 9\frac{1}{5} & = & 9\frac{5}{25} & = & 8\frac{30}{25} \\ -\ 8\frac{6}{25} & = & -\ 8\frac{6}{25} & = & -\ 8\frac{6}{25} \\ \hline & & & & \frac{24}{25} \end{array}$$

41.
$$\begin{array}{r} 2\frac{3}{4} \\ +\ 1\frac{1}{4} \\ \hline 3\frac{4}{4} = 3+1 = 4 \end{array}$$

43.
$$\begin{array}{rcrcr} 15\frac{4}{7} & = & 15\frac{8}{14} & = & 14\frac{22}{14} \\ -\ 9\frac{11}{14} & = & -\ 9\frac{11}{14} & = & -\ 9\frac{11}{14} \\ \hline & & & & 5\frac{11}{14} \end{array}$$

45. $3\frac{1}{9}\cdot 2 = \frac{28}{9}\cdot\frac{2}{1} = \frac{56}{9} = 6\frac{2}{9}$

47. $1\frac{2}{3} \div 2\frac{1}{5} = \frac{5}{3} \div \frac{11}{5} = \frac{5}{3}\cdot\frac{5}{11} = \frac{25}{33}$

49. $22\frac{4}{9} + 13\frac{5}{18} = 22\frac{8}{18} + 13\frac{5}{18} = 35\frac{13}{18}$

51.
$$\begin{array}{rcl} 5\frac{2}{3} & = & 5\frac{4}{6} \\ -\ 3\frac{1}{6} & = & -\ 3\frac{1}{6} \\ \hline & & 2\frac{3}{6} = 2\frac{1}{2} \end{array}$$

53.
$$\begin{array}{rcl} 15\frac{1}{5} & = & 15\frac{6}{30} \\ 20\frac{3}{10} & = & 20\frac{9}{30} \\ +\ 37\frac{2}{15} & = & +\ 37\frac{4}{30} \\ \hline & & 72\frac{19}{30} \end{array}$$

55.
$$\begin{array}{rcrcr} 6\frac{4}{7} & = & 6\frac{8}{14} & = & 6\frac{22}{14} \\ -\ 5\frac{11}{14} & = & -\ 5\frac{11}{14} & = & -\ 5\frac{11}{14} \\ \hline & & & & \frac{11}{14} \end{array}$$

57. $4\frac{2}{7}\cdot 1\frac{3}{10} = \frac{30}{7}\cdot\frac{13}{10} = \frac{3\cdot 10\cdot 13}{7\cdot 10} = \frac{39}{7} = 5\frac{4}{7}$

59.

$$\begin{array}{rl} 6\frac{2}{11} & = 6\frac{6}{33} \\ 3 & = 3 \\ +\ 4\frac{10}{33} & = 4\frac{10}{33} \\ \hline & 13\frac{16}{33} \end{array}$$

61. The phrase "total duration" tells us to add. Find the sum of the three durations.

$$\begin{array}{rcl} 4\frac{14}{15} & = & 4\frac{56}{60} \\ 4\frac{7}{60} & = & 4\frac{7}{60} \\ +1\frac{2}{3} & = & +1\frac{40}{60} \\ \hline & & 9\frac{103}{60} = 10\frac{43}{60} \end{array}$$

The total duration is $1\frac{43}{60}$ minutes.

63. The phrase "how much longer" tells us to subtract. Subtract $4\frac{7}{60}$ minutes from $4\frac{14}{15}$ minutes.

$$\begin{array}{rcl} 4\frac{14}{15} & = & 4\frac{56}{60} \\ -\ 4\frac{7}{60} & = & -\ 4\frac{7}{60} \\ \hline & & \frac{49}{60} \end{array}$$

The June 21, 2001 eclipse will be $\frac{49}{60}$ minute longer than the March 29, 2006 eclipse.

65.

$$\begin{aligned} 6\cdot 3\frac{1}{4} &= \frac{6}{1}\cdot\frac{13}{4} \\ &= \frac{2\cdot 3\cdot 13}{1\cdot 2\cdot 2} \\ &= \frac{39}{2} \\ &= 19\frac{1}{2} \end{aligned}$$

The sidewalk is $19\frac{1}{2}$ inches wide.

67.

$$\begin{aligned} 525\div 43\frac{3}{4} &= \frac{525}{1}\div\frac{175}{4} \\ &= \frac{525}{1}\cdot\frac{4}{175} \\ &= \frac{175\cdot 3\cdot 4}{1\cdot 175} \\ &= \frac{12}{1} \\ &= 12 \end{aligned}$$

He purchased 12 shares.

69. Subtract $1\frac{1}{2}$ inches from $1\frac{9}{16}$ inches.

$$\begin{array}{rcl} 1\frac{9}{16} & = & 1\frac{9}{16} \\ -\ 1\frac{1}{2} & = & -\ 1\frac{8}{16} \\ \hline & & \frac{1}{16} \end{array}$$

The entrance holes for Mountain Bluebirds should be $\frac{1}{16}$ inch wider than the entrance for Eastern Bluebirds.

71. The phrase "cuts off" tells us to subtract. First subtract $2\frac{1}{2}$ feet from $15\frac{2}{3}$ feet. Then subtract $3\frac{1}{4}$ feet from that result.

$$\begin{array}{rcr} 15\frac{2}{3} & = & 15\frac{4}{6} \\ -2\frac{1}{2} & = & -2\frac{3}{6} \\ \hline & & 13\frac{1}{6} \end{array}$$

$$\begin{array}{rcrcr} 13\frac{1}{6} & = & 13\frac{2}{12} & = & 12\frac{14}{12} \\ -3\frac{1}{4} & = & -3\frac{3}{12} & = & -3\frac{3}{12} \\ \hline & & & & 9\frac{11}{12} \end{array}$$

No, the pipe will be $\frac{1}{12}$ of a foot short.

73. $58\frac{3}{4} \div 7\frac{1}{2} = \frac{235}{4} \div \frac{15}{2}$

$= \frac{235}{4} \cdot \frac{2}{15}$

$= \frac{5 \cdot 47 \cdot 2}{2 \cdot 2 \cdot 3 \cdot 5}$

$= \frac{47}{6}$ or $7\frac{5}{6}$

$7\frac{5}{6}$ gallons were used each hour.

75. The phrase "how much more" tells us to subtract. Subtract $3\frac{3}{5}$ inches from $11\frac{1}{4}$ inches.

$$\begin{array}{rcrcr} 11\frac{1}{4} & = & 11\frac{5}{20} & = & 10\frac{25}{20} \\ -3\frac{3}{5} & = & -3\frac{12}{20} & = & -3\frac{12}{20} \\ \hline & & & & 7\frac{13}{20} \end{array}$$

Tucson gets $7\frac{13}{20}$ inches more, on average, than Yuma.

77. $2 \cdot 1\frac{3}{4} = \frac{2}{1} \cdot \frac{7}{4}$

$= \frac{2 \cdot 7}{1 \cdot 4}$

$= \frac{2 \cdot 7}{1 \cdot 2 \cdot 2}$

$= \frac{7}{1 \cdot 2}$

$= \frac{7}{2}$ or $3\frac{1}{2}$

The area is $\frac{7}{2}$ or $3\frac{1}{2}$ square yards.

79. $5\frac{1}{2} \cdot 5\frac{1}{2} = \frac{11}{2} \cdot \frac{11}{2}$

$= \frac{11 \cdot 11}{2 \cdot 2}$

$= \frac{121}{4}$ or $30\frac{1}{4}$

Its area is $\frac{121}{4}$ or $30\frac{1}{4}$ square inches.

81. Find the distance around. Add the lengths of the three sides.

$$\begin{array}{r} 2\frac{1}{3} \\ 2\frac{1}{3} \\ +\ 2\frac{1}{3} \\ \hline 6\frac{3}{3} = 7 \end{array}$$

The perimeter is 7 miles.

83. Find the distance around. Add the lengths of the four sides.

$$\begin{array}{rcl} 3 & = & 3 \\ 5\frac{1}{3} & = & 5\frac{8}{24} \\ 5 & = & 5 \\ +\ 7\frac{7}{8} & = & 7\frac{21}{24} \\ \hline & & 20\frac{29}{24} = 21\frac{5}{24} \end{array}$$

The perimeter is $21\frac{5}{24}$ meters.

85. Subtract $1\frac{1}{3}$ hours from $2\frac{3}{4}$ hours.

$$\begin{array}{rcl} 2\frac{3}{4} & = & 2\frac{9}{12} \\ -1\frac{1}{3} & = & -1\frac{4}{12} \\ \hline & & 1\frac{5}{12} \end{array}$$

He will have to wait $1\frac{5}{12}$ hours.

87. The phrase "overall height' tells us to add. Add the two heights.

$$\begin{array}{rcl} 152\frac{1}{6} & = & 152\frac{1}{6} \\ +\ 154\frac{1}{2} & = & 154\frac{3}{6} \\ \hline & & 306\frac{4}{6} = 306\frac{2}{3} \end{array}$$

The overall height is $306\frac{2}{3}$ feet.

89.

$$\begin{aligned} 15\frac{1}{5} \div 24 &= \frac{76}{5} \div \frac{24}{1} \\ &= \frac{76}{5} \cdot \frac{1}{24} \\ &= \frac{76 \cdot 1}{5 \cdot 24} \\ &= \frac{4 \cdot 19 \cdot 1}{5 \cdot 4 \cdot 6} \\ &= \frac{19 \cdot 1}{5 \cdot 6} \\ &= \frac{19}{30} \end{aligned}$$

On average $\frac{19}{30}$ inch fell each hour.

91. $2x - 5 + 7x - 8 = 2x + 7x - 5 - 8$
$= 9x - 13$

93. $3(y - 2) - 6y = 3y - 6 - 6y$
$= 3y - 6y - 6$
$= -3y - 6$

95. $2^3 + 3^2 = 2 \cdot 2 \cdot 2 + 3 \cdot 3 = 8 + 9 = 17$

97. $\frac{7-3}{2^2} = \frac{4}{4} = 1$

99. Weight of Supreme box:

$$\begin{array}{rcl} 2\frac{1}{4}\text{ lb} & = & 2\frac{1}{4}\text{ lb} \\ +\ 3\frac{1}{2}\text{ lb} & = & +\ 3\frac{2}{4}\text{ lb} \\ \hline & & 5\frac{3}{4}\text{ lb} = 5\frac{6}{8}\text{ lb} \end{array}$$

Weight of Deluxe box:

$$\begin{array}{rcl} 1\frac{3}{8}\text{ lb} & = & 1\frac{3}{8}\text{ lb} \\ +\ 4\frac{1}{4}\text{ lb} & = & +\ 4\frac{2}{8}\text{ lb} \\ \hline & & 5\frac{5}{8}\text{ lb} \end{array}$$

The Supreme box is heavier.

$$\begin{array}{r} 5\frac{6}{8}\text{ lb} \\ -\ 5\frac{5}{8}\text{ lb} \\ \hline \frac{1}{8}\text{ lb} \end{array}$$

It is $\frac{1}{8}$ pound heavier.

101. Answers may vary.

103. Answers may vary.

Chapter 4 Review

1. $\frac{3}{4}$ ← Three parts shaded / ← Four equal parts

So the answer expressed as a fraction is $\frac{3}{4}$.

2. $\frac{2}{5}$ ← Two parts shaded / ← Five equal parts

So the answer expressed as a fraction is $\frac{2}{5}$.

3. To graph $\frac{7}{9}$ on a number line, divide the distance from 0 to 1 into 9 equal parts. Then start at 0 and count over 7 parts.

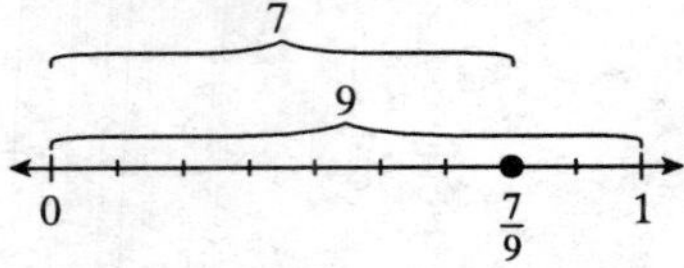

4. To graph $\frac{4}{7}$ on a number line, divide the distance from 0 to 1 into 7 equal parts. Then start at 0 and count over 4 parts.

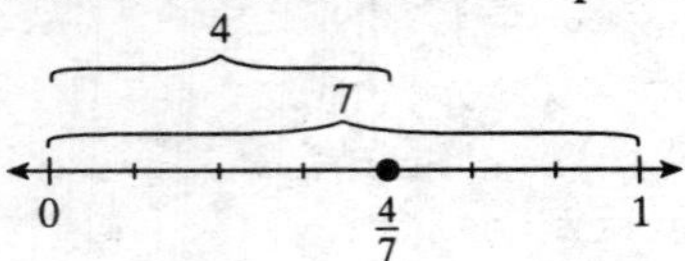

5. To graph $\frac{5}{4}$ on a number line, divide the distances from 0 to 1 and 1 to 2 into 4 equal parts. Then start at 0 and count over 5 parts.

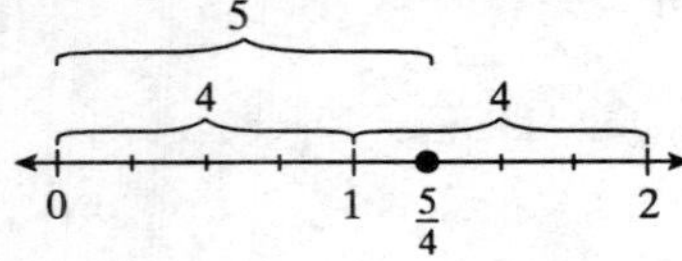

6. To graph $\frac{7}{5}$ on a number line, divide the distances from 0 to 1 and from 1 to 2 into 5 equal parts. Then start at 0 and count over 7 parts.

7. $\frac{\text{Number closed}}{\text{Total number}} = \frac{35}{242}$
$\frac{35}{242}$ of job specialties are closed to women.

8. $\frac{17}{20}$ of U.S. Army soldiers are men.

9. $\frac{2}{3} = \frac{2 \cdot 10}{3 \cdot 10} = \frac{20}{30}$

10. $\frac{5}{8} = \frac{5 \cdot 7}{8 \cdot 7} = \frac{35}{56}$

11. $\frac{7a}{6} = \frac{7a \cdot 7}{6 \cdot 7} = \frac{49a}{42}$

12. $\frac{9b}{4} = \frac{9b \cdot 5}{4 \cdot 5} = \frac{45b}{20}$

13. $\frac{4}{5x} = \frac{4 \cdot 10}{5x \cdot 10} = \frac{40}{50x}$

14. $\frac{5}{9y} = \frac{5 \cdot 2}{9y \cdot 2} = \frac{10}{18y}$

15. $\frac{12}{28} = \frac{2 \cdot 2 \cdot 3}{2 \cdot 2 \cdot 7} = \frac{3}{7}$

16. $\frac{15}{27} = \frac{3 \cdot 5}{3 \cdot 9} = \frac{5}{9}$

17. $\frac{25x}{75x^2} = \frac{5 \cdot 5 \cdot x}{3 \cdot 5 \cdot 5 \cdot x \cdot x} = \frac{1}{3x}$

18. $\frac{36y^3}{72y} = \frac{36y \cdot y^2}{2 \cdot 36y} = \frac{y^2}{2}$

19. $\frac{29ab}{32abc} = \frac{29 \cdot a \cdot b}{32 \cdot a \cdot b \cdot c} = \frac{29}{32c}$

20. $\frac{18xyz}{23xy} = \frac{18z}{23}$

21. $\frac{45x^2y}{27xy^3} = \frac{3 \cdot 3 \cdot 5 \cdot x \cdot x \cdot y}{3 \cdot 3 \cdot 3 \cdot x \cdot y \cdot y^2} = \frac{5x}{3y^2}$

22. $\frac{42ab^2c}{30abc^3} = \frac{2 \cdot 3 \cdot 7a \cdot b \cdot b \cdot c}{2 \cdot 3 \cdot 5a \cdot b \cdot c \cdot c^2} = \frac{7b}{5c^2}$

23. $\frac{\text{inches in the part}}{\text{inches in a foot}} = \frac{8}{12} = \frac{4 \cdot 2}{4 \cdot 3} = \frac{2}{3}$

8 inches is $\frac{2}{3}$ of a foot.

24. $1 - \frac{6}{15} = \frac{15-6}{15} = \frac{9}{15} = \frac{3 \cdot 3}{3 \cdot 5} = \frac{3}{5}$

$\frac{3}{5}$ of the cars are not white.

25. $\frac{3}{5} \cdot \frac{1}{2} = \frac{3 \cdot 1}{5 \cdot 2} = \frac{3}{10}$

26. $-\frac{6}{7} \cdot \frac{5}{12} = -\frac{6 \cdot 5}{7 \cdot 2 \cdot 6} = -\frac{5}{14}$

27. $\frac{7}{8x} \cdot -\frac{2}{3} = -\frac{7 \cdot 2}{4 \cdot 2 \cdot x \cdot 3} = -\frac{7}{12x}$

28. $\frac{6}{15} \cdot \frac{5y}{8} = \frac{2 \cdot 3 \cdot 5y}{3 \cdot 5 \cdot 2 \cdot 4} = \frac{y}{4}$

29. $-\frac{24x}{5} \cdot -\frac{15}{8x^3} = \frac{3 \cdot 8 \cdot x \cdot 3 \cdot 5}{5 \cdot 8 \cdot x \cdot x \cdot x} = \frac{9}{x^2}$

30. $\frac{27y^3}{21} \cdot \frac{7}{18y^2} = \frac{3 \cdot 9 \cdot y \cdot y^2 \cdot 7}{3 \cdot 7 \cdot 2 \cdot 9 \cdot y^2} = \frac{y}{2}$

31. $\left(-\frac{1}{3}\right)^3 = -\frac{1}{3} \cdot -\frac{1}{3} \cdot -\frac{1}{3} = -\frac{1 \cdot 1 \cdot 1}{3 \cdot 3 \cdot 3} = -\frac{1}{27}$

32. $\left(-\frac{5}{12}\right)^2 = -\frac{5}{12} \cdot -\frac{5}{12} = \frac{5 \cdot 5}{12 \cdot 12} = \frac{25}{144}$

33. $\frac{x^3}{y} \cdot \frac{y^3}{x} = \frac{x \cdot x^2 \cdot y \cdot y^2}{y \cdot x} = x^2y^2$

34. $\frac{ac}{b} \cdot \frac{b^2}{a^3c} = \frac{a \cdot c \cdot b \cdot b}{b \cdot a \cdot a^2 \cdot c} = \frac{b}{a^2}$

35. $xy = \frac{2}{3} \cdot \frac{1}{5} = \frac{2 \cdot 1}{3 \cdot 5} = \frac{2}{15}$

36. $ab = -7 \cdot \frac{9}{10} = -\frac{7}{1} \cdot \frac{9}{10} = -\frac{7 \cdot 9}{1 \cdot 10} = -\frac{63}{10}$

37. $-\frac{3}{4} \div \frac{3}{8} = -\frac{3}{4} \cdot \frac{8}{3} = -\frac{3 \cdot 2 \cdot 4}{4 \cdot 3} = -2$

38. $\frac{21a}{4} \div \frac{7a}{5} = \frac{21a}{4} \cdot \frac{5}{7a} = \frac{3 \cdot 7 \cdot a \cdot 5}{4 \cdot 7 \cdot a} = \frac{15}{4}$

39. $\frac{18x}{5} \div \frac{2}{5x} = \frac{18x}{5} \cdot \frac{5x}{2}$

$= \frac{2 \cdot 9 \cdot x \cdot 5 \cdot x}{5 \cdot 2}$

$= 9x^2$

40. $-\frac{9}{2} \div -\frac{1}{3} = \frac{9}{2} \cdot \frac{3}{1} = \frac{9 \cdot 3}{2 \cdot 1} = \frac{27}{2}$

41. $-\frac{5}{3} \div 2y = -\frac{5}{3} \cdot \frac{1}{2y} = -\frac{5 \cdot 1}{3 \cdot 2y} = -\frac{5}{6y}$

42. $\frac{5x^2}{y} \div \frac{10x^3}{y^3} = \frac{5x^2}{y} \cdot \frac{y^3}{10x^3}$

$= \frac{5x^2y \cdot y^2}{y \cdot 2 \cdot 5 \cdot x \cdot x^2}$

$= \frac{y^2}{2x}$

43. $x \div y = \frac{9}{7} \div \frac{3}{4} = \frac{9}{7} \cdot \frac{4}{3} = \frac{3 \cdot 3 \cdot 4}{7 \cdot 3} = \frac{12}{7}$

44. $a \div b = -5 \div \frac{2}{3} = -\frac{5}{1} \cdot \frac{3}{2} = -\frac{5 \cdot 3}{1 \cdot 2} = -\frac{15}{2}$

45. $A = \frac{11}{6} \cdot \frac{7}{8} = \frac{77}{48}$ or $1\frac{29}{48}$ sq. ft.

46. $A = \left(\frac{2}{3}\right)^2 = \frac{2}{3} \cdot \frac{2}{3} = \frac{4}{9}$ sq. m

47. $\frac{7}{11} + \frac{3}{11} = \frac{7+3}{11} = \frac{10}{11}$

48. $\frac{4}{9} + \frac{2}{9} = \frac{6}{9} = \frac{2}{3}$

49. $\frac{1}{12} - \frac{5}{12} = \frac{1-5}{12} = \frac{-4}{12} = -\frac{4}{4 \cdot 3} = -\frac{1}{3}$

50. $\frac{3}{y}-\frac{1}{y}=\frac{2}{y}$

51. $\frac{11x}{15}+\frac{x}{15}=\frac{11x+x}{15}$
$=\frac{12x}{15}$
$=\frac{3\cdot 4\cdot x}{3\cdot 5}$
$=\frac{4x}{5}$

52. $\frac{4y}{21}-\frac{3}{21}=\frac{4y-3}{21}$

53. $\frac{4}{15}+\frac{3}{15}-\frac{2}{15}=\frac{4+3-2}{15}=\frac{5}{15}=\frac{5}{3\cdot 5}=\frac{1}{3}$

54. $\frac{4}{15}-\frac{3}{15}-\frac{2}{15}=\frac{4-3-2}{15}=\frac{-1}{15}=-\frac{1}{15}$

55. $3=3$
$x=x$
$\text{LCD}=3x$

56. $4=2\cdot 2$
$8=2\cdot 2\cdot 2$
$12=2\cdot 2\cdot 3$
$\text{LCD}=2\cdot 2\cdot 2\cdot 3=24$

57. $z+\frac{1}{5}=1$
$\frac{4}{5}+\frac{1}{5}=1$
$\frac{5}{5}=1$
$1=1$
Yes, it is a solution.

58. $x-\frac{2}{4}=\frac{1}{4}$
$\frac{3}{4}-\frac{2}{4}=\frac{1}{4}$
$\frac{1}{4}=\frac{1}{4}$
Yes, it is a solution.

59. $\frac{3}{8}+\frac{1}{8}=\frac{4}{8}=\frac{4\cdot 1}{4\cdot 2}=\frac{1}{2}$
She studied $\frac{1}{2}$ hour.

60. $\frac{5}{8}+\frac{1}{8}-\frac{3}{8}=\frac{3}{8}$
$\frac{3}{8}$ gallon is left.

61. $\frac{3}{8}+\frac{2}{8}+\frac{1}{8}=\frac{6}{8}=\frac{2\cdot 3}{2\cdot 4}=\frac{3}{4}$
He did $\frac{3}{4}$ of his homework.

62. Recall that perimeter means distance around. Find the sum of the lengths of the four sides.
$\frac{3}{16}+\frac{9}{16}+\frac{3}{16}+\frac{9}{16}=\frac{3+9+3+9}{16}$
$=\frac{24}{16}$
$=\frac{3\cdot 8}{2\cdot 8}$
$=\frac{3}{2}$ or $1\frac{1}{2}$
The perimeter is $\frac{3}{2}$ or $1\frac{1}{2}$ miles.

63. $\frac{7}{18}+\frac{2}{9}=\frac{7}{18}+\frac{4}{18}=\frac{11}{18}$

64. $\frac{4}{13}-\frac{1}{26}=\frac{4\cdot 2}{13\cdot 2}-\frac{1}{26}=\frac{8}{26}-\frac{1}{26}=\frac{7}{26}$

65. $-\frac{1}{3}+\frac{1}{4}=-\frac{4}{12}+\frac{3}{12}=-\frac{1}{12}$

66. $-\frac{2}{3}+\frac{1}{4}=\frac{-2\cdot 4}{3\cdot 4}+\frac{1\cdot 3}{4\cdot 3}$
$=\frac{-8}{12}+\frac{3}{12}$
$=\frac{-5}{12}$
$=-\frac{5}{12}$

67. $\frac{5x}{11}+\frac{2}{55}=\frac{25x}{55}+\frac{2}{55}=\frac{25x+2}{55}$

68. $\frac{4}{15}+\frac{b}{5}=\frac{4}{15}+\frac{b\cdot 3}{5\cdot 3}=\frac{4+3b}{15}$

69. $\frac{5y}{12}-\frac{2y}{9}=\frac{5y\cdot 3}{12\cdot 3}-\frac{2y\cdot 4}{9\cdot 4}=\frac{15y}{36}-\frac{8y}{36}=\frac{7y}{36}$

70. $\frac{7x}{18}+\frac{2x}{9}=\frac{7x}{18}+\frac{2x\cdot 2}{9\cdot 2}=\frac{7x+4x}{18}=\frac{11x}{18}$

71. $\frac{4}{9}+\frac{5}{y}=\frac{4y}{9y}+\frac{45}{9y}=\frac{4y+45}{9y}$

72. $-\frac{9}{14}-\frac{3}{7}=-\frac{9}{14}-\frac{3\cdot 2}{7\cdot 2}=-\frac{9}{14}-\frac{6}{14}=-\frac{15}{14}$

73. $\frac{4}{25}+\frac{23}{75}+\frac{7}{50}=\frac{24}{150}+\frac{46}{150}+\frac{21}{150}=\frac{91}{150}$

74.
$$\begin{aligned}\frac{2}{3}-\frac{2}{9}-\frac{1}{6}&=\frac{2\cdot 6}{3\cdot 6}-\frac{2\cdot 2}{9\cdot 2}-\frac{1\cdot 3}{6\cdot 3}\\&=\frac{12}{18}-\frac{4}{18}-\frac{3}{18}\\&=\frac{5}{18}\end{aligned}$$

75.
$$\begin{aligned}a-\frac{2}{3}&=\frac{1}{6}\\a-\frac{2}{3}+\frac{2}{3}&=\frac{1}{6}+\frac{2}{3}\\a&=\frac{1}{6}+\frac{4}{6}\\a&=\frac{5}{6}\end{aligned}$$

76.
$$\begin{aligned}9x+\frac{1}{5}-8x&=-\frac{7}{10}\\(9x-8x)+\frac{1}{5}&=-\frac{7}{10}\\x+\frac{1}{5}-\frac{1}{5}&=-\frac{7}{10}-\frac{1}{5}\\x&=-\frac{7}{10}-\frac{1\cdot 2}{5\cdot 2}\\x&=\frac{-7-2}{10}\\&=-\frac{9}{10}\end{aligned}$$

77. Find the sum of the lengths of the four sides.
$$\begin{aligned}\frac{2}{9}+\frac{5}{6}+\frac{2}{9}+\frac{5}{6}&=\frac{4}{9}+\frac{10}{6}\\&=\frac{4\cdot 2}{9\cdot 2}+\frac{10\cdot 3}{6\cdot 3}\\&=\frac{8}{18}+\frac{30}{18}\\&=\frac{38}{18}\\&=\frac{19}{9}\end{aligned}$$
The perimeter is $\frac{19}{9}$ meters.

78. Find the sum of the three sides.
$$\begin{aligned}\frac{1}{5}+\frac{3}{5}+\frac{7}{10}&=\frac{2\cdot 1}{2\cdot 5}+\frac{2\cdot 3}{2\cdot 5}+\frac{7}{10}\\&=\frac{2+6+7}{10}\\&=\frac{15}{10}\\&=\frac{3\cdot 5}{2\cdot 5}\\&=\frac{3}{2}\end{aligned}$$
The perimeter is $\frac{3}{2}$ feet.

79.
$$x+\frac{1}{3}=\frac{35}{11}$$
$$\frac{8}{11}+\frac{1}{3}=\frac{35}{11}$$
$$\frac{24}{33}+\frac{11}{33}=\frac{105}{33}$$
$$\frac{35}{33}\neq\frac{105}{33}$$
No, it is not a solution.

80.
$$\frac{1}{9}y-\frac{1}{4}=\frac{1}{4}$$
$$\frac{1}{9}\left(\frac{9}{2}\right)-\frac{1}{4}=\frac{1}{4}$$
$$\frac{1}{2}-\frac{1}{4}=\frac{1}{4}$$
$$\frac{2}{4}-\frac{1}{4}=\frac{1}{4}$$
$$\frac{1}{4}=\frac{1}{4}$$
Yes, it is a solution.

81. $\frac{9}{25}+\frac{3}{50}=\frac{9\cdot 2}{25\cdot 2}+\frac{3}{50}=\frac{18+3}{50}=\frac{21}{50}$

$\frac{21}{50}$ have type A blood.

82. $\frac{2}{3}-\frac{5}{12}=\frac{8}{12}-\frac{5}{12}=\frac{3}{12}=\frac{1}{4}$

The difference is $\frac{1}{4}$ yard.

83. $\frac{\frac{2x}{5}}{\frac{7}{10}}=\frac{2x}{5}\div\frac{7}{10}=\frac{2x}{5}\cdot\frac{10}{7}=\frac{4x}{7}$

84. $\frac{\frac{3y}{7}}{\frac{11}{7}}=\frac{7\cdot\frac{3y}{7}}{7\cdot\frac{11}{7}}=\frac{3y}{11}$

85. The LCD of 4 and 8 is 8.
$$\frac{2+\frac{3}{4}}{1-\frac{1}{8}}=\frac{8\left(2+\frac{3}{4}\right)}{8\left(1-\frac{1}{8}\right)}$$
$$=\frac{8\cdot 2+8\cdot\frac{3}{4}}{8\cdot 1-8\cdot\frac{1}{8}}$$
$$=\frac{16+6}{8-1}$$
$$=\frac{22}{7}$$

86. The LCD of 3 and 6 is 6.
$$\frac{\frac{5}{6}+2}{\frac{11}{3}-1}=\frac{6\left(\frac{5}{6}+2\right)}{6\left(\frac{11}{3}-1\right)}$$
$$=\frac{6\cdot\frac{5}{6}+6\cdot 2}{6\cdot\frac{11}{3}-6\cdot 1}$$
$$=\frac{5+12}{22-6}$$
$$=\frac{17}{16}$$

87. The LCD of 5, 2, 4, and 10 is 20.
$$\frac{\frac{2}{5}-\frac{1}{2}}{\frac{3}{4}-\frac{7}{10}}=\frac{20\left(\frac{2}{5}-\frac{1}{2}\right)}{20\left(\frac{3}{4}-\frac{7}{10}\right)}$$
$$=\frac{20\cdot\frac{2}{5}-20\cdot\frac{1}{2}}{20\cdot\frac{3}{4}-20\cdot\frac{7}{10}}$$
$$=\frac{8-10}{15-14}$$
$$=\frac{-2}{1}$$
$$=-2$$

88. The LCD of 6, 4, and $12y$ is $12y$.

$$\frac{\frac{5}{6}-\frac{1}{4}}{\frac{-1}{12y}} = \frac{12y\left(\frac{5}{6}-\frac{1}{4}\right)}{12y\left(\frac{-1}{12y}\right)}$$
$$= \frac{12y\left(\frac{5}{6}\right)-12y\left(\frac{1}{4}\right)}{12y\left(\frac{-1}{12y}\right)}$$
$$= \frac{10y-3y}{-1}$$
$$= \frac{7y}{-1}$$
$$= -7y$$

89. $2x+y = 2\cdot\frac{1}{2}+\left(-\frac{2}{3}\right) = 1-\frac{2}{3} = \frac{3}{3}-\frac{2}{3} = \frac{1}{3}$

90.
$$\frac{x}{y+z} = \frac{\frac{1}{2}}{-\frac{2}{3}+\frac{4}{5}}$$
$$= \frac{30\left(\frac{1}{2}\right)}{30\left(-\frac{2}{3}+\frac{4}{5}\right)}$$
$$= \frac{15}{30\left(-\frac{2}{3}\right)+30\left(\frac{4}{5}\right)}$$
$$= \frac{15}{-20+24}$$
$$= \frac{15}{4}$$

91.
$$\frac{x+y}{z} = \frac{\frac{1}{2}+\left(-\frac{2}{3}\right)}{\frac{4}{5}}$$
$$\frac{\frac{1}{2}-\frac{2}{3}}{\frac{4}{5}} = \frac{30\left(\frac{1}{2}-\frac{2}{3}\right)}{30\left(\frac{4}{5}\right)}$$
$$= \frac{30\cdot\frac{1}{2}-30\cdot\frac{2}{3}}{24}$$
$$= \frac{15-20}{24}$$
$$= -\frac{5}{24}$$

92.
$$x+y+z = \frac{1}{2}+\left(-\frac{2}{3}\right)+\frac{4}{5}$$
$$= \frac{15}{30}+\frac{-20}{30}+\frac{24}{30}$$
$$= \frac{19}{30}$$

93. $y^2 = \left(-\frac{2}{3}\right)^2 = -\frac{2}{3}\cdot -\frac{2}{3} = \frac{2\cdot 2}{3\cdot 3} = \frac{4}{9}$

94. $x-z = \frac{1}{2}-\frac{4}{5} = \frac{5}{10}-\frac{8}{10} = -\frac{3}{10}$

95.
$$-\frac{3}{5}x = 6$$
$$-\frac{5}{3}\cdot-\frac{3}{5}x = -\frac{5}{3}\cdot 6$$
$$x = -10$$

96.
$$\frac{2}{9}y = -\frac{4}{3}$$
$$\frac{9}{2}\cdot\frac{2}{9}y = \frac{9}{2}\cdot\left(-\frac{4}{3}\right)$$
$$y = -\frac{9\cdot 4}{2\cdot 3}$$
$$= -\frac{3\cdot 3\cdot 2\cdot 2}{2\cdot 3}$$
$$= -\frac{3\cdot 2}{1}$$
$$= -6$$

97.
$$\frac{x}{7}-3 = -\frac{6}{7}$$
$$7\left(\frac{x}{7}-3\right) = 7\cdot -\frac{6}{7}$$
$$7\cdot\frac{x}{7}-7\cdot 3 = -6$$
$$x-21 = -6$$
$$x-21+21 = -6+21$$
$$x = 15$$

98.
$$\begin{aligned}\frac{y}{5}+2&=\frac{11}{5}\\ 5\left(\frac{y}{5}+2\right)&=5\cdot\frac{11}{5}\\ 5\cdot\frac{y}{5}+5\cdot 2&=11\\ y+10&=11\\ y+10-10&=11-10\\ y&=1\end{aligned}$$

99.
$$\begin{aligned}\frac{1}{6}+\frac{x}{4}&=\frac{17}{12}\\ 12\left(\frac{1}{6}+\frac{x}{4}\right)&=12\cdot\frac{17}{12}\\ 12\cdot\frac{1}{6}+12\cdot\frac{x}{4}&=17\\ 2+3x&=17\\ 2+3x-2&=17-2\\ 3x&=5\\ \frac{3x}{3}&=\frac{15}{3}\\ x&=5\end{aligned}$$

100.
$$\begin{aligned}\frac{x}{5}-\frac{5}{4}&=\frac{x}{2}-\frac{1}{20}\\ 20\left(\frac{x}{5}-\frac{5}{4}\right)&=20\left(\frac{x}{2}-\frac{1}{20}\right)\\ 20\cdot\frac{x}{5}-20\cdot\frac{5}{4}&=20\cdot\frac{x}{2}-20\cdot\frac{1}{20}\\ 4x-25&=10x-1\\ 4x-25+25&=10x-1+25\\ 4x&=10x+24\\ 4x-10x&=10x+24-10x\\ -6x&=24\\ \frac{-6x}{-6}&=\frac{24}{-6}\\ x&=-4\end{aligned}$$

101.
$$\begin{array}{r}3\\ 4\overline{)\ 15}\\ \underline{-12}\\ 3\end{array}$$
$$\frac{15}{4}=3\frac{3}{4}$$

102. $\frac{39}{13}=\frac{3\cdot 13}{13}=3$

103. $\frac{7}{7}=1$

104.
$$\begin{array}{r}31\\ 4\overline{)\ 125}\\ \underline{-12}\\ 5\\ \underline{-4}\\ 1\end{array}$$
$$\frac{125}{4}=31\frac{1}{4}$$

105. $2\frac{1}{5}=\frac{2\cdot 5+1}{5}=\frac{11}{5}$

106. $5=\frac{5}{1}$

107. $3\frac{8}{9}=\frac{3\cdot 9+8}{9}=\frac{35}{9}$

108. $3=\frac{3}{1}$

109.
$$\begin{array}{rcr}31\frac{2}{7} & = & 31\frac{6}{21}\\ +\ 14\frac{10}{21} & = & 14\frac{10}{21}\\ \hline & & 45\frac{16}{21}\end{array}$$

110.
$$\begin{array}{r}24\frac{4}{5}\\ +\ 35\frac{1}{5}\\ \hline 59\frac{5}{5}\end{array}=59+1=60$$

111.
$$\begin{array}{rcrcr}69\frac{5}{22} & = & 69\frac{5}{22} & = & 68\frac{27}{22}\\ -\ 36\frac{7}{11} & = & -\ 36\frac{14}{22} & = & -\ 36\frac{14}{22}\\ \hline & & & & 32\frac{13}{22}\end{array}$$

112.
$$\begin{array}{rcrcr} 36\frac{3}{20} & = & 36\frac{9}{60} & = & 35\frac{69}{60} \\ -\ 32\frac{5}{6} & = & -\ 32\frac{50}{60} & = & -\ 32\frac{50}{60} \\ \hline & & & & 3\frac{19}{60} \end{array}$$

113.
$$\begin{array}{rl} 29\frac{2}{9} & = 29\frac{4}{18} \\ 27\frac{7}{18} & = 27\frac{7}{18} \\ +\ 54\frac{2}{3} & = 54\frac{12}{18} \\ \hline & 110\frac{23}{18} = 110 + 1\frac{5}{8} = 111\frac{5}{18} \end{array}$$

114.
$$\begin{array}{rl} 7\frac{3}{8} & = 7\frac{9}{24} \\ 9\frac{5}{6} & = 9\frac{20}{24} \\ +\ 3\frac{1}{12} & = 3\frac{2}{24} \\ \hline & 19\frac{31}{24} = 19 + 1\frac{7}{24} = 20\frac{7}{24} \end{array}$$

115. $1\frac{5}{8} \cdot \frac{2}{3} = \frac{13}{8} \cdot \frac{2}{3} = \frac{13 \cdot 2}{4 \cdot 2 \cdot 3} = \frac{13}{12} = 1\frac{1}{12}$

116. $3\frac{6}{11} \cdot \frac{5}{13} = \frac{39}{11} \cdot \frac{5}{13} = \frac{3 \cdot 13 \cdot 5}{11 \cdot 13} = \frac{15}{11} = 1\frac{4}{11}$

117. $4\frac{1}{6} \cdot 2\frac{2}{5} = \frac{25}{6} \cdot \frac{12}{5} = \frac{5 \cdot 5 \cdot 6 \cdot 2}{6 \cdot 5} = \frac{10}{1} = 10$

118. $5\frac{2}{3} \cdot 2\frac{1}{4} = \frac{17}{3} \cdot \frac{9}{4} = \frac{17 \cdot 3 \cdot 3}{3 \cdot 4} = \frac{51}{4} = 12\frac{3}{4}$

119.
$$\begin{aligned} 6\frac{3}{4} \div 1\frac{2}{7} &= \frac{27}{4} \div \frac{9}{7} \\ &= \frac{27}{4} \cdot \frac{7}{9} \\ &= \frac{3 \cdot 9 \cdot 7}{4 \cdot 9} \\ &= \frac{21}{4} \\ &= 5\frac{1}{4} \end{aligned}$$

120.
$$\begin{aligned} 5\frac{1}{2} \div 2\frac{1}{11} &= \frac{11}{2} \div \frac{23}{11} \\ &= \frac{11}{2} \cdot \frac{11}{23} \\ &= \frac{11 \cdot 11}{2 \cdot 23} \\ &= \frac{121}{46} \\ &= 2\frac{29}{46} \end{aligned}$$

121. $\frac{7}{2} \div 1\frac{1}{2} = \frac{7}{2} \div \frac{3}{2} = \frac{7}{2} \cdot \frac{2}{3} = \frac{7 \cdot 2}{2 \cdot 3} = \frac{7}{3}$ or $2\frac{1}{3}$

122. $1\frac{3}{5} \div \frac{1}{4} = \frac{8}{5} \cdot \frac{4}{1} = \frac{8 \cdot 4}{5 \cdot 1} = \frac{32}{5} = 6\frac{2}{5}$

123.
$$\begin{array}{rcl} 3\frac{3}{4} & = & 3\frac{15}{20} \\ +\ 2\frac{3}{5} & = & 2\frac{12}{20} \\ \hline & & 5\frac{27}{20} = 5 + 1\frac{7}{20} = 6\frac{7}{20} \end{array}$$

Their combined weight is $6\frac{7}{20}$ pounds.

124.
$$\begin{array}{rcr} 50 & = & 49\frac{2}{2} \\ -\ 5\frac{1}{2} & = & -\ 5\frac{1}{2} \\ \hline & & 44\frac{1}{2} \end{array}$$

The amount remaining on the reel is $44\frac{1}{2}$ yards.

125. $$\begin{array}{rcrcr} 62\frac{3}{10} & = & 62\frac{3}{10} & = & 61\frac{13}{10} \\ -\ 54\frac{1}{2} & = & -\ 54\frac{5}{10} & = & -\ 54\frac{5}{10} \\ \hline & & & & 7\frac{8}{10} = 7\frac{4}{5} \end{array}$$

The annual snowfall was $7\frac{4}{5}$ inches below normal.

126. $2\frac{1}{4}\cdot 3\frac{1}{3} = \frac{9}{4}\cdot\frac{10}{3}$
$= \frac{9\cdot 10}{4\cdot 3}$
$= \frac{90}{12}$
$= \frac{2\cdot 3\cdot 3\cdot 5}{2\cdot 2\cdot 3}$
$= \frac{15}{2}$
$= 7\frac{1}{2}$

The area is $7\frac{1}{2}$ square feet.

127. Find the sum of 4 sides.
$1\frac{1}{4}+1\frac{1}{4}+1\frac{1}{4}+1\frac{1}{4} = \frac{5}{4}+\frac{5}{4}+\frac{5}{4}+\frac{5}{4}$
$= \frac{20}{4}$
$= 5$
The perimeter is 5 feet.

128. $\frac{7}{10}\cdot 2\frac{1}{8} = \frac{7}{10}\cdot\frac{17}{8}$
$= \frac{7\cdot 17}{10\cdot 8}$
$= \frac{119}{80}$ or $1\frac{39}{80}$

The area is $\frac{119}{80}$ or $1\frac{39}{80}$ square inches.

129. Find the sum of the eight sides of his flower beds.
$\frac{2}{3}+\frac{2}{3}+\frac{2}{3}+\frac{2}{3}+\frac{7}{8}+\frac{1}{3}+\frac{7}{8}+\frac{1}{3}$
$= \frac{2+2+2+2+1+1}{3}+\frac{7+7}{8}$
$= \frac{10}{3}+\frac{14}{8}$
$= \frac{10\cdot 8}{3\cdot 8}+\frac{14\cdot 3}{8\cdot 3}$
$= \frac{80}{24}+\frac{42}{24}$
$= \frac{122}{24}$
$= \frac{2\cdot 61}{2\cdot 12}$
$= \frac{61}{12}$ or $5\frac{1}{12}$
The total perimeter of his flower beds is $5\frac{1}{12}$ meters.

130. $58\cdot 3\frac{1}{2} = \frac{58}{1}\cdot\frac{7}{2} = 203$ calories

131. $3\frac{1}{3}\cdot 4 = \frac{10}{3}\cdot\frac{4}{1} = \frac{40}{3}$ grams or $13\frac{1}{3}$ grams

132. $5\frac{1}{4}\div 5 = \frac{21}{4}\cdot\frac{1}{5} = \frac{21\cdot 1}{4\cdot 5} = \frac{21}{20}$ or $1\frac{1}{20}$
He walks $\frac{21}{20}$ or $1\frac{1}{20}$ mile each day.

Chapter 4 Test

1. $7\frac{2}{3} = \frac{3\cdot 7+2}{3} = \frac{23}{3}$

2. $3\frac{6}{11} = \frac{11\cdot 3+6}{11} = \frac{39}{11}$

3. $$\begin{array}{r} 4 \\ 5\overline{)\ 23} \\ \underline{-20} \\ 3 \end{array}$$

$\frac{23}{5} = 4\frac{3}{5}$

4.
$$\begin{array}{r} 18 \\ 4\overline{)\,75} \\ \underline{-4} \\ 35 \\ \underline{-32} \\ 3 \end{array}$$
$$\frac{75}{4} = 18\frac{3}{4}$$

5. $\frac{54}{210} = \frac{9 \cdot 6}{35 \cdot 6} = \frac{9}{35}$

6. $\frac{42}{70} = \frac{3 \cdot 14}{5 \cdot 14} = \frac{3}{5}$

7. $\frac{4}{4} \div \frac{3}{4} = \frac{4}{4} \cdot \frac{4}{3} = \frac{4}{3}$

8. $-\frac{4}{3} \cdot \frac{4}{4} = -\frac{4}{3}$

9. $\frac{7x}{9} + \frac{x}{9} = \frac{7x + x}{9} = \frac{8x}{9}$

10.
$$\begin{aligned} \frac{1}{7} - \frac{3}{x} &= \frac{1 \cdot x}{7 \cdot x} - \frac{3 \cdot 7}{x \cdot 7} \\ &= \frac{x}{7x} - \frac{21}{7x} \\ &= \frac{x - 21}{7x} \end{aligned}$$

11. $\frac{xy^3}{z} \cdot \frac{z}{xy} = \frac{x \cdot y \cdot y^2 \cdot z}{z \cdot x \cdot y} = \frac{y^2}{1} = y^2$

12. $-\frac{2}{3} \cdot -\frac{8}{15} = \frac{2 \cdot 8}{3 \cdot 15} = \frac{16}{45}$

13. $\frac{9a}{10} + \frac{2}{5} = \frac{9a}{10} + \frac{2 \cdot 2}{5 \cdot 2} = \frac{9a + 4}{10}$

14. $-\frac{8}{15y} - \frac{2}{15y} = \frac{-8 - 2}{15y} = \frac{-10}{15y} = -\frac{2}{3y}$

15.
$$\begin{aligned} 8y^3 \div \frac{y}{3} &= \frac{8y^3}{1} \div \frac{y}{3} \\ &= \frac{8y^3}{1} \cdot \frac{3}{y} \\ &= \frac{8 \cdot 3 \cdot y \cdot y^2}{y} \\ &= \frac{24y^2}{1} \\ &= 24y^2 \end{aligned}$$

16. $5\frac{1}{4} \div \frac{7}{12} = \frac{21}{4} \cdot \frac{12}{7} = \frac{3 \cdot 7 \cdot 3 \cdot 4}{4 \cdot 7} = 9$

17.
$$\begin{aligned} 3\frac{7}{8} &= 3\frac{7 \cdot 5}{8 \cdot 5} = 3\frac{35}{40} \\ 7\frac{2}{5} &= 7\frac{2 \cdot 8}{5 \cdot 8} = 7\frac{16}{40} \\ \underline{2\frac{3}{4}} &= \underline{2\frac{3 \cdot 10}{4 \cdot 10}} = \underline{2\frac{30}{40}} \\ & \quad 12\frac{81}{40} = 12 + 2\frac{1}{40} = 14\frac{1}{40} \end{aligned}$$

18.
$$\begin{aligned} \frac{3a}{8} \cdot \frac{16}{6a^3} &= \frac{3a \cdot 2 \cdot 8}{8 \cdot 3 \cdot 2a \cdot a^2} \\ &= \frac{2 \cdot 3 \cdot 8 \cdot a}{2 \cdot 3 \cdot 8 \cdot a \cdot a^2} \\ &= \frac{1}{a^2} \end{aligned}$$

19. $-\frac{16}{3} \div -\frac{3}{12} = \frac{16}{3} \cdot \frac{12}{3} = \frac{16 \cdot 3 \cdot 4}{3 \cdot 3} = \frac{64}{3}$

20. $3\frac{1}{3} \cdot 6\frac{3}{4} = \frac{10}{3} \cdot \frac{27}{4} = \frac{5 \cdot 9}{2} = \frac{45}{2}$ or $22\frac{1}{2}$

21.
$$\begin{aligned} 12 \div 3\frac{1}{3} &= 12 \div \frac{10}{3} \\ &= 12 \cdot \frac{3}{10} \\ &= \frac{2 \cdot 6 \cdot 3}{2 \cdot 5} \\ &= \frac{18}{5} \\ &= 3\frac{3}{5} \end{aligned}$$

22. $\left(\frac{14}{5}\cdot\frac{25}{21}\right)\div 10 = \frac{14}{5}\cdot\frac{25}{21}\cdot\frac{1}{10}$
$= \frac{2\cdot 7\cdot 5\cdot 5}{5\cdot 3\cdot 7\cdot 2\cdot 5}$
$= \frac{1}{3}$

23. $\frac{11}{12}-\frac{3}{8}+\frac{5}{24} = \frac{11\cdot 2}{12\cdot 2}-\frac{3\cdot 3}{8\cdot 3}+\frac{5}{24}$
$= \frac{22}{24}-\frac{9}{24}+\frac{5}{24}$
$= \frac{22-9+5}{24}$
$= \frac{18}{24}$
$= \frac{3\cdot 6}{4\cdot 6}$
$= \frac{3}{4}$

24. $\frac{\frac{5x}{7}}{\frac{20x^2}{21}} = \frac{21\cdot\frac{5x}{7}}{21\cdot\frac{20x^2}{21}} = \frac{15x}{20x^2} = \frac{3\cdot 5x}{4\cdot 5\cdot x\cdot x} = \frac{3}{4x}$

25. $\frac{5+\frac{3}{7}}{2-\frac{1}{2}} = \frac{14\left(5+\frac{3}{7}\right)}{14\left(2-\frac{1}{2}\right)}$
$= \frac{14\cdot 5+14\cdot\frac{3}{7}}{14\cdot 2-14\cdot\frac{1}{2}}$
$= \frac{70+6}{28-7}$
$= \frac{76}{21}$

26. $-\frac{3}{8}x = \frac{3}{4}$
$-\frac{8}{3}\cdot -\frac{3}{8}x = -\frac{8}{3}\cdot\frac{3}{4}$
$x = -\frac{8\cdot 3}{3\cdot 4}$
$x = -2$

27. $\frac{x}{5}+x = -\frac{24}{5}$
$5\cdot\left(\frac{x}{5}+x\right) = 5\cdot -\frac{24}{5}$
$5\cdot\frac{x}{5}+5x = -24$
$x+5x = -24$
$6x = -24$
$\frac{6x}{6} = -\frac{24}{6}$
$x = -4$

28. $\frac{2}{3}+\frac{x}{4} = \frac{5}{12}+\frac{x}{2}$
$12\left(\frac{2}{3}+\frac{x}{4}\right) = 12\left(\frac{5}{12}+\frac{x}{2}\right)$
$12\left(\frac{2}{3}\right)+12\left(\frac{x}{4}\right) = 12\left(\frac{5}{12}\right)+12\left(\frac{x}{2}\right)$
$8+3x = 5+6x$
$8+3x-6x = 5+6x-6x$
$8-3x = 5$
$8-3x-8 = 5-8$
$-3x = -3$
$\frac{-3x}{-3} = \frac{-3}{-3}$
$x = 1$

29. $-5x = -5\left(-\frac{1}{2}\right) = \frac{5}{2}$

30. $x\div y = \frac{1}{2}\div 3\frac{7}{8}$
$= \frac{1}{2}\div\frac{31}{8}$
$= \frac{1}{2}\cdot\frac{8}{31}$
$= \frac{1\cdot 2\cdot 4}{2\cdot 31}$
$= \frac{4}{31}$

31. $\frac{280 \text{ calories}}{560 \text{ calories}} = \frac{28\cdot 10}{28\cdot 2\cdot 10} = \frac{1}{2}$
$\frac{1}{2}$ of the calories are from fat.

32. The phrase "How long is the remaining" tells us to subtract. Subtract $2\frac{3}{4}$ feet from $6\frac{1}{2}$ feet.

$$\begin{array}{rcrcr} 6\frac{1}{2} & = & 6\frac{2}{4} & = & 5\frac{6}{4} \\ -\ 2\frac{3}{4} & = & -\ 2\frac{3}{4} & = & -\ 2\frac{3}{4} \\ \hline & & & & 3\frac{3}{4} \end{array}$$

The remaining piece is $3\frac{3}{4}$ feet.

33. Find the sum of the fractions representing Back Woods and Westward.

$$\begin{aligned} \frac{3}{16}+\frac{1}{8} &= \frac{3}{16}+\frac{1\cdot 2}{8\cdot 2} \\ &= \frac{3}{16}+\frac{2}{16} \\ &= \frac{5}{16} \end{aligned}$$

$\frac{5}{16}$ of backpack sales go to Back Woods and Westward.

34. Multiply the fraction representing Wilderness, Inc. $\left(\frac{1}{4}\right)$ by 500,000.

$$\begin{aligned} \frac{1}{4}\cdot 500{,}000 &= \frac{1}{4}\cdot\frac{500{,}000}{1} \\ &= \frac{1\cdot 500{,}000}{4\cdot 1} \\ &= \frac{1\cdot 4\cdot 125{,}000}{4\cdot 1} \\ &= \frac{1\cdot 125{,}000}{1} \\ &= \frac{125{,}000}{1} \\ &= 125{,}000 \end{aligned}$$

Wilderness, Inc. sells 125,000 backpacks each year.

35.

$$\begin{aligned} \text{Area} &= \text{length}\cdot\text{width} \\ &= (100+10+10)\cdot\left(53\frac{1}{3}+10+10\right) \\ &= 120\cdot 73\frac{1}{3} \\ &= \frac{120}{1}\cdot\frac{220}{3} \\ &= \frac{120\cdot 220}{1\cdot 3} \\ &= \frac{40\cdot 220}{1} \\ &= \frac{8800}{1} \\ &= 8800 \end{aligned}$$

8800 square yards of turf are needed.

36. $1\frac{8}{9}\cdot\frac{2}{3}=\frac{17}{9}\cdot\frac{2}{3}=\frac{17\cdot 2}{9\cdot 3}=\frac{34}{27}$ or $1\frac{7}{27}$

The area of the figure is $\frac{34}{27}$ or $1\frac{7}{27}$ square miles.

37.

$$\begin{aligned} 258\div 10\frac{3}{4} &= \frac{258}{1}\div\frac{43}{4} \\ &= \frac{258}{1}\cdot\frac{4}{43} \\ &= \frac{258\cdot 4}{1\cdot 43} \\ &= \frac{43\cdot 6\cdot 4}{1\cdot 43} \\ &= \frac{24}{1} \\ &= 24 \end{aligned}$$

We can expect the car to travel 24 miles.

38.

$$\begin{aligned} 120\cdot\frac{3}{4} &= \frac{120}{1}\cdot\frac{3}{4} \\ &= \frac{120\cdot 3}{1\cdot 4} \\ &= \frac{4\cdot 30\cdot 3}{4} \\ &= \frac{30\cdot 3}{1} \\ &= 90 \end{aligned}$$

The stock sold for $90 after the spill.

39. $$\frac{3}{5}x = 120$$
$$\frac{5}{3}\cdot\frac{3}{5}x = \frac{5}{3}\cdot\frac{120}{1}$$
$$x = \frac{5\cdot 40\cdot 3}{3\cdot 1}$$
$$x = \frac{5\cdot 40}{1}$$
$$x = 200$$
The regular price is $200.

Chapter 5

Pretest

1. $0.27 = \frac{27}{100}$

2. Since $0 < 1$, $0.205 < 0.213$.

3. 54.651 rounded to the nearest tenth is 54.7. We round up because the digit in the hundredths place is greater than or equal to 5.

4. $$\begin{array}{r} {}^{2\,1} \\ 38.410 \\ 14.032 \\ +\ 7.600 \\ \hline 60.042 \end{array}$$

5. $$\begin{array}{rl} 3.4 & \text{1 decimal} \\ \times\ 2.1 & \text{1 decimal} \\ \hline 34 & \\ 680 & \\ \hline 7.14 & \text{2 decimal places} \end{array}$$

6. $(2.016)(100) = 201.6$

7. $0.4\overline{)16.24}$

   Move decimal points 1 place.

   $$\begin{array}{r} 40.6 \\ 4.\overline{)\ 162.4} \\ \underline{-16} \\ 02 \\ \underline{-0} \\ 24 \\ \underline{-24} \\ 0 \end{array}$$

   $16.24 \div 0.4 = 40.6$

8. $\frac{891}{10,000} = 0.0891$

9. $x - y = 12.3 - 0.61 = 11.69$

   $$\begin{array}{r} 12.30 \\ -\ 0.61 \\ \hline 11.69 \end{array}$$

10. $$\begin{aligned} &-9.8 - 6.2x - 7.9 + 1.4x \\ &= -9.8 - 7.9 - 6.2x + 1.4x \\ &= -17.7 - 4.8x \end{aligned}$$

11. $xy = (4.2)(0.03) = 0.126$

    $$\begin{array}{rl} 4.2 & \text{1 decimal place} \\ \times\ 0.03 & \text{2 decimal places} \\ \hline 0.126 & \text{3 decimal places} \end{array}$$

12. $$\begin{aligned} \text{Circumference} &= 2\pi r \\ &= 2\pi \cdot 6 \\ &= 12\pi \\ &\approx 12(3.14) \\ &= 37.68 \end{aligned}$$

    The circumference is 12π inches which is approximately 37.68 inches.

13. $\frac{\$576}{20} = \$28.80$

    $$\begin{array}{r} 28.8 \\ 20\overline{)\ 576.0} \\ \underline{-40} \\ 176 \\ \underline{-160} \\ 160 \\ \underline{-160} \\ 0 \end{array}$$

14. $0.2(6.9 - 3.01) = 0.2(3.89) = 0.778$

15. $\frac{3}{8} = 0.375$

    $$\begin{array}{r} 0.375 \\ 8\overline{)\ 3.000} \\ \underline{-2\,4} \\ 60 \\ \underline{-56} \\ 40 \\ \underline{-40} \\ 0 \end{array}$$

16. $\frac{9}{11} \approx 0.81818 < 0.8182$

$\frac{9}{11} < 0.8182$

17.
$$\begin{aligned} 4(x+0.22) &= 2x-3.4 \\ 4x+0.88 &= 2x-3.4 \\ 4x+0.88-0.88 &= 2x-3.4-0.88 \\ 4x &= 2x-4.28 \\ 4x-2x &= 2x-4.28-2x \\ 2x &= -4.28 \\ \frac{2x}{2} &= -\frac{4.28}{2} \\ x &= -2.14 \end{aligned}$$

18. $\sqrt{\frac{36}{49}} = \frac{6}{7}$ since $\frac{6}{7} \cdot \frac{6}{7} = \frac{36}{49}$.

19. $\sqrt{46} \approx 6.78$

20.
$$\begin{aligned} a^2+b^2 &= c^2 \\ 9^2+12^2 &= c^2 \\ 81+144 &= c^2 \\ 225 &= c^2 \\ \sqrt{225} &= c \\ 15 &= c \end{aligned}$$

The length of the hypotenuse is 15 inches.

Section 5.1

Practice Problems

1. a. 0.08 is eight hundredths

   b. 500.025 is five hundred and twenty-five thousandths.

   c. 0.0329 is three hundred twenty-nine ten-thousandths.

2. Ninety-seven and twenty-eight hundredths

3. Seventy-two and one thousand eighty-five ten-thousandths.

4. 300.96

5. 39.042

6. $0.037 = \frac{37}{1000}$

7. $14.97 = 14\frac{97}{100}$

8. $0.12 = \frac{12}{100} = \frac{3}{25}$

9. $57.8 = 57\frac{8}{10} = 57\frac{4}{5}$

10. $209.986 = 209\frac{986}{1000} = 209\frac{493}{500}$

11. The tens, ones, and tenths places are all the same. The hundredths places are different. Since $0 < 8$, then $13.208 < 13.28$.

12. In the tenths place, 1 is greater than 0 so $0.12 > 0.086$.

13. To round 123.7817 to the nearest thousandth, observe that the digit in the ten-thousandths place is 7. Since this digit is at least 5, we need to add 1 to the digit in the thousandths place. The number 123.7817 rounded to the nearest thousandth is 123.782.

14. To round 1.2789 to the nearest hundredth, observe that the digit in the thousandths place is 8. Since this digit is at least 5, we need to add 1 to the digit in the hundredths place. The number 1.0789 rounded to the nearest hundredth is 1.28. The price is $1.28.

15. To round 24.43 to the nearest tenth, observe that the digit in the hundredths place is 3. Since this digit is less than 5, we do not add 1 to the digit in the tenths place. The number 24.43 rounded to the nearest tenth is 24.4. The bill is $24.40.

Mental Math

1. 70
↑ tens

2. 700
   ↑ hundreds

3. 0.7
   ↑ tenths

4. 0.07
   ↑ hundredths

Exercise Set 5.1

1. 6.52 is six and fifty-two hundredths.

3. 16.23 is sixteen and twenty-three hundredths.

5. 3.205 is three and two hundred five thousandths.

7. 167.009 is one hundred sixth-seven and nine thousanths.

9. Six and five-tenths is 6.5.

11. Nine and eight-hundredths is 9.08.

13. Five and six hundred twenty-five thousandths is 5.625.

15. Sixty-four ten-thousandths is 0.0064.

17. Twenty and thirty-three hundredths is 20.33.

19. Five and nine tenths is 5.9.

21. $0.3 = \frac{3}{10}$

23. $0.27 = \frac{27}{100}$

25. $5.47 = 5\frac{47}{100}$

27. $0.048 = \frac{48}{1000} = \frac{6}{125}$

29. $7.07 = 7\frac{7}{100}$

31. $15.802 = 15\frac{802}{1000} = 15\frac{401}{500}$

33. $0.3005 = \frac{3005}{10,000} = \frac{601}{2000}$

35. $487.32 = 487\frac{32}{100} = 487\frac{8}{25}$

37. 0.15 0.16
    ↑ ↑
    $5 < 6$ so
    $0.15 < 0.16$

39. 0.57 0.54
    ↑ ↑
    $7 > 4$ so
    $0.57 > 0.54$

41. 0.098 0.1
    ↑ ↑
    $0 < 1$ so
    $0.098 < 0.1$

43. 0.54900 0.549
    ↑ ↑
    $0 = 0$ so
    $0.54900 = 0.549$

45. 167.908 167.980
    ↑ ↑
    $0 < 8$ so
    $167.908 < 167.980$

47. 420,000 0.000042
    ↑ ↑
    $4 > 0$ so
    $420,000 > 0.000042$

49. To round 0.57 to the nearest tenth, observe that the digit in the hundredths place is 7. Since this digit is at least 5, we need to add 1 to the digit in the tenths place. The number 0.57 rounded to the nearest tenth is 0.6.

51. To round 0.234 to the nearest hundredth, observe that the digit in the thousandths place is 4. Since this digit is less than 5, we do not add 1 to the digit in the hundredths place. The number 0.234 rounded to the nearest hundredth is 0.23.

53. To round 0.5942 to the nearest thousandth, observe that the digit in the ten-thousandths place is 2. Since this digit is less than 5, we do not add 1 to the digit in the thousandths place. The number 0.5942 rounded to the nearest thousandth is 0.594.

55. To round 98,207.23 to the nearest ten, observe that the digit in the ones place is 7. Since this digit is at least 5, we need to add 1 to the digit in the tens place. The number 98,207.32 rounded to the nearest ten is 98,210.

57. To round 12,342 to the nearest tenth, observe that the digit in the hundredths place is 4. Since this digit is less than 5, we do not need to add 1 to the digit in the tenths place. The number 12.342 rounded to the nearest tenth is 12.3.

59. To round 17.667 to the nearest hundredth, observe that the digit in the thousandths place is 7. Since this digit is less than 5, we do not need to add 1 to the digit in the tenths place. The number 17.667 rounded to the nearest hundredth is 17.67.

61. To round 0.501 to the nearest tenth, observe that the digit in the hundredths place is 0. Since the digit is less than 5, we do not add 1 to the digit in the tenths place. The number 0.501 rounded to the nearest tenth is 0.5.

63. To round 0.067 to the nearest hundredth, observe that the digit in the thousandths place is 7. Since this digit is at least 5, we need to add 1 to the digit in the hundredths place. The number 0.067 rounded to the nearest hundredth is 0.07. The amount is $0.07.

65. To round 26.95 to the nearest one, observe that the digit in the tenths place is 9. Since this digit is at least 5, we need to add 1 to the digit in the ones place. The number 26.95 rounded to the nearest one is 27. The amount is $27.

67. To round 0.1992 to the nearest hundredth, observe that the digit in the thousandths place is 9. Since the digit is at least 5, we need to add 1 the digit in the hundredths place. The number 0.1992 rounded to the nearest hundredth is 0.2. The amount is $0.20.

69. 0.26559 rounds to 0.27
0.26499 rounds to 0.26
0.25786 rounds to 0.26
0.25186 rounds to 0.25
Therefore, 0.26499 and 0.25786 round to 0.26.

71. To round 39,867 to the nearest thousand, observe that the digit in the hundreds place is 8. Since this digit is at least 5, we need to add 1 to the digit in the thousands place. The number 39,867 rounded to the nearest thousand is 40,000.

73. To round 2.39027 to the nearest hundredth, observe that the digit in the thousandths place is 0. Since this digit is less than 5, we do not add 1 to the digit in the hundredths place. The number 2.39027 rounded to the nearest hundredth is 2.39. The time is 2.39 hours.

75. To round 24.6229 to the nearest thousandth, observe that the digit in the ten-thousandths place is 9. Since this digit is at least 5, we need to add 1 to the digit in the thousandths place. The number 24.6229 rounded to the nearest thousandth is 24.623. The length is 24.623 hours.

77. To round 135.74 to the nearest one, observe that the digit in the tenths place is 7. Since this digit is at least 5, we need to add 1 to the digit in the ones place. The number 135.74 rounded to the nearest one is 136. The record is 136 mph.

79. To round 26.75 to the nearest whole point, observe that the digit in the tenths place is 7. Since this digit is at least 5, we need to add 1 to the digit in the ones place. The number 26.75 rounded to the nearest one is 27. The average number of points he scored per game is 27.

81. Since 28.21 > 27.91 and 28.21 < 28.49, the hurricane is a Category 3.

83. $\begin{array}{r} 3452 \\ +2314 \\ \hline 5766 \end{array}$

85. $\begin{array}{r} 94 \\ -23 \\ \hline 71 \end{array}$

87. $\begin{array}{r} 482 \\ -239 \\ \hline 243 \end{array}$

89. All the hundreds places are the same. When comparing the tens places, 6 of the values have 2 and 2 > 1. So continue to compare the ones places for these 6 numbers. Notice then that 6 > 5. So the highest average score is 226.130 achieved by Walter Ray Williams Jr. in 1998.

91. Compare each place value and list from greatest to least.
226.130, 225.490, 225.370, 222.980, 222.830, 222.008, 219.702, 218.158, 215.432.

93. Answers may vary.

Section 5.2

Practice Problems

1. a. $\begin{array}{r} 15.520 \\ +2.371 \\ \hline 17.891 \end{array}$

b. $\begin{array}{r} 20.060 \\ +\ 17.612 \\ \hline 37.672 \end{array}$

c. $\begin{array}{r} 0.125 \\ +\ 122.800 \\ \hline 122.925 \end{array}$

2. a. $\begin{array}{r} {\scriptstyle 1\,1\;\;1\;\;\;\;\;\;\;} \\ 34.5670 \\ 129.4300 \\ +\ 2.8903 \\ \hline 166.8873 \end{array}$

b. $\begin{array}{r} {\scriptstyle 1\;\;\;\;\;\;\;\;\;\;} \\ 11.210 \\ 46.013 \\ +\ 362.526 \\ \hline 419.749 \end{array}$

3. $\begin{array}{r} 27.00000 \\ +\ 0.00043 \\ \hline 27.00043 \end{array}$

4. Subtract the absolute values.
$\begin{array}{r} 99.2 \\ -\ 8.1 \\ \hline 91.1 \end{array}$
Thus, 8.1 + (−99.2) = −91.1.

5. $\begin{array}{r} 5.80 \\ -3.92 \\ \hline 1.88 \end{array}$
Check:
$\begin{array}{r} {\scriptstyle 1\;1\;\;\;} \\ 1.88 \\ +\ 3.92 \\ \hline 5.80 \end{array}$ or 5.8

6. $\begin{array}{r} 53.00 \\ -29.31 \\ \hline 23.69 \end{array}$

Check:

$\begin{array}{r} {}^{1\,1\,1} \\ 23.69 \\ +29.31 \\ \hline 53.00 \end{array}$ or 53

7. $\begin{array}{r} {}^{1\ 16} \\ 26.99 \\ -18.00 \\ \hline 8.99 \end{array}$

Check:

$\begin{array}{r} 8.99 \\ +18.00 \\ \hline 26.99 \end{array}$

8. To subtract 9.6, add the opposite of 9.6, or −9.6.
$-3.4 - 9.6 = -3.4 + (-9.6)$
To add two numbers with the same sign, add their absolute values.

$\begin{array}{r} 3.4 \\ +9.6 \\ \hline 13.0 \end{array}$

The sign in the answer is the common sign.
Thus, $(-3.4) + (-9.6) = -13$.

9. $y - z = 11.6 - 10.8$
$= 0.8$

$\begin{array}{r} 11.6 \\ -10.8 \\ \hline 0.8 \end{array}$

10. $y - 4.3 = 7.8$
$12.1 - 4.3 \stackrel{?}{=} 7.8$
$7.8 = 7.8$ True
Yes, 12.1 is a solution.

11. $-4.3y + 7.8 - 20.1y + 14.6$
$= (-4.3y) + (-20.1y) + 7.8 + 14.6$
$= (-24.4y) + 22.4$
$= -24.4y + 22.4$

12. The phrase "total monthly cost" tells us to add.
Find the sum of the three expenses.

$\begin{array}{r} {}^{111} \\ 52.70 \\ 536.50 \\ +\ 87.50 \\ \hline 676.70 \end{array}$

The total monthly cost is \$676.70.

13. The phrase "how much taller" tells us to subtract.
Subtract 70.8 from 72.6.

$\begin{array}{r} 72.6 \\ -70.8 \\ \hline 1.8 \end{array}$

Check:

$\begin{array}{r} 1.8 \\ +70.8 \\ \hline 72.6 \end{array}$

The difference in average height is 1.8 inches.

Calculator Explorations

1. $315.782 + 12.96 = 328.742$
2. $29.68 + 85.902 = 115.582$
3. $6.249 - 1.0076 = 5.2414$
4. $5.238 - 0.682 = 4.556$
5. $\begin{array}{r} 12.555 \\ 224.987 \\ 5.2 \\ +\ 622.65 \\ \hline 865.392 \end{array}$

6.
$$\begin{array}{r} 47.006 \\ 0.17 \\ 313.259 \\ +\ 139.088 \\ \hline 499.523 \end{array}$$

Mental Math

1. $\begin{array}{r} 0.3 \\ +\ 0.2 \\ \hline 0.5 \end{array}$

2. $\begin{array}{r} 0.4 \\ +0.5 \\ \hline 0.9 \end{array}$

3. $\begin{array}{r} 1.00 \\ +0.26 \\ \hline 1.26 \end{array}$

4. $\begin{array}{r} 3.00 \\ +0.19 \\ \hline 3.19 \end{array}$

5. $\begin{array}{r} 7.6 \\ +\ 1.3 \\ \hline 8.9 \end{array}$

6. $\begin{array}{r} 4.5 \\ +3.2 \\ \hline 7.7 \end{array}$

7. $\begin{array}{r} 0.9 \\ -0.3 \\ \hline 0.6 \end{array}$

8. $\begin{array}{r} 0.6 \\ -0.2 \\ \hline 0.4 \end{array}$

Exercise Set 5.2

1. $\begin{array}{r} 1.3 \\ +2.2 \\ \hline 3.5 \end{array}$

3. $5.7 = 5.70$
$\begin{array}{r} 5.70 \\ +\ 1.13 \\ \hline 6.83 \end{array}$

5. $\begin{array}{r} 24.6000 \\ 2.3900 \\ +0.0678 \\ \hline 27.0578 \end{array}$

7. $\begin{array}{r} 45.023 \\ 3.006 \\ +8.403 \\ \hline 56.432 \end{array}$

9. $-2.6 + (-5.97)$
Add the absolute values.
$\begin{array}{r} 2.60 \\ +\ 5.97 \\ \hline 8.57 \end{array}$
Attach the common sign.
-8.57

11. $15.78 + (-4.62)$
Subtract the absolute values.
$\begin{array}{r} 15.78 \\ -4.62 \\ \hline 11.16 \end{array}$
Attach the sign of the number with the larger absolute value.
11.16

13. $\begin{array}{r} 8.8 \\ -2.3 \\ \hline 6.5 \end{array}$

15. $\begin{array}{r} 18.0 \\ -2.7 \\ \hline 15.3 \end{array}$

17. $\begin{array}{r} 654.90 \\ -56.67 \\ \hline 598.23 \end{array}$

19. $\begin{array}{r} 23.0 \\ -6.7 \\ \hline 16.3 \end{array}$

21. $-1.12 - 5.2 = -1.12 + (-5.2)$
Add the absolute values.
$\begin{array}{r} 1.2 \\ +\ 5.20 \\ \hline 6.32 \end{array}$
Attach the common sign.
–6.32

23. $7.7 - 14.1 = 7.7 + (-14.1)$
Subtract the absolute values.
$\begin{array}{r} 14.1 \\ -\ 7.7 \\ \hline 6.4 \end{array}$
Attach the sign of the number with the larger absolute value.
–6.4

25. $\begin{array}{r} 0.9 \\ +\ 2.2 \\ \hline 3.1 \end{array}$

27. $\begin{array}{r} 362.300 \\ 1.098 \\ +\ 87.340 \\ \hline 450.738 \end{array}$

29. $-5.9 - 4 = -5.9 + (-4)$
Add the absolute values.
$\begin{array}{r} 5.9 \\ +4.0 \\ \hline 9.9 \end{array}$
Attach the common sign.
–9.9

31. $\begin{array}{r} 45.67 \\ -20.00 \\ \hline 25.67 \end{array}$

33. $-6.06 + 0.44$
Subtract the absolute values.
$\begin{array}{r} 6.06 \\ -\ 0.44 \\ \hline 5.62 \end{array}$
Attach the sign of the number with the largest absolute value.
–5.62

35. $\begin{array}{r} 900.34 \\ -123.45 \\ \hline 776.89 \end{array}$

37. $\begin{array}{r} 3490.23 \\ +8493.09 \\ \hline 11,983.32 \end{array}$

39. $\begin{array}{r} 234.89 \\ +230.67 \\ \hline 465.56 \end{array}$

41. $50.2 - 600 = 50.2 + (-600)$
Subtract the absolute values.
$\begin{array}{r} 600.0 \\ -50.2 \\ \hline 549.8 \end{array}$
Attach the sign of the number with the larger absolute value.
–549.8

43. $\begin{array}{r} 923.5 \\ -\ 61.9 \\ \hline 861.6 \end{array}$

45. $\begin{array}{r} 100.009 \\ 6.080 \\ +9.034 \\ \hline 115.123 \end{array}$

47. $\begin{array}{r} 1000.0 \\ -123.4 \\ \hline 876.6 \end{array}$

49. $-0.003 + 0.091$
Subtract the absolute values.
$\begin{array}{r} 0.091 \\ -0.003 \\ \hline 0.088 \end{array}$
Attach the sign of the number with the larger absolute value.
0.088

51. $\begin{array}{r} 500.000 \\ -\ 34.098 \\ \hline 465.902 \end{array}$

53. $x + z = 3.6 + 0.21 = 3.81$

55. $x - z = 3.6 - 0.21 = 3.39$

57. $\begin{aligned} y - xz &= 5 - 3.6 + 0.217 \\ &= 5.00 - 3.60 \\ &= 1.40 + 0.21 \\ &= 1.61 \end{aligned}$

59. $x + 2.7 = 9.3$
$7 + 2.7 = 9.3$
$9.7 = 9.3$ False
No, it is not a solution.

61. $27.4 - y = 16$
$27.4 - 11.4 = 16$
$16 = 16$
Yes, it is a solution.

63. $2.3 + x = 5.3 - x$
$2.3 + 1 = 5.3 - 1$
$3.3 \neq 4.3$ False
No, it is not a solution.

65. $30.7x + 17.6 - 23.8x - 10.7$
$= 30.7x + (-23.8x) + 17.6 + (-10.7)$
$= 6.9x + 6.9$

67. $-8.61 + 4.23y - 2.36 - 0.76y$
$= -8.61 + (-2.36) + 4.23y + (-0.76y)$
$= -10.97 + 3.47y$

69. The phrase "total monthly cost" tells us to add. Find the sum of the four expenses.
$\begin{array}{r} {\scriptstyle 2\,1\,1\,1} \\ 275.36 \\ 83.00 \\ 81.60 \\ +\ 14.75 \\ \hline 454.71 \end{array}$
The total monthly cost is $454.71.

71. The phrase "By how much did the price change" tells us to subtract. Subtract 0.979 from 1.039.
$\begin{array}{r} 1.039 \\ -\ 0.979 \\ \hline 0.060 \end{array}$
Check:
$\begin{array}{r} {\scriptstyle 1\ 1} \\ 0.060 \\ +\ 0.979 \\ \hline 1.039 \end{array}$
The price changed by $0.06.

73. Subtract the cost of the book from what she paid ($20 + $20 = $40).
$\begin{array}{r} 40.00 \\ -\ 32.48 \\ \hline 7.52 \end{array}$
Check:
$\begin{array}{r} {\scriptstyle 11\ 1} \\ 7.52 \\ +\ 32.48 \\ \hline 40.00 \end{array}$ or 40
Her change was $7.52.

75. The phrase "How much more" tells us to subtract. Subtract 175.9 from 373.7.

$$\begin{array}{r} 373.7 \\ -\ 175.9 \\ \hline 197.8 \end{array}$$

Check:

$$\begin{array}{r} {\scriptstyle 1\,1\,1} \\ 197.8 \\ +\ 175.9 \\ \hline 373.7 \end{array}$$

The difference in consumption is 197.8 pounds per person.

77. The phrase "How much more" tells us to subtract. Subtract 46.07 from 61.88.

$$\begin{array}{r} 61.88 \\ -\ 46.07 \\ \hline 15.81 \end{array}$$

Check:

$$\begin{array}{r} {\scriptstyle 1} \\ 15.81 \\ +\ 46.07 \\ \hline 61.88 \end{array}$$

New Orleans receives 15.81 more inches of rain annually than Houston.

79. To find Green's speed, we must add the old record and the increase.

$$\begin{array}{r} {\scriptstyle 1\,1\,1\,1} \\ 633.468 \\ +\ 129.567 \\ \hline 763.035 \end{array}$$

The new record by Green is 763.035 mph.

81. The phrase "total amount" tells us to add. Find the sum of the 3 concert's earnings.

$$\begin{array}{r} {\scriptstyle 1\,1} \\ 121.2 \\ 103.5 \\ +\ 98.0 \\ \hline 322.7 \end{array}$$

The total amount of money these three concert's have earned is \$322.7 million.

83. To find the total amount of snow Blue Canyon receives we must add Marquette's snow fall to that of the stated increase.

$$\begin{array}{r} {\scriptstyle 1} \\ 129.2 \\ +\ 111.6 \\ \hline 240.8 \end{array}$$

Blue Canyon receives an average of 240.8 inches of snow annually.

85. Find the sum of the lengths of the three sides.

$$\begin{array}{r} 12.40 \\ 29.34 \\ +\ 25.70 \\ \hline 67.44 \end{array}$$

The architect needs 67.44 feet of border material.

87. Subtract 124.002 from 153.616.

$$\begin{array}{r} 153.616 \\ -\ 124.002 \\ \hline 29.614 \end{array}$$

Check:

$$\begin{array}{r} {\scriptstyle 1} \\ 29.614 \\ +\ 124.002 \\ \hline 153.616 \end{array}$$

The average Indianapolis 500 speed was 29.614 mph faster in 1995 than 1950.

89. Add the durations of all four missions.

$$\begin{array}{r} {\scriptstyle 212\ \ 11} \\ 330.583 \\ 94.567 \\ 147.000 \\ +\ 142.900 \\ \hline 715.050 \text{ or } 715.05 \end{array}$$

James A. Lovell has spent 715.05 hours in spaceflight.

91. Compare the five numbers, finding the largest. Switzerland has the largest number of 22 pounds. Switzerland has the greatest chocolate consumption per person.

93. Subtract 13.9 from 22.

$$\begin{array}{r} 22.0 \\ -\ 13.9 \\ \hline 8.1 \end{array}$$

Check:

$$\begin{array}{r} {}^{11} \\ 8.1 \\ +\ 13.9 \\ \hline 22.0 \text{ or } 22 \end{array}$$

The difference in consumption is 8.1 pounds per person.

95.

Country	Pounds of Chocolate per person
Switzerland	22.0
Norway	16.0
Germany	15.8
United Kingdom	14.5
Belgium	13.9

97. $\left(\frac{1}{5}\right)^3 = \frac{1}{5} \cdot \frac{1}{5} \cdot \frac{1}{5} = \frac{1}{125}$

99.

$$\begin{aligned} \frac{25}{36} \cdot \frac{24}{40} &= \frac{25}{36} \cdot \frac{3 \cdot 8}{5 \cdot 8} \\ &= \frac{25 \cdot 3}{36 \cdot 5} \\ &= \frac{5 \cdot 5 \cdot 3}{3 \cdot 12 \cdot 5} \\ &= \frac{5 \cdot 1}{12 \cdot 1} \\ &= \frac{5}{12} \end{aligned}$$

101. Answers may vary.

103. $14.271 - 8.968x + 1.333 - 201.815x + 101.239x = -109.544x + 15.604$

Section 5.3

Practice Problems

1.
$$\begin{array}{r} 45.9 \\ \times\ 0.42 \\ \hline 918 \\ 18360 \\ 000 \\ \hline 19.278 \end{array}$$

2.
$$\begin{array}{r} 0.112 \\ \times\ \ 0.6 \\ \hline 672 \\ 00 \\ \hline 0.0672 \end{array}$$

3.
$$\begin{array}{r} 0.0721 \\ \times\ \ \ 48 \\ \hline 5768 \\ 28840 \\ \hline 3.4608 \end{array}$$

4. The product of a positive number and a negative number is a negative number.
$(5.4)(-1.3) = -7.02$

5. To find 23.7×10, note that 10 has 1 zero. Therefore, we move the decimal point of 23.7 to the right 1 place. The product is 237.

6. To find 203.004×100, note that 100 has 2 zeros. Therefore, we move the decimal point of 203.004 to the right 2 places. The product is 20,300.4.

7. To find 1.15×1000, note that 1000 has 3 zeros. Therefore, we move the decimal point of 1.15 to the right 3 places. The product is 1150.

8. To find 7.62×0.1, note that 0.1 has 1 decimal place. Therefore, we move the decimal point of 7.62 to the left 1 place. The product is 0.762.

9. To find 1.9×0.01, note that 0.01 has 2 decimal places. Therefore, we move the decimal point of 1.9 to the left 2 places. The product is 0.019.

10. To find 7682×0.001, note that 0.001 has 3 decimal places. Therefore, we move the decimal point of 7682 to the left 3 places. The product is 7.682.

11.
$$\begin{aligned} 2192 \text{ thousand} &= 2192 \times 1000 \\ &= 2{,}192{,}000 \end{aligned}$$
There are 2,192,000 farms in the U.S.

12. Recall that $7y$ means $7 \cdot y$.
$$\begin{aligned} 7y &= (7)(0.028) \\ &= 0.196 \leftarrow \end{aligned} \qquad \begin{array}{r} 0.028 \\ \times\ \ \ \ 7 \\ \hline 0.196 \end{array}$$

13.
$$\begin{aligned} 4x &= 22 \\ 4(5.5) &\stackrel{?}{=} 22 \\ 22 &= 22 \end{aligned}$$
True. Since $22 = 22$ is a true statement, 5.5 is a solution.

14.
$$\begin{aligned} \text{Circumference} &= 2 \cdot \pi \cdot \text{ radius} \\ &= 2 \cdot \pi \cdot 11 \\ &= 22\pi \\ &= 22(3.14) \\ &= 69.08 \end{aligned}$$
The circumference is $22\pi \approx 69.08$ meters.

15. Multiply 60.5 by 5.6.
$$\begin{array}{r} 60.5 \\ \times\ \ \ 5.6 \\ \hline 3630 \\ 3025\ \ \\ \hline 338.80 \end{array} \text{ or } 338.8$$
She needs 338.8 ounces of fertilizer.

Exercise Set 5.3

1.
$$\begin{array}{rl} 0.2 & 1 \text{ decimal place} \\ \times\ 0.6 & 1 \text{ decimal place} \\ \hline 0.12 & 2 \text{ decimal places} \end{array}$$

3. $\begin{array}{r} 1.2 \\ \times\ 0.5 \\ \hline 0.60 \end{array}$ 1 decimal place; 2 decimal places; 2 decimal places

0.6 The trailing 0 can be dropped.

5. $(-2.3)(7.65)$

$\begin{array}{r} 7.65 \\ \times\ -2.3 \\ \hline 2295 \\ 15300 \\ \hline -17.595 \end{array}$ 2 decimal places; 1 decimal place; 3 decimal places

7. $(-6.89)(-5.7)$

$\begin{array}{r} 6.89 \\ \times\ 5.7 \\ \hline 4823 \\ 34450 \\ \hline 39.273 \end{array}$ 2 decimal places; 1 decimal place; 3 decimal places

9. $6.5 \times 10 = 65$

11. $6.5 \times 0.1 = 0.65$

13. $(-7.093)(1000) = -7093$

15. $(-9.83)(-0.01) = (9.83)(0.01) = 0.0983$

17. $\begin{array}{r} 5.62 \\ \times\ 7.7 \\ \hline 3934 \\ 39340 \\ \hline 43.274 \end{array}$ 2 decimal places; 1 decimal place; 3 decimal places

19. $\begin{array}{r} 1.0047 \\ \times\ 8.2 \\ \hline 20094 \\ 80376 \\ \hline 8.23854 \end{array}$ 4 decimal places; 1 decimal place; 5 decimal places

21. $(147.9)(100) = 14{,}790$

23. $(937.62)(-0.01) = -9.3762$

25. $\begin{array}{r} 49.02 \\ \times\ 0.023 \\ \hline 14706 \\ 98080 \\ \hline 1.12746 \end{array}$ 3 decimals; 3 decimal places; 5 decimal places

27. $\begin{array}{r} -0.023 \\ \times\ 6.28 \\ \hline 184 \\ 0460 \\ 13800 \\ \hline -0.14444 \end{array}$ 3 decimal places; 2 decimal places; 5 decimal places

29. $5.5 \text{ billion} = 5.5 \times 1{,}000{,}000{,}000$
$= 5{,}500{,}000{,}000$

31. $36.4 \text{ million} = 36.4 \times 1{,}000{,}000$
$= 36{,}400{,}000$

33. $1.6 \text{ million} = 1.6 \times 1{,}000{,}000$
$= 1{,}600{,}000$

35. Recall that xy means $x \cdot y$.

$xy = (3)(-0.2)$ $\qquad \begin{array}{r} 3.0 \\ \times\ -0.2 \\ \hline -0.60 \end{array}$

$= -0.6 \quad \leftarrow$

37. Recall that xz means $x \cdot z$.

$xz = (3)(5.7)$ $\qquad \begin{array}{r} 5.7 \\ \times\ 3 \\ \hline 17.1 \end{array}$

$= 17.1 \quad \leftarrow$

39. Recall that $20z$ means $20 \cdot z$.

$20z = 20(5.7)$ $\qquad \begin{array}{r} 20 \\ \times\ 5.7 \\ \hline 14\ 0 \\ 100 \\ \hline 114.0 \end{array}$

$= 114 \quad \leftarrow$

41. $0.6x = 4.92$ Show substitution exactly.
$0.6(14.2) = 4.92$
$8.52 = 4.92$ False
No, it is not a solution.

43. $-3x = -2.4$
$-3(0.08) = -2.4$
$-0.24 = -2.4$ False
No, it is not a solution.

45. $3.5y = -14$
$3.5(-4) = -14$
$-14.0 = -14$
$-14 = -14$ True
Yes, it is a solution.

47. Circumference $= 2 \cdot \pi \cdot$ radius
$C = 2 \cdot \pi \cdot 4 = 8\pi$
$C \approx 8(3.14) = 25.12$
The circumference is 8π meters, which is approximately 25.12 meters.

49. Circumference $= \pi \cdot$ diameter
$C = \pi \cdot 10 = 10\pi$
$C \approx 10(3.14) = 31.4$
The circumference is 10π centimeters, which is approximately 31.4 centimeters.

51. Circumference $= 2 \cdot \pi \cdot$ radius
$C = 2 \cdot \pi \cdot 9.1 = 18.2\pi$
$C \approx 57.148$
The circumference is 18.2π yards, which is approximately 57.148 yards.

53. Circumference $= \pi \cdot$ diameter
$C = \pi \cdot 250 = 250\pi$
$C \approx 250(3.14) = 785$
The circumference of the ferris wheel is 250π feet which is approximately 785 feet.

55. a. $C = 2 \cdot \pi \cdot 10 = 20\pi \approx 20(3.14) = 62.8$ m

$$\begin{aligned} C &= 2 \cdot \pi \cdot 20 \\ &= 40\pi \\ &\approx 40(3.14) \\ &= 125.6 \text{ m} \end{aligned}$$

b. yes

57. Multiply the number of ounces by the number of grams of fat in 1 ounce to get the total amount of fat.

$$\begin{array}{r} 6.2 \\ \times\ 4 \\ \hline 24.8 \end{array}$$

There are 24.8 grams of fat in a 4-ounce serving of cream cheese.

59. To find 2.7×1000, note that 1000 has 3 zeros. Therefore, we move the decimal point of 2.7 to the right 3 places. The product is 2700. The farmer received $2700.

61. Multiply 39.37 by 1.65.

$$\begin{array}{r} 39.37 \\ \times\ 1.65 \\ \hline 19685 \\ 236220 \\ 393700 \\ \hline 64.9605 \end{array}$$

She is about 64.9605 inches tall.

63. Multiply the number of hours by the hourly wage.

$$\begin{array}{r} 13.88 \\ \times\ 40 \\ \hline 0 \\ 55520 \\ \hline 555.20 \end{array}$$

His pay was $555.20

65. $675 \times 108 = 72{,}900$ yen

67. $350 \times 1.48 = 518$ Canadian dollars

69.

$$\begin{array}{r} 26 \\ 5 \overline{)\ 130} \\ \underline{-10} \\ 30 \\ \underline{-30} \\ 0 \end{array}$$

71.

$$\begin{array}{r} 36 \\ 56 \overline{)\ 2016} \\ \underline{-168} \\ 336 \\ \underline{-336} \\ 0 \end{array}$$

73.

$$\begin{array}{r} 8 \\ 365 \overline{)\ 2920} \\ \underline{-2920} \\ 0 \end{array}$$

75. $\frac{24}{7} \div \frac{8}{21} = \frac{24}{7} \cdot \frac{21}{8}$
$= \frac{24 \cdot 21}{7 \cdot 8}$
$= \frac{8 \cdot 3 \cdot 7 \cdot 3}{7 \cdot 8}$
$= \frac{3 \cdot 3}{1 \cdot 1}$
$= \frac{9}{1}$
$= 9$

77. (20.6)(1.86)(100,000) = 3,831,600 miles

79. Answers may vary.

Section 5.4

Practice Problems

1. $5.6\overline{)166.88}$ becomes

$$\begin{array}{r} 29.8 \\ 56\overline{)\ 1668.8} \\ \underline{-112} \\ 548 \\ \underline{-504} \\ 448 \\ \underline{-448} \\ 0 \end{array}$$

2. Recall that a negative number divided by a negative number gives a positive quotient.

$$\begin{array}{r} 0.027 \\ 104\overline{)\ 2.808} \\ \underline{-2\,08} \\ 728 \\ \underline{-728} \\ 0 \end{array}$$

Thus −2.808 ÷ (−104) = 0.027.

3. $0.57\overline{)23.4}$ becomes

$$\begin{array}{r} 41.052 \approx 41.05 \\ 57.\overline{)\ 2340.000} \\ \underline{-228} \\ 60 \\ \underline{-57} \\ 30 \\ \underline{-0} \\ 300 \\ \underline{-285} \\ 150 \\ \underline{-114} \\ 36 \end{array}$$

4. To find 28 ÷ 1000, note that 1000 has 3 zeros. Therefore, we move the decimal point of 28 to
the left 3 places. The quotient is 0.028.

5. To find 8.56 ÷ 100, note that 100 has 2 zeros. Therefore, we move the decimal point of 8.56 to the left 2 places. The quotient is 0.0856.

6. $x \div y = 0.035 \div 0.02$

$0.02\overline{)0.035}$ becomes $\begin{array}{r} 1.75 \\ 2\overline{)3.50} \end{array}$
$= 1.75$

7. $\frac{x}{100} = 3.9$
$\frac{39}{100} \stackrel{?}{=} 3.9$
$0.39 \stackrel{?}{=} 3.9$ False
Since 0.39 = 3.9 is a false statement, 39 is not a solution.

8.
$$\begin{array}{r} 11.84 \\ 1250\overline{)\ 14,800.00} \\ -1250 \\ \hline 2300 \\ -1250 \\ \hline 10500 \\ -10000 \\ \hline 5000 \\ -5000 \\ \hline 0 \end{array}$$
He needs 12 bags.

Exercise Set 5.4

1.
$$\begin{array}{r} 0.094 \\ 5\overline{)0.470} \\ -45 \\ \hline 20 \\ -20 \\ \hline 0 \end{array}$$
$0.47 \div 5 = 0.094$

3. $0.06\overline{)18.00}$
Move the decimal points 2 places.
$$\begin{array}{r} 300 \\ 6.\overline{)\ 1800.} \\ -18 \\ \hline 00 \\ -0 \\ \hline 00 \\ -0 \\ \hline 0 \end{array}$$
$-18 \div 0.06 = -300$

5. $0.82\overline{)4.756}$
Move the decimal points 2 places.
$$\begin{array}{r} 5.8 \\ 82.\overline{)\ 475.6} \\ -410 \\ \hline 656 \\ -656 \\ \hline 0 \end{array}$$
$4.756 \div 0.82 = 5.8$

7. $5.5\overline{)36.3}$
Move the decimal points 1 place.
$$\begin{array}{r} 6.6 \\ 55.\overline{)\ 363.0} \\ -330 \\ \hline 330 \\ -330 \\ \hline 0 \end{array}$$
The answer is negative.
$-36.3 \div 5.5 = -6.6$

9. $2.4\overline{)429.34}$
Move the decimal points one place.
$$\begin{array}{r} 178.... \\ 24.\overline{)4293.4} \\ -24 \\ \hline 189 \\ -168 \\ \hline 213 \\ -192 \\ \hline 21 \end{array}$$
178 rounded to the nearest hundred is 200.

11. $0.023\overline{)0.549}$
Move the decimal points 3 places.
$$\begin{array}{r} 23.869 \\ 23.\overline{)\ 549.00} \\ -46 \\ \hline 89 \\ -69 \\ \hline 200 \\ -184 \\ \hline 160 \\ -138 \\ \hline 220 \\ -207 \\ \hline 13 \end{array}$$
23.869 rounded to the nearest hundredth is 23.87.

13. $0.4\overline{)45.23}$

Move the decimal points one place.

$$\begin{array}{r} 113.0 \\ 4.\overline{)\ 452.300} \\ \underline{-4} \\ 05 \\ \underline{-4} \\ 12 \\ \underline{-12} \\ 03 \\ \underline{-0} \\ 3 \end{array}$$

113.0 rounded to the nearest ten is 110.

15. $54.982 \div 100$ Move decimal point 2 places to the left.

$= 0.54982$

17. $12.9 \div (-1000)$ Move the decimal point 3 places to the left and attach the negative sign.

$= -0.0129$

19. $87 \div 10$ Move the decimal point 1 place to the left.

$= 8.7$

21.

$$\begin{array}{r} 0.413 \\ 3\overline{)\ 1.239} \\ \underline{-1\ 2} \\ 03 \\ \underline{-3} \\ 09 \\ \underline{-9} \\ 0 \end{array}$$

$1.239 \div 3 = 0.413$

23. $0.6\overline{)4.2}$

Move the decimal points 1 place.

$$\begin{array}{r} 7. \\ 6.\overline{)\ 42.} \\ \underline{-42} \\ 0 \end{array}$$

The answer is negative.

$-4.2 \div 0.6 = -7$

25. $0.27\overline{)1.296}$

Move the decimal points 2 places.

$$\begin{array}{r} 4.8 \\ 27.\overline{)\ 129.6} \\ \underline{-108} \\ 21\ 6 \\ \underline{-21\ 6} \\ 0 \end{array}$$

$1.296 \div 0.27 = 4.8$

27. $0.02\overline{)42.}$

Move the decimal points 2 places.

$$\begin{array}{r} 2100. \\ 2.\overline{)\ 4200.} \\ \underline{-4200} \\ 0 \end{array}$$

29. $-18 \div -0.6 = 180 \div -6 = 30$

31. $35 \div 0.005 = 35{,}000 \div 5 = 7000$

33. $1.6\overline{)1.104}$

Move the decimal points 1 place.

$$\begin{array}{r} 0.69 \\ 16.\overline{)\ 11.04} \\ \underline{-96} \\ 1\ 44 \\ \underline{-1\ 44} \\ 0 \end{array}$$

The answer is negative.

$-1.104 \div 1.6 = -0.69$

35. $-2.4 \div -100$ Move the decimal point 2 places to the left. The answer is positive.

$= 0.024$

37. $\dfrac{4.615}{0.071} = \dfrac{4615}{71} = 65$

39. $8.9\overline{)0.00263}$

Move the decimal points 1 place.

$$\begin{array}{r} 0.00029\ldots \\ 89.\overline{)0.02630} \end{array}$$

0.00029 rounded to the nearest ten-thousandth is 0.0003.

41. $0.0043\overline{)500}$

Move the decimal points 4 places.

$$\begin{array}{r} 116279. \\ 43.\overline{)\ 5000000.} \\ \underline{-43} \\ 70 \\ \underline{-43} \\ 270 \\ \underline{-258} \\ 120 \\ \underline{-86} \\ 340 \\ \underline{-301} \\ 390 \\ \underline{-387} \\ 3 \end{array}$$

116,279 rounded to the nearest ten-thousand is 120,000.

43. $z \div y = 4.52 \div 0.8$

$0.8\overline{)4.52}$

Move the decimal points one place.

$$\begin{array}{r} 5.65 \\ 8.\overline{)\ 45.20} \\ \underline{-40} \\ 52 \\ \underline{-48} \\ 40 \\ \underline{-40} \\ 0 \end{array}$$

45. $x \div y = 5.65 \div 0.8 = 7.0625$

$0.8\overline{)5.65}$

Move the decimal point 1 place.

$$\begin{array}{r} 7.0625 \\ 8.\overline{)\ 56.5000} \\ \underline{-56} \\ 0\ 5 \\ \underline{-\ 0} \\ 50 \\ \underline{-48} \\ 20 \\ \underline{-16} \\ 40 \\ \underline{-40} \\ 0 \end{array}$$

47. $y \div 2 = 0.8 \div 2 = 0.4$

$$\begin{array}{r} 0.4 \\ 2\overline{)0.8} \\ \underline{-8} \\ 0 \end{array}$$

49. $\frac{x}{4} = 3.04$

$\frac{12.16}{4} = 3.04$

$3.04 = 3.04$ True

Yes, it is a solution.

51. $\frac{x}{4.3} = 2$

$\frac{0.86}{4.3} = 2$

$0.2 = 2$ False

Not, it is not a solution.

53. $\frac{z}{10} = 0.8$

$\frac{8}{10} = 0.8$

$08. = 0.8$ True

Yes, it is a solution.

55. Divide the total amount by the amount per month to get the number of months.

$73.86\overline{)1772.64}$

Move the decimal points 2 places.

$$\begin{array}{r} 24 \\ 7386.\overline{)\ 177{,}264.} \\ -147\ 72 \\ \hline 29\ 544 \\ -29\ 544 \\ \hline 0 \end{array}$$

It will be paid off in 24 months.

57. There are 52 weeks per year and 40 hours per week. Therefore, there are $52 \times 40 = 2080$ hours per year.

$$\begin{array}{r} 1245.687 \\ 2080\overline{)\ 2{,}591{,}031.000} \\ -2\ 080 \\ \hline 511\ 0 \\ -416\ 0 \\ \hline 95\ 03 \\ -83\ 20 \\ \hline 11\ 831 \\ -10\ 400 \\ \hline 1\ 4310 \\ -1\ 2480 \\ \hline 18300 \\ -16640 \\ \hline 16600 \\ -14560 \\ \hline 2040 \end{array}$$

His hourly wage was \$1245.69.

59.

$$\begin{array}{r} 202.14 \approx 202.1 \\ 39.\overline{)\ 7883.50} \\ -78 \\ \hline 08 \\ -0 \\ \hline 83 \\ -78 \\ \hline 55 \\ -39 \\ \hline 160 \\ -156 \\ \hline 4 \end{array}$$

She needs to buy 202.1 pounds.

61. $39.37\overline{)200}$

Move the decimal points 2 places.

$$\begin{array}{r} 5.08 \approx 5.1 \\ 3937.\overline{)\ 20{,}000.00} \\ -19\ 685 \\ \hline 3150 \\ -0 \\ \hline 31500 \\ -31496 \\ \hline 4 \end{array}$$

There are 5.1 meters in 200 inches.

63. To find $59.6 \div 100$, note that 100 has 2 zeros. Therefore, we move the decimal point of 59.6 to the left 2 places. The quotient is 0.596. The average price was \$0.596 per pound.

65.

$$\begin{array}{r} 128.63 \approx 128.6 \\ 24\overline{)\ 3087.12} \\ -24 \\ \hline 68 \\ -48 \\ \hline 207 \\ -192 \\ \hline 151 \\ -144 \\ \hline 72 \\ -72 \\ \hline 0 \end{array}$$

Their average speed was about 128.6 mph.

67.
$$\begin{array}{r} 22.12 \approx 22.1 \\ 31\overline{)\ 686.00} \\ \underline{-62} \\ 66 \\ \underline{-62} \\ 40 \\ \underline{-31} \\ 90 \\ \underline{-62} \\ 28 \end{array}$$
Her average was 22.1 points/game.

69. 75 mph ÷ 1.15
$1.15\overline{)75}$
Move the decimal points 2 places.
$$\begin{array}{r} 65.21 \approx 65.2 \\ 115.\overline{)\ 7500.00} \\ \underline{-690} \\ 600 \\ \underline{-575} \\ 250 \\ \underline{-230} \\ 200 \\ \underline{-115} \\ 85 \end{array}$$
95 mph ÷ 1.15
$1.15\overline{)95}$
Move the decimal points 2 places.
$$\begin{array}{r} 82.60 \approx 82.6 \\ 115.\overline{)\ 9500.00} \\ \underline{-920} \\ 300 \\ \underline{-230} \\ 70\ 0 \\ \underline{-69\ 0} \\ 1\ 00 \\ \underline{-\ 0} \\ 1\ 00 \end{array}$$
The wind speed range for a Category 1 hurricane is from 65.2 knots to 82.6 knots.

71. To round 345.219 to the nearest hundredth, observe that the digit in the thousandths place is 9. Since this digit is at least 5, we need to add 1 to the digit in the hundredths place. The number 345.219 rounded to the nearest hundredth is 345.22.

73. To round 1000.994 to the nearest tenth, observe that the digit in the hundredths place is 9. Since this digit is at least 5, we need to add 1 to the digit in the tenths place. The number 1000.994 rounded to the nearest tenth is 1001.0.

75. $2 + 3 \cdot 6 = 2 + 18 = 20$

77. $20 - 10 \div 5 = 20 - 2 = 18$

79. $(86 + 78 + 91 + 85) \div 4 = 340 \div 4$
$$\begin{array}{r} 85 \\ 4\overline{)\ 340} \\ \underline{-32} \\ 20 \\ \underline{-20} \\ 0 \end{array}$$
The average is 85.

81. $4.5\overline{)38.7}$
Move the decimal points 1 place.
$$\begin{array}{r} 8.6 \\ 45.\overline{)\ 387.0} \\ \underline{-360} \\ 27\ 0 \\ \underline{-27\ 0} \\ 0 \end{array}$$
The length is 8.6 feet.

83. Answers may vary.

Integrated Review

1.
$$\begin{array}{r} 1.60 \\ \underline{+0.97} \\ 2.57 \end{array}$$

2. $$\begin{array}{r} 3.20 \\ +0.85 \\ \hline 4.05 \end{array}$$

3. $$\begin{array}{r} 9.8 \\ -0.9 \\ \hline 8.9 \end{array}$$

4. $$\begin{array}{r} 10.2 \\ -6.7 \\ \hline 3.5 \end{array}$$

5. $$\begin{array}{r} 0.8 \\ \times 0.2 \\ \hline 0.16 \end{array}$$

6. $$\begin{array}{r} 0.6 \\ \times 0.4 \\ \hline 0.24 \end{array}$$

7. $$\begin{array}{r} 0.27 \\ 8\overline{)\ 2.16} \\ -1\ 6 \\ \hline 56 \\ -56 \\ \hline 0 \end{array}$$

8. $$\begin{array}{r} 0.52 \\ 6\overline{)\ 3.12} \\ -30 \\ \hline 12 \\ -12 \\ \hline 0 \end{array}$$

9. $$\begin{array}{r} 9.6 \\ \times 0.5 \\ \hline 4.80 \end{array}$$ or $(9.6)(-0.5) = -4.8$

10. $$\begin{array}{r} 8.7 \\ \times 0.7 \\ \hline 6.09 \end{array}$$ $(-8.7)(-0.7) = 6.09$

11. $$\begin{array}{r} 123.60 \\ -48.04 \\ \hline 75.56 \end{array}$$

12. $$\begin{array}{r} 325.20 \\ -36.08 \\ \hline 289.12 \end{array}$$

13. $-25 + 0.026 = -24.974$

14. $$\begin{array}{r} 44.000 \\ -0.125 \\ \hline 43.875 \end{array}$$ $0.125 + (-44) = -43.875$

15. $3.4\overline{)29.24}$ becomes

$$\begin{array}{r} 8.6 \\ 34\overline{)\ 292.4} \\ -272 \\ \hline 204 \\ -204 \\ \hline 0 \end{array}$$

-8.6

The answer is negative.

16. $1.9\overline{)10.26}$ becomes

$$\begin{array}{r} 5.4 \\ 19\overline{)102.6} \\ 95 \\ \hline 76 \\ 76 \\ \hline 0 \end{array}$$ $-10.26 \div (-1.9) = 5.4$

17. $2.8 \times 100 = 280$

18. $1.6 \times 1000 = 1600$

19. $$\begin{array}{r} 96210 \\ 7.028 \\ +121.700 \\ \hline 224.938 \end{array}$$

20. $$\begin{array}{r} 0.268 \\ 1.940 \\ +142.881 \\ \hline 145.079 \end{array}$$

21.
$$\begin{array}{r} 0.56 \\ 46\overline{)\ 25.76} \\ \underline{-23\ 0} \\ 2\ 76 \\ \underline{-2\ 76} \\ 0 \end{array}$$

22.
$$\begin{array}{r} 0.63 \\ 43\overline{)\ 27.09} \\ \underline{-25\ 8} \\ 1\ 29 \\ \underline{-1\ 29} \\ 0 \end{array}$$

23.
$$\begin{array}{r} 12.004 \\ \underline{\times\ 2.3} \\ 36012 \\ \underline{240080} \\ 26.6092 \end{array}$$

24.
$$\begin{array}{r} 28.006 \\ \underline{\times\ 5.2} \\ 56012 \\ \underline{1400300} \\ 145.6312 \end{array}$$

25.
$$\begin{array}{r} 10.0 \\ \underline{-\ 4.6} \\ 5.4 \end{array}$$

26.
$$\begin{array}{r} 18.00 \\ \underline{-\ 0.26} \\ 17.74 \end{array}$$
$0.26 - 18 = -17.74$

27.
$$\begin{array}{r} 268.19 \\ \underline{+\ 146.25} \\ 414.44 \end{array}$$
$-268.19 - 414.44 = -414.44$

28.
$$\begin{array}{r} 860.18 \\ \underline{+\ 434.85} \\ 1295.03 \end{array}$$
$-860.18 - 434.85 = -1295.03$

29. $0.087\overline{)2.958}$ becomes
$$\begin{array}{r} 34 \\ 87.\overline{)\ 2958} \\ \underline{-261} \\ 348 \\ \underline{-348} \\ 0 \end{array}$$
-34
The answer is negative.

30. $0.061\overline{)1.708}$ becomes
$$\begin{array}{r} 28 \\ 61.\overline{)\ 1708} \\ \underline{-122} \\ 488 \\ \underline{-488} \\ 0 \end{array}$$
-28
The answer is negative.

31.
$$\begin{array}{r} 160.00 \\ \underline{-43.19} \\ 116.81 \end{array}$$

32.
$$\begin{array}{r} 120.00 \\ \underline{-\ 101.21} \\ 18.79 \end{array}$$

33. The tenths place in 0.38 is less than the tenths place in 0.5. Thus, the cookie can be labeled
"Fat Free."

34.
$$\begin{array}{r} 8020.78 \\ \underline{-\ 380.53} \\ 7640.25 \end{array}$$
The DJIA opened at 7640.25 points.

35.
$$\begin{array}{r} 26.30 \\ 49\overline{)1289.00} \\ \underline{98} \\ 309 \\ \underline{-294} \\ 15\ 0 \\ \underline{14\ 7} \\ 30 \end{array}$$
His average was 26.3 points/game.

Section 5.5

Practice Problems

1.
$$\begin{array}{r} 65.34 \\ \underline{-14.68} \\ 50.66 \end{array} \quad \text{Estimate} \quad \begin{array}{r} 65 \\ \underline{-15} \\ 50 \end{array}$$
The estimated difference is \$50, so \$50.66 is reasonable.

2.
$$\begin{array}{r} 30.26 \\ \underline{\times\ 2.98} \\ 24208 \\ 272340 \\ \underline{605200} \\ 90.1748 \end{array} \quad \text{Estimate} \quad \begin{array}{r} 30 \\ \underline{\times\ 3} \\ 90 \end{array}$$
The answer 90.1748 is reasonable.

3. $91.5\overline{)713.7}$ Estimate $\begin{array}{r} 8 \\ 90\overline{)720} \end{array}$

becomes
$$\begin{array}{r} 7.8 \\ 915.\overline{)\ 7137.0} \\ \underline{-6405} \\ 732\ 0 \\ \underline{-732\ 0} \\ 0 \end{array}$$
The estimate is 8, so 7.8 is reasonable.

4.
$$\begin{array}{r} 79.2 \\ \underline{-53.7} \end{array} \quad \text{Estimate} \quad \begin{array}{r} 79 \\ \underline{-\ 54} \\ 25 \end{array}$$
The estimate is 25 miles.

5. $-8.6(3.2-1.8) = -8.6(1.4)$
$= -12.04$

6. $(0.7)^2 = (0.7)(0.7)$
$= 0.49$

7. $\dfrac{8.78-2.8}{20} = \dfrac{5.98}{20} = 0.299$

8. $1.7y - 2 = 1.7(2.3) - 2$
$= 3.91 - 2$
$= 1.91$

9. $3x + 7.5 = 1.2$
$3(-2.1) + 7.5 \stackrel{?}{=} 1.2$
$-6.3 + 7.5 \stackrel{?}{=} 1.2$
$1.2 = 1.2$ True
Since $1.2 = 1.2$ is a true statement, -2.1 is a solution.

Exercise Set 5.5

1. $4.9 - 2.1 = 2.8$
$5 - 2 = 3$
3 is close to 2.8, so the answer is reasonable.

3.
$$\begin{array}{r} 6 \\ \underline{\times\ 483.11} \\ 2898.66 \end{array} \quad \text{Estimate:} \quad \begin{array}{r} 6 \\ \underline{\times\ 500} \\ 3000 \end{array}$$
3000 is close to 2898.66, so the answer is reasonable.

5. $62.16 \div 14.8 = 4.2$
$60 \div 15 = 4$
4 is close to 4.2 so the answer is reasonable.

7.
$$\begin{array}{r} 69.2 \\ 32.1 \\ \underline{+\ 48.5} \\ 149.7 \end{array} \quad \begin{array}{r} 70 \\ 30 \\ \underline{+\ 50} \\ 150 \end{array}$$
149.8 is close to 150, so the answer is reasonable.

9.
$$\begin{array}{r} 34.92 \\ \underline{-12.03} \\ 22.89 \end{array} \quad \begin{array}{r} 35 \\ \underline{-12} \\ 23 \end{array}$$
23 is close to 22.89 so the answer is reasonable.

11. $2(12.2) + 2(5.9) \approx 2(12) + 2(6)$
$= 24 + 12$
$= 36$ inches

13. $11.8 + 12.9 + 14.2 \approx 12 + 13 + 14 = 39$ ft

15. $3.14(7)(2) = 43.96$ meters

17. $$\begin{array}{r} 51.6 \\ 30\overline{)\ 1550.0} \\ \underline{-150} \\ 50 \\ \underline{-30} \\ 200 \\ \underline{-180} \\ 20 \end{array}$$
Their car will use about 52 gallons.

19. 198.79 is approximately 200
5 years is $5 \times 12 = 60$ months.
$60 \times 200 = 12{,}000$
They will pay about $12,000.

21. $19.9 + 15.1 + 10.9 + 6.7 = ?$
Estimate:
$20 + 15 + 11 + 7 = 53$ miles
The distance is about 53 miles.

23. $600.8 million → $600 million
$461.0 million → $500 million
$421.4 million → $400 million
$399.8 million → $400 million
$357.1 million → $400 million
$329.7 million → $300 million
The estimate is $2600 million

25. 485 rounded to the nearest 10 is 490.
271 rounded to the nearest 10 is 270.
$$\begin{array}{r} 490 \\ \underline{\times\ \ 270} \\ 0 \\ 34300 \\ \underline{98000} \\ 132{,}300 \end{array}$$
The population is about 132,300 people.

27. $(0.4)^2 = (0.4)(0.4) = 0.16$

29. $\frac{1+0.8}{-0.6} = \frac{1.8}{-0.6} = \frac{18}{-6} = -3$

31. $1.4(2 - 1.8) = 1.4(0.2) = 0.28$

33. $4.83 \div 2.1 = 2.3$

35. $(-2.3)^2 = (-2.3)(-2.3) = 5.29$

37. $(3.1 + 0.7)(2.9 - 0.9) = (3.8)(2.0) = 7.6$

39. $\frac{(4.5)^2}{100} = \frac{20.25}{100} = 0.2025$

41. $\frac{7+0.74}{-6} = \frac{7.74}{-6} = -1.29$

43. z^2
$(-2.4)^2 = 5.76$

45. $x - y$
$6 - (0.3) = 5.7$

47. $4y - z$
$4(0.3) - (-2.4) = 1.2 + 2.4 = 3.6$

49. $$\begin{aligned} 7x + 2.1 &= -7 \\ 7(-1.3) + 2.1 &= -7 \\ -9.1 + 2.1 &= -7 \\ -7 &= -7 \quad \text{True} \end{aligned}$$
Yes, –1.3 is a solution.

51. $$\begin{aligned} x - 6.5 &= 2x + 1.8 \\ -4.7 - 6.5 &= 2(-4.7) + 1.8 \\ -11.2 &= -9.4 + 1.8 \\ -11.2 &= -7.6 \quad \text{False} \end{aligned}$$
No, –4.7 is not a solution.

53. $\frac{3}{4} \cdot \frac{5}{12} = \frac{3 \cdot 5}{4 \cdot 3 \cdot 4} = \frac{5}{16}$

55. $$\begin{aligned} \frac{36}{56} \div \frac{30}{35} &= \frac{36}{56} \cdot \frac{35}{30} \\ &= \frac{2 \cdot 2 \cdot 3 \cdot 3 \cdot 5 \cdot 7}{2 \cdot 2 \cdot 2 \cdot 7 \cdot 2 \cdot 3 \cdot 5} \\ &= \frac{3}{4} \end{aligned}$$

57. $\frac{5}{12} - \frac{1}{3} = \frac{5}{12} - \frac{4}{12} = \frac{1}{12}$

59. $$\begin{aligned} 1.96(7.852 - 3.147)^2 &= 1.96(4.705)^2 \\ &= 1.96(22.137025) \\ &= 43.388569 \end{aligned}$$
Estimate: $2(8 - 3)^2 = 2(5)^2 = 2(25) = 50$
The result is reasonable.

61. Answers may vary.

Section 5.6

Practice Problems

1. a. $\frac{2}{5} = 2 \div 5$

$$\begin{array}{r} 0.4 \\ 5\overline{)\,2.0} \\ \underline{-2\,0} \\ 0 \end{array}$$

$\frac{2}{5} = 0.4$

b. $\frac{9}{40} = 9 \div 40$

$$\begin{array}{r} 0.225 \\ 40\overline{)\,9.000} \\ \underline{-8\,0} \\ 100 \\ \underline{-\,80} \\ 200 \\ \underline{-200} \\ 0 \end{array}$$

$\frac{9}{40} = 0.225$

2. a.

$$\begin{array}{r} 0.833\ldots \\ 6\overline{)\,5.000} \\ \underline{-4\,8} \\ 20 \\ \underline{-18} \\ 20 \\ \underline{-18} \\ 2 \end{array}$$

$\frac{5}{6} = 0.8\overline{3} \approx 0.83$

b.

$$\begin{array}{r} 0.22\ldots \\ 9\overline{)\,2.00} \\ \underline{-1\,8} \\ 20 \\ \underline{-18} \\ 2 \end{array}$$

$\frac{2}{9} = 0.\overline{2} \approx 0.2$

3.

$$\begin{array}{r} 0.1111\ldots \\ 9\overline{)1.0000} \\ \underline{-9} \\ 10 \\ \underline{-9} \\ 10 \end{array}$$

$\frac{1}{9} = 0.\overline{1} \approx 0.111$

4. $\frac{1}{5} = 1 \div 5$

$$\begin{array}{r} 0.2 \\ 5\overline{)\,1.0} \\ \underline{-1\,0} \\ 0 \end{array}$$

$\frac{1}{5} = 0.20$

Since $0.20 < 0.25$, $\frac{1}{5} < 0.25$.

5. a.

$$\begin{array}{r} 0.5 \\ 2\overline{)\,1.0} \\ \underline{-1\,0} \\ 0 \end{array}$$

$\frac{1}{2} = 0.50$

Since $0.50 < 0.54$, $\frac{1}{2} < 0.54$.

b. Since $\frac{2}{9} = 0.\overline{2}$, $0.\overline{2} = \frac{2}{9}$.

c.

$$\begin{array}{r} 0.714 \\ 7\overline{)\,5.000} \\ \underline{-4\,9} \\ 10 \\ \underline{-7} \\ 30 \\ \underline{-28} \\ 2 \end{array}$$

Since $\frac{5}{7} \approx 0.714$ and $0.714 < 0.72$,

$\frac{5}{7} < 0.72$.

6. a.

Original number	$\frac{1}{3}$	0.302	$\frac{3}{8}$
Decimals	0.333…	0.302	0.375
Compare in order	2nd	1st	3rd

Then, written in order: 0.302, $\frac{1}{3}$, $\frac{3}{8}$

b.

Original number	1.26	$1\frac{1}{4}$	$1\frac{2}{5}$
Decimals	1.26	1.25	1.40
Compare in order	2nd	1st	3rd

Then, written in order: $1\frac{1}{4}$, 1.26, $1\frac{2}{5}$

c.

Original number	0.4	0.41	$\frac{5}{7}$
Decimals	0.40	0.41	≈0.714
Compare in order	1st	2nd	3rd

Then, written in order: 0.4, 0.41, $\frac{5}{7}$

7. $\text{area} = \frac{1}{2} \cdot \text{base} \cdot \text{height}$

$= \frac{1}{2} \cdot 7 \cdot 2.1$

$= 0.5 \cdot 7 \cdot 2.1$

$= 7.35$

The area is 7.35 square meters.

Exercise Set 5.6

1. $\frac{1}{5} = 0.2$

$$\begin{array}{r} 0.2 \\ 5\overline{)\ 1.0} \\ \underline{-1\ 0} \\ 0 \end{array}$$

3. $\frac{4}{8} = 0.5$

$$\begin{array}{r} 0.5 \\ 8\overline{)\ 4.0} \\ \underline{-4\,0} \\ 0 \end{array}$$

5. $\frac{3}{4} = 0.75$

$$\begin{array}{r} 0.75 \\ 4\overline{)\ 3.00} \\ \underline{-2\,8} \\ 20 \\ \underline{-20} \\ 0 \end{array}$$

7. $\frac{2}{25} = 0.08$

$$\begin{array}{r} 0.08 \\ 25\overline{)\ 2.00} \\ \underline{-2\,00} \\ 0 \end{array}$$

9. $\frac{3}{8} = 0.375$

$$\begin{array}{r} 0.375 \\ 8\overline{)\ 3.000} \\ \underline{-2\ 4} \\ 60 \\ \underline{-56} \\ 40 \\ \underline{-40} \\ 0 \end{array}$$

11. $\frac{11}{12} = 0.91\overline{6}$

$$\begin{array}{r} 0.9166\ldots \\ 12\overline{)\ 11.0000} \\ \underline{-10\ 8} \\ 20 \\ \underline{-12} \\ 80 \\ \underline{-72} \\ 80 \\ \underline{-72} \\ 8 \end{array}$$

13. $\frac{17}{40} = 0.425$

$$\begin{array}{r} 0.425 \\ 40\overline{)\ 17.000} \\ \underline{-16\ 0} \\ 1\ 00 \\ \underline{-80} \\ 200 \\ \underline{-200} \\ 0 \end{array}$$

15. $\frac{9}{20} = 0.45$

$$\begin{array}{r} 0.45 \\ 20\overline{)\ 9.00} \\ \underline{-8\ 0} \\ 1\ 00 \\ \underline{-1\ 00} \\ 0 \end{array}$$

17. $\frac{1}{3} = 0.\overline{3}$

$$\begin{array}{r} 0.33\ldots \\ 3\overline{)1.00} \\ \underline{-9} \\ 10 \\ \underline{-9} \\ 1 \end{array}$$

19. $\frac{7}{16} = 0.4375$

$$\begin{array}{r} 0.4375 \\ 16\overline{)\ 7.0000} \\ \underline{-6\ 4} \\ 60 \\ \underline{-48} \\ 120 \\ \underline{-112} \\ 80 \\ \underline{-80} \\ 0 \end{array}$$

21. $\frac{2}{9} = 0.\overline{2}$

$$\begin{array}{r} 0.22\ldots \\ 9\overline{)\ 2.00} \\ \underline{-18} \\ 20 \\ \underline{-18} \\ 2 \end{array}$$

23. $\frac{5}{3} = 1.\overline{6}$

$$\begin{array}{r} 1.66\ldots \\ 3\overline{)\ 5.00} \\ \underline{-3} \\ 2\ 0 \\ \underline{-1\ 8} \\ 20 \\ \underline{-18} \\ 2 \end{array}$$

25. $0.\overline{3} \approx 0.333\ldots \approx 0.33$

27. $0.4375 \approx 0.44$

29. $0.\overline{2} = 0.222\ldots \approx 0.2$

31. $1.\overline{6} = 1.666\ldots \approx 1.7$

33.

$$\begin{array}{r} 0.1941\ldots \\ 376\overline{)\ 73.0000} \\ \underline{-37\ 6} \\ 35\ 40 \\ \underline{-33\ 84} \\ 1\ 560 \\ \underline{-1\ 504} \\ 560 \\ \underline{-376} \\ 184 \end{array}$$

$\frac{73}{376} \approx 0.194$

35.

$$\begin{array}{r} 0.44 \\ 25\overline{)\ 11.00} \\ \underline{-10\ 0} \\ 1\ 00 \\ \underline{-1\ 00} \\ 0 \end{array}$$

$\frac{11}{25} = 0.44$

37.

$$\begin{array}{r} 0.8 \\ 5\overline{)\ 4.0} \\ \underline{-4\ 0} \\ 0 \end{array}$$

$\frac{4}{5} = 0.8$

39. $2 < 9$, so $0.562 < 0.569$

41. $2 > 1$, so $0.823 > 0.813$

43. $2 < 3$, so $0.0923 < 0.0932$

45. $\frac{2}{3} = \frac{4}{6}$ and $\frac{4}{6} < \frac{5}{6}$, so $\frac{2}{3} < \frac{5}{6}$

47. $\frac{5}{9} = 0.\overline{5} = 0.555\ldots$

$\frac{51}{91} \approx 0.5604$

$0.5\overline{5} < 0.5604$, so $\frac{5}{9} < \frac{51}{91}$.

49. $\frac{4}{7} \approx 0.5714$ and $0.5714 > 0.14$,
so $\frac{4}{7} > 0.14$.

51. $\frac{18}{13} \approx 1.3846$
$1.38 < 1.3846$, so $1.38 < \frac{18}{13}$.

53. $\frac{456}{64} = 7.125$
$7.123 < 7.125$, so $7.123 < \frac{456}{64}$

55. 0.32, 0.34, 0.35

57. $0.49 = 0.490$
0.49, 0.491, 0.498

59. $\frac{3}{4} = 0.75$
0.73, $\frac{3}{4}$, 0.78

61. $\frac{4}{7} \approx 0.571$
0.412, 0.453, $\frac{4}{7}$

63. $\frac{42}{8} = 5.25$
5.23, $\frac{42}{8}$, 5.34

65. $\frac{12}{5} = 2.4$, $\frac{17}{8} = 2.125$
$\frac{17}{8}$, 2.37, $\frac{12}{5}$

67.
$$\begin{aligned} \text{Area} &= \frac{1}{2} \cdot \text{base} \cdot \text{height} \\ &= \frac{1}{2} \times 5.7 \times 9 \\ &= 0.5 \times 5.7 \times 9 \\ &= 25.65 \end{aligned}$$
The area is 25.65 square inches.

69.
$$\begin{aligned} \text{Area} &= \frac{1}{2} \cdot \text{base} \cdot \text{height} \\ &= \frac{1}{2} \cdot 5.2 \cdot 3.6 \\ &= 0.5 \cdot 5.2 \cdot 3.6 \\ &= 9.36 \end{aligned}$$
The area is 9.36 square centimeters.

71.
$$\begin{aligned} \text{Area} &= \text{base} \cdot \text{height} \\ &= 0.62 \cdot \frac{2}{5} \\ &= 0.62 \cdot 0.4 \\ &= 0.248 \end{aligned}$$
The area is 0.248 square yards.

73. $2^3 = 2 \cdot 2 \cdot 2 = 8$

75. $6^2 \cdot 2 = 6 \cdot 6 \cdot 2 = 72$

77. $\left(\frac{1}{3}\right)^4 = \frac{1}{3} \cdot \frac{1}{3} \cdot \frac{1}{3} \cdot \frac{1}{3} = \frac{1}{81}$

79. $\left(\frac{3}{5}\right)^2 = \frac{3}{5} \cdot \frac{3}{5} = \frac{9}{25}$

81. $\left(\frac{2}{5}\right)\left(\frac{5}{2}\right)^2 = \frac{2}{5} \cdot \frac{5}{2} \cdot \frac{5}{2} = \frac{5}{2}$

83. $\frac{2321}{10,506} \approx 0.221$

$$\begin{array}{r} 0.2209 \approx 0.221 \\ 10,506 \overline{)\ 2321.0000} \\ \underline{-2101\ 2} \\ 219\ 80 \\ \underline{-210\ 12} \\ 9\ 680 \\ \underline{-\quad 0} \\ 9\ 6800 \\ \underline{-9\ 4554} \\ 2246 \end{array}$$

85.

	Estimate
2321	2300
1576	1600
1396	1400
1109	1100
1088	1100
+ 803	+ 800
	8300

87.

$$\begin{array}{r} 0.625 \\ 8\overline{)\ 5.000} \\ \underline{-4\ 8} \\ 20 \\ \underline{-16} \\ 40 \\ \underline{-40} \\ 0 \end{array}$$

You should enter 0.625 as the margin width.

89. Answers may vary.

91. $\frac{3}{4}-(9.6)(5)=0.75-48$
$=-47.25$

93. $\left(\frac{1}{10}\right)^2+(1.6)(2.1)=\frac{1}{100}+(1.6)(2.1)$
$=0.01+3.36$
$=3.37$

95. $\frac{3}{8}(4.7-5.9)=0.375(4.7-5.9)$
$=0.375\cdot -12$
$=-0.45$

Section 5.7

Practice Problems

1. $z+0.9=1.3$
$z+0.9-0.9=1.3-0.9$
$z=0.4$

2. $0.17x=-0.34$
$\frac{0.17x}{0.17}=\frac{-0.34}{0.17}$
$x=-2$

3. $6.3-5x=3(x+2.9)$
$6.3-5x=3x+8.7$
$6.3-5x-6.3=3x+8.7-6.3$
$-5x=3x+2.4$
$-5x-3x=3x+2.4-3x$
$-8x=2.4$
$\frac{-8x}{-8}=\frac{2.4}{-8}$
$x=-0.3$

4. $0.2y+2.6=4$
$10(0.2y+2.6)=10(4)$
$2y+26=40$
$2y+26-26=40-26$
$2y=14$
$\frac{2y}{2}=\frac{14}{2}$
$y=7$

Exercise Set 5.7

1. $x+1.2=7.1$
$x+1.2-1.2=7.1-1.2$
$x=5.9$

3. $5y=2.15$
$\frac{5y}{5}=\frac{2.15}{5}$
$y=0.43$

5. $6x+8.65=3x+10$
$6x+8.65-8.65=3x+10-8.65$
$6x=3x+1.35$
$6x-3x=3x-3x+1.35$
$3x=1.35$
$\frac{3x}{3}=\frac{1.35}{3}$
$x=0.45$

7. $2(x-1.3)=5.8$
$2x-2.6=5.8$
$2x-2.6+2.6=5.8+2.6$
$2x=8.4$
$\frac{2x}{2}=\frac{8.4}{2}$
$x=4.2$

9.
$$\begin{aligned} 0.4x+0.7&=-0.9\\ 4x+7&=-9\\ 4x+7-7&=-9-7\\ 4x&=-16\\ \frac{4x}{4}&=\frac{-16}{4}\\ x&=-4 \end{aligned}$$

11.
$$\begin{aligned} 7x-10.8&=x\\ 70x-108&=10x\\ 70x-108+108&=10x+108\\ 70x&=10x+108\\ 70x-10x&=10x-10x+108\\ 60x&=108\\ \frac{60x}{60}&=\frac{108}{60}\\ x&=1.8 \end{aligned}$$

13.
$$\begin{aligned} 2.1x+5-1.6x&=10\\ 21x+50-16x&=100\\ 5x+50&=100\\ 5x+50-50&=100-50\\ 5x&=50\\ \frac{5x}{5}&=\frac{50}{5}\\ x&=10 \end{aligned}$$

15.
$$\begin{aligned} y-3.6&=4\\ y-3.6+3.6&=4+3.6\\ y&=7.6 \end{aligned}$$

17.
$$\begin{aligned} -0.02x&=-1.2\\ -2x&=-120\\ \frac{-2x}{-2}&=\frac{-120}{-2}\\ x&=60 \end{aligned}$$

19.
$$\begin{aligned} 6.5&=10x+7.2\\ 65&=100x+72\\ 65-72&=100x+72-72\\ -7&=100x\\ -0.07&=x \end{aligned}$$

21.
$$\begin{aligned} 200x-0.67&=100x+0.81\\ 200x-0.67+0.67&=100x+0.81+0.67\\ 200x&=100x+1.48\\ 200x-100x&=100x-100x+1.48\\ 100x&=1.48\\ x&=0.0148 \end{aligned}$$

23.
$$\begin{aligned} 3(x+2.71)&=2x\\ 3x+8.13&=2x\\ 3x-2x&=-8.13\\ x&=-8.13 \end{aligned}$$

25.
$$\begin{aligned} 1.2+0.3x&=0.9\\ 12+3x&=9\\ 12-12+3x&=9-12\\ 3x&=-3\\ \frac{3x}{3}&=\frac{-3}{3}\\ x&=-1 \end{aligned}$$

27.
$$\begin{aligned} 0.9x+2.65&=0.5x+5.45\\ 90x+265&=50x+545\\ 90x+265-265&=50x+545-265\\ 90x-50x&=50x-50x+280\\ 40x&=280\\ x&=7 \end{aligned}$$

29.
$$\begin{aligned} 4x+7.6&=2(3x-3.2)\\ 4x+7.6&=6x-6.4\\ 4x-6x+7.6-7.6&=6x-6x-6.4-7.6\\ -2x&=-14\\ x&=7 \end{aligned}$$

31. $2x-6+4x-10=(2x+4x)+(-6-10)=6x-16$

33. $3(x-5)+10=3x-15+10=3x-5$

35.
$$\begin{aligned} 5y-1.2-7y+8&=(5y-7y)+(8-1.2)\\ &=-2y+6.8 \end{aligned}$$

37. Answers may vary.

39.
$$\begin{aligned} -5.25x&=-40.33575\\ \frac{-5.25x}{-5.25}&=\frac{-40.33575}{-5.25}\\ x&=7.683 \end{aligned}$$

41.
$$\begin{aligned} 1.95y + 6.834 &= 7.65y - 19.8591 \\ 19.8591 + 6.834 &= 7.65y - 1.95y \\ 26.6931 &= 5.7y \\ \frac{26.6931}{5.7} &= \frac{5.7y}{5.7} \\ 4.683 &= y \end{aligned}$$

Section 5.8

Practice Problems

1. a. $\sqrt{100} = 10$ because $10^2 = 100$ and 10 is positive.

   b. $\sqrt{64} = 8$ because $8^2 = 64$ and 8 is positive.

   c. $\sqrt{121} = 11$ because $11^2 = 121$ and 11 is positive.

   d. $\sqrt{0} = 0$ because $0^2 = 0$.

2. $\sqrt{\frac{1}{4}} = \frac{1}{2}$ because $\frac{1}{2} \cdot \frac{1}{2} = \frac{1}{4}$.

3. $\sqrt{\frac{9}{16}} = \frac{3}{4}$

4. $\sqrt{11} \approx 3.317$

5. $\sqrt{29} \approx 5.385$

6.
$$\begin{aligned} a^2 + b^2 &= c^2 \\ 12^2 + 16^2 &= c^2 \\ 144 + 256 &= c^2 \\ 400 &= c^2 \\ c &= \sqrt{400} \\ c &= 20 \end{aligned}$$

The hypotenuse is 20 feet.

7.
$$\begin{aligned} a^2 + b^2 &= c^2 \\ 9^2 + 7^2 &= c^2 \\ 81 + 49 &= c^2 \\ 130 &= c^2 \\ c &= \sqrt{130} \\ c &\approx 11 \end{aligned}$$

The length of the hypotenuse is exactly $\sqrt{130}$ kilometers, which is approximately 11 kilometers.

8.
$$\begin{aligned} a^2 + b^2 &= c^2 \\ a^2 + 7^2 &= 11^2 \\ a^2 + 49 &= 121 \\ a^2 &= 72 \\ a &= \sqrt{72} \\ a &\approx 8.49 \end{aligned}$$

The length of the leg is exactly $\sqrt{72}$ feet, which is approximately 8.49 feet.

9.

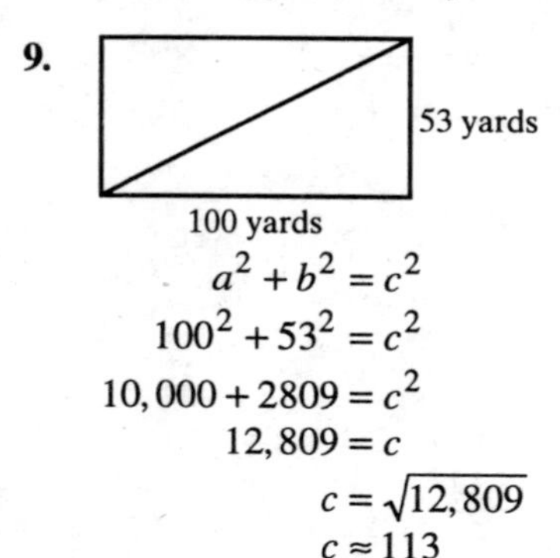

$$\begin{aligned} a^2 + b^2 &= c^2 \\ 100^2 + 53^2 &= c^2 \\ 10,000 + 2809 &= c^2 \\ 12,809 &= c \\ c &= \sqrt{12,809} \\ c &\approx 113 \end{aligned}$$

The diagonal of the football field is exactly $\sqrt{12,809}$ yards, which is approximately 113 yards.

Calculator Explorations

1. $\sqrt{1024} = 32$

2. $\sqrt{676} = 26$

3. $\sqrt{15} \approx 3.872983346 \approx 3.873$

4. $\sqrt{19} \approx 4.358898944 \approx 4.359$

5. $\sqrt{97} \approx 9.848857802 \approx 9.849$

6. $\sqrt{56} \approx 7.483314774 \approx 7.483$

Exercise Set 5.8

1. $\sqrt{4} = 2$ because $2^2 = 4$.

3. $\sqrt{625} = 25$ because $25^2 = 625$.

5. $\sqrt{\frac{1}{81}} = \frac{1}{9}$ because $\frac{1}{9} \cdot \frac{1}{9} = \frac{1}{81}$.

7. $\sqrt{\frac{144}{64}} = \frac{12}{8} = \frac{3}{2}$ because $\frac{12}{8} \cdot \frac{12}{8} = \frac{144}{64}$.

9. $\sqrt{256} = 16$ because $16^2 = 256$.

11. $\sqrt{\frac{9}{4}} = \frac{3}{2}$ because $\frac{3}{2} \cdot \frac{3}{2} = \frac{9}{4}$.

13. $\sqrt{3} \approx 1.732$

15. $\sqrt{15} \approx 3.873$

17. $\sqrt{14} \approx 3.742$

19. $\sqrt{47} \approx 6.856$

21. $\sqrt{8} \approx 2.828$

23. $\sqrt{26} \approx 5.099$

25. $\sqrt{71} \approx 8.426$

27. $\sqrt{7} \approx 2.646$

29.
$$\begin{aligned} a^2 + b^2 &= c^2 \\ 5^2 + 12^2 &= c^2 \\ 25 + 144 &= c^2 \\ 169 &= c^2 \\ c &= \sqrt{169} \\ c &= 13 \end{aligned}$$

The length of the hypotenuse is 13 inches.

31.
$$\begin{aligned} a^2 + b^2 &= c^2 \\ 10^2 + b^2 &= 12^2 \\ 100 + b^2 &= 144 \\ b^2 &= 44 \\ b &= \sqrt{44} \\ b &\approx 6.633 \end{aligned}$$

The length of the leg is approximately 6.633 centimeters.

33.

c, 4, 3

$$\begin{aligned} a^2 + b^2 &= c^2 \\ 3^2 + 4^2 &= c^2 \\ 9 + 16 &= c^2 \\ 25 &= c^2 \\ c &= \sqrt{25} \\ c &= 5 \end{aligned}$$

35.

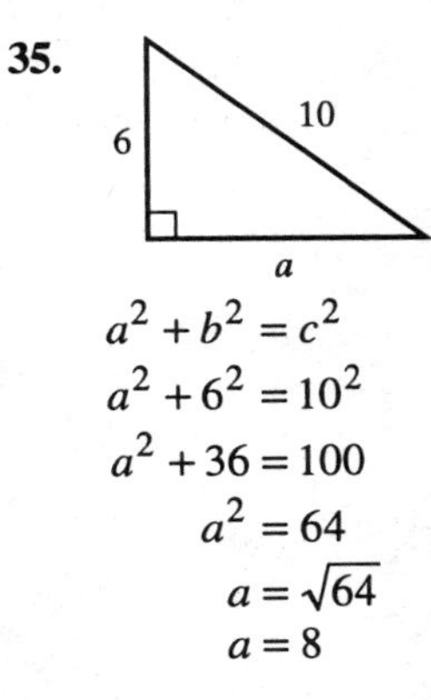

$$\begin{aligned} a^2 + b^2 &= c^2 \\ a^2 + 6^2 &= 10^2 \\ a^2 + 36 &= 100 \\ a^2 &= 64 \\ a &= \sqrt{64} \\ a &= 8 \end{aligned}$$

37.

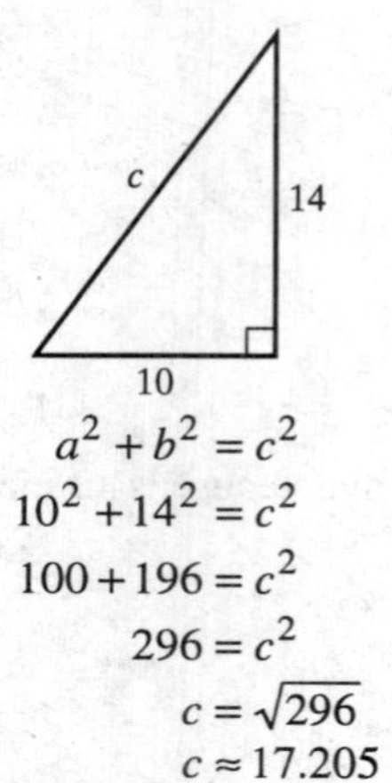

$$a^2 + b^2 = c^2$$
$$10^2 + 14^2 = c^2$$
$$100 + 196 = c^2$$
$$296 = c^2$$
$$c = \sqrt{296}$$
$$c \approx 17.205$$

39.

2 c 16

$$a^2 + b^2 = c^2$$
$$2^2 + 16^2 = c^2$$
$$4 + 256 = c^2$$
$$260 = c^2$$
$$c = \sqrt{260}$$
$$c \approx 16.125$$

41.

13 5 b

$$a^2 + b^2 = c^2$$
$$5^2 + b^2 = 13^2$$
$$25 + b^2 = 169$$
$$b^2 = 144$$
$$b = 12$$

43.

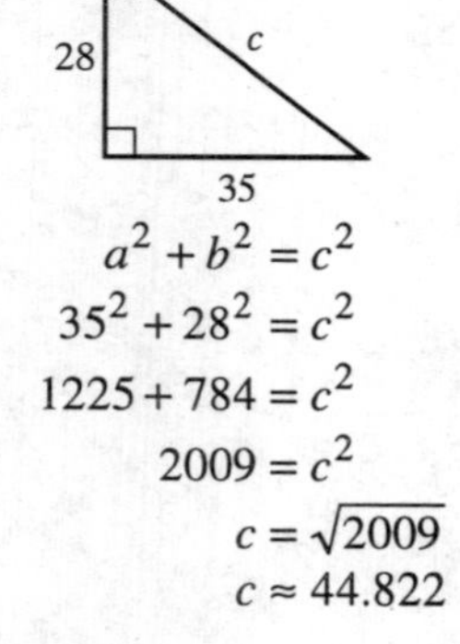

$$a^2 + b^2 = c^2$$
$$35^2 + 28^2 = c^2$$
$$1225 + 784 = c^2$$
$$2009 = c^2$$
$$c = \sqrt{2009}$$
$$c \approx 44.822$$

45.

30 c 30

$$a^2 + b^2 = c^2$$
$$30^2 + 30^2 = c^2$$
$$900 + 900 = c^2$$
$$1800 = c^2$$
$$c = \sqrt{1800}$$
$$c \approx 42.426$$

47.

1 2 a

$$a^2 + b^2 = c^2$$
$$a^2 + 1^2 = 2^2$$
$$a^2 + 1 = 4$$
$$a^2 = 3$$
$$a = \sqrt{3}$$
$$a \approx 1.732$$

49.

$$a^2 + b^2 = c^2$$
$$100^2 + 100^2 = c^2$$
$$10{,}000 + 10{,}000 = c^2$$
$$20{,}000 = c^2$$
$$c = \sqrt{20{,}000}$$
$$c \approx 141.42$$

The length is approximately 141.42 yards.

51.

$$a^2 + b^2 = c^2$$
$$20^2 + b^2 = 32^2$$
$$400 + b^2 = 1024$$
$$b^2 = 624$$
$$b = \sqrt{624}$$
$$b \approx 25.0$$

The height is approximately 25.0 feet.

53.
$$\begin{aligned} a^2 + b^2 &= c^2 \\ 300^2 + 160^2 &= c^2 \\ 90{,}000 + 25{,}600 &= c^2 \\ 115{,}600 &= c^2 \\ c &= \sqrt{115{,}600} \\ c &= 340 \end{aligned}$$

The length is 340 feet.

55. $\frac{10}{12} = \frac{2 \cdot 5}{2 \cdot 6} = \frac{5}{6}$

57. $\frac{24}{60} = \frac{2 \cdot 2 \cdot 2 \cdot 3}{2 \cdot 3 \cdot 2 \cdot 5} = \frac{2}{5}$

59. $\frac{30}{72} = \frac{2 \cdot 3 \cdot 5}{2 \cdot 36} = \frac{2 \cdot 3 \cdot 5}{2 \cdot 6 \cdot 6} = \frac{2 \cdot 3 \cdot 5}{2 \cdot 2 \cdot 3 \cdot 2 \cdot 3} = \frac{5}{12}$

61. $\sqrt{38}$ is between 6 and 7.
$\sqrt{38} \approx 6.164$

63. $\sqrt{101}$ is between 10 and 11.
$\sqrt{101} \approx 10.050$

65. Answers may vary.

Chapter 5 Review

1. 23.45
↑ (4)
tenths

2. 0.000345
↑ (4)
hundred thousandths

3. 23.45 is twenty-three and forty-five hundredths.

4. 0.00345 is three hundred forty-five hundred thousandths.

5. 109.23 is one hundred nine and twenty-three hundredths.

6. 200.000032 is two hundred and thirty-two millionths

7. 2.15

8. 503.102

9. 16,025.0014

10. $0.16 = \frac{16}{100} = \frac{4}{25}$

11. $12.023 = 12\frac{23}{1000}$

12. $1.0045 = 1\frac{45}{10{,}000} = 1\frac{9}{2000}$

13. $0.00231 = \frac{231}{100{,}000}$

14. $25.25 = 25\frac{25}{100} = 25\frac{1}{4}$

15. 0.49　　0.43
↑　　↑
9　>　3
$0.49 > 0.43$

16. $0.973 = 0.9730$

17. 402.00032　　402.000032
↑　　↑
3　>　0
$402.00032 > 402.000032$

18. 0.230505　　0.23505
↑　　↑
0　<　5
$0.230505 < 0.23505$

19. 0.623 to the nearest tenth is 0.6.

20. 0.9384 to the nearest hundredth is 0.94.

21. 42.895 to the nearest hundredth is 42.90.

22. 16.34925 to the nearest thousandth is 16.349.

23. 13,490.5 people rounded to the nearest hundred is 13,500 people.

24. $10.75 = 10\frac{75}{100} = 10\frac{3}{4}$ teaspoons

25.
$$\begin{array}{r} 2.4 \\ +\ 7.1 \\ \hline 9.5 \end{array}$$

26.
$$\begin{array}{r} 3.9 \\ +\ 1.2 \\ \hline 5.1 \end{array}$$

27. $-6.4 + (-0.88)$

Add the absolute values.
$$\begin{array}{r} 6.40 \\ +\ 0.88 \\ \hline 7.28 \end{array}$$
Attach the common sign.

-7.28

28. Subtract the absolute values.
$$\begin{array}{r} 19.02 \\ -6.98 \\ \hline 12.04 \end{array}$$
Since $19.02 > 6.98$, the answer is negative.

The answer is -12.04.

29.
$$\begin{array}{r} 200.490 \\ 16.820 \\ +\ 103.002 \\ \hline 320.312 \end{array}$$

30.
$$\begin{array}{r} 100.45000 \\ 48.29000 \\ +\ 0.00236 \\ \hline 148.74236 \end{array}$$

31.
$$\begin{array}{r} 4.9 \\ -3.2 \\ \hline 1.7 \end{array}$$

32.
$$\begin{array}{r} 5.23 \\ -2.74 \\ \hline 2.49 \end{array}$$

33. Add the absolute values.
$$\begin{array}{r} 892.1 \\ +432.4 \\ \hline 1324.5 \end{array}$$
Attach the common sign.

-1324.5

34. Subtract the absolute values.
$$\begin{array}{r} 10.200 \\ -\ 0.064 \\ \hline 10.136 \end{array}$$
Since $10.2 > 0.064$, the answer is negative.

The answer is -10.136.

35.
$$\begin{array}{r} 100.00 \\ -34.98 \\ \hline 65.02 \end{array}$$

36.
$$\begin{array}{r} 200.00000 \\ -\ 0.00198 \\ \hline 199.99802 \end{array}$$

37.
$$\begin{array}{r} 10,305.77 \\ +\ 617.78 \\ \hline 10,923.55 \end{array}$$
The DJIA opened at 10,923.55.

38. $x - y = 1.2 - 6.9 = -5.7$

39. $2.3x + 6.5 + 1.9x + 6.3 = 2.3x + 1.9x + 6.5 + 6.3$

$= 4.2x + 12.8$

40. $8.6y - 7.61 + 1.29y + 3.44$

$= 8.6y + 1.29y - 7.61 + 3.44$

$= 9.89y - 4.17$

41. $7.2 \times 10 = 72$

42. $9.345 \times 1000 = 9345$

43.
$$\begin{array}{rl} 34.02 & \text{2 decimal places} \\ \times\ 2.3 & \text{1 decimal place} \\ \hline 10206 & \\ 68040 & \\ \hline 78.246 & \text{3 decimal places} \end{array}$$
The answer is negative.

$-34.02 \times 2.3 = -78.246$

44.
$$\begin{array}{rl} -839.02 & \text{2 decimal places} \\ \times\,(-87.3) & \text{1 decimal place} \\ \hline 251706 & \\ 5873140 & \\ +67121600 & \\ \hline 73246446 & \text{3 decimal places} \end{array}$$

45. $C = 2 \cdot \pi \cdot r = 2 \cdot \pi \cdot 7 = 14\pi$ meters
$C \approx 14(3.14) = 43.96$ meters

46.
$$\begin{array}{r} 0.625 \\ \times\ 102 \\ \hline 1250 \\ +\ 62\,500 \\ \hline \end{array}$$
63.750 rounded to the nearest tenth is 63.8 miles.

47. $21 \div 0.3$
$$\frac{21}{0.3} = \frac{210}{3} = 70$$

48. $0.03\overline{)0.0063}$
Move decimal points 2 places.
$$\begin{array}{r} 0.21 \\ 3.\overline{)0.63} \\ -6 \\ \hline 03 \\ -\ 3 \\ \hline 0 \end{array}$$
The answer is negative.
$-0.0063 \div 0.03 = -0.21$

49. $0.005\overline{)24.5}$
Move decimal point 3 places.
$$\begin{array}{r} 4900 \\ 5.\overline{)\ 24500.} \\ -20 \\ \hline 45 \\ -45 \\ \hline 00 \\ -0 \\ \hline 00 \\ -0 \\ \hline 0 \end{array}$$
The answer is negative.
$24.5 \div (-0.005) = -4900$

50. $2.3\overline{)54.98}$
Move decimal points 1 place.
$$\begin{array}{r} 23.9043 \approx 23.904 \\ 23.\overline{)\ 549.8000} \\ -46 \\ \hline 89 \\ -69 \\ \hline 208 \\ -207 \\ \hline 10 \\ -0 \\ \hline 100 \\ -92 \\ \hline 80 \\ -69 \\ \hline 11 \end{array}$$
$54.98 \div 2.3 \approx 23.904$

51.
$$\begin{array}{r} 8.0588 \approx 8.059 \\ 34\overline{)\ 274.0000} \\ -272 \\ \hline 20 \\ -0 \\ \hline 200 \\ -170 \\ \hline 300 \\ -272 \\ \hline 280 \\ -272 \\ \hline 8 \end{array}$$

52.
$$\begin{array}{r} 158.25 \\ 20\overline{)\ 3165.00} \\ -20 \\ \hline 116 \\ -100 \\ \hline 165 \\ -160 \\ \hline 50 \\ -\ 40 \\ \hline 100 \\ -100 \\ \hline 0 \end{array}$$
$-3165 \div (-20) = 158.25$

53. $3.28\overline{)24.00}$

Move decimal points 2 places.

$$\begin{array}{r} 7.31 \approx 7.3 \\ 328.\overline{)\ 2400.00} \\ -2296 \\ \hline 1040 \\ -984 \\ \hline 560 \\ -328 \\ \hline 232 \end{array}$$

There are about 7.3 meters in 24 feet.

54. $69.71\overline{)3136.95}$

Move decimal points 2 places.

$$\begin{array}{r} 45 \\ 6971\overline{)\ 313695} \\ -27884 \\ \hline 34855 \\ -34855 \\ \hline 0 \end{array}$$

The loan will be paid off in 45 months.

55.
$$\begin{array}{r} 2.4 \\ 6.7 \\ +9.1 \\ \hline 18.2 \end{array} \qquad \begin{array}{r} 2 \\ 7 \\ +9 \\ \hline 18 \end{array}$$

56.
$$\begin{array}{r} 15.9 \\ +34.1 \\ \hline 50.0 \end{array} \qquad \begin{array}{r} 16 \\ +34 \\ \hline 50 \end{array}$$

57.
$$\begin{array}{r} 340.03 \\ -240.98 \\ \hline 99.05 \end{array} \qquad \begin{array}{r} 340 \\ -241 \\ \hline 99 \end{array}$$

58.
$$\begin{array}{r} 100.0 \\ -\ 45.9 \\ \hline 54.1 \end{array} \qquad \begin{array}{r} 100 \\ -46 \\ \hline 54 \end{array}$$

59.
$$\begin{array}{r} 6.02 \\ \times\ 5.91 \\ \hline 602 \\ 5\ 4180 \\ 30\ 1000 \\ \hline 35.5782 \end{array} \qquad \begin{array}{r} 6 \\ \times\ 6 \\ \hline 36 \end{array}$$

60.
$$\begin{array}{r} 0.205 \\ \times\ 1.72 \\ \hline 410 \\ 14350 \\ 20500 \\ \hline 0.35260 \end{array} \qquad \begin{array}{r} 0.2 \\ \times\ 1.7 \\ \hline 0.34 \end{array}$$

61. $1.9\overline{)62.13}$

becomes

$$\begin{array}{r} 32.7 \\ 19.\overline{)\ 621.3} \\ -57 \\ \hline 51 \\ -38 \\ \hline 13\ 3 \\ -13\ 3 \\ \hline 0 \end{array} \qquad \begin{array}{r} 31 \\ 2\overline{)\ 62} \\ -6 \\ \hline 02 \\ -2 \\ \hline 0 \end{array}$$

62. $19.8\overline{)601.92}$

becomes

$$\begin{array}{r} 30.4 \\ 198.\overline{)\ 6019.2} \\ -594 \\ \hline 79 \\ -0 \\ \hline 79\ 2 \\ -79\ 2 \\ \hline 0 \end{array} \qquad \begin{array}{r} 30.1 \\ 20\overline{)\ 602.0} \\ -60 \\ \hline 02 \\ -0 \\ \hline 2\ 0 \\ -2\ 0 \\ \hline 0 \end{array}$$

63.
$$\begin{array}{lr} 1994 & 220 \\ 1995 & 230 \\ 1996 & 220 \\ 1997 & 210 \\ 1998 & +\ 190 \\ \hline & 1070 \end{array}$$

The estimated total expenditures was \$1070 million.

64. $77.3 \times 115.9 \approx 77 \times 116 = 8932$

The area is about 8932 square feet.

65.

$$\begin{array}{r} 1.07 \\ 1.89 \\ +\ 0.99 \\ \hline \end{array} \quad \text{Estimate} \quad \begin{array}{r} 1 \\ 2 \\ +\ 1 \\ \hline 4 \end{array}$$

Yes, \$5 is enough.

66. $(-7.6)(1.9) + 2.5 = -14.44 + 2.5 = -11.94$

67. $2.3^2 - 1.4 = 5.29 - 1.4 = 3.89$

68. $\frac{(-3.2)^2}{100} = \frac{10.24}{100} = 0.1024$

69. $(2.6 + 1.4)(4.5 - 3.6) = 4(0.9) = 3.6$

70.

$$\begin{array}{r} 0.8 \\ 5 \overline{)\ 4.0} \\ -4\,0 \\ \hline 0 \end{array}$$

$\frac{4}{5} = 0.8$

71.

$$\begin{array}{r} 0.9230 \approx 0.923 \\ 13 \overline{)\ 12.0000} \\ -11\,7 \\ \hline 30 \\ -26 \\ \hline 40 \\ -39 \\ \hline 10 \\ -0 \\ \hline 10 \end{array}$$

$\frac{12}{13} \approx 0.923$

72.

$$\begin{array}{r} 0.4285 \approx 0.429 \\ 7 \overline{)\ 3.0000} \\ -2\,8 \\ \hline 20 \\ -14 \\ \hline 60 \\ -56 \\ \hline 40 \\ -35 \\ \hline 5 \end{array}$$

$\frac{3}{7} \approx 0.429$

73.

$$\begin{array}{r} 0.2166\ldots = 0.21\overline{6} \\ 60 \overline{)\ 13.0000} \\ -12\,0 \\ \hline 1\,00 \\ -60 \\ \hline 400 \\ -360 \\ \hline 400 \\ -360 \\ \hline 40 \end{array}$$

$\frac{13}{60} = 0.21\overline{6} \approx 0.217$

74.

$$\begin{array}{r} 0.1125 \approx 0.113 \\ 80 \overline{)\ 9.0000} \\ -8\,0 \\ \hline 1\,00 \\ -80 \\ \hline 200 \\ -160 \\ \hline 400 \\ -400 \\ \hline 0 \end{array}$$

$\frac{9}{80} = 0.1125 \approx 0.113$

75.
$$\begin{array}{r} 51.0571 \approx 51.057 \\ 175\overline{)\;8935.0000} \\ \underline{-875} \\ 185 \\ \underline{-175} \\ 100 \\ \underline{-0} \\ 1000 \\ \underline{-875} \\ 1250 \\ \underline{-1225} \\ 250 \\ \underline{-175} \\ 75 \end{array}$$

$$\frac{8935}{175} \approx 51.057$$

76. $0.3920 = 0.392$

77. $\frac{4}{7} \approx 0.5714$

$\frac{5}{8} = 0.625$

$\frac{4}{7} < \frac{5}{8}$

78. $\frac{5}{17} \approx 0.294$

$0.293 < \frac{5}{17}$

79. $\frac{6}{11} \approx 0.545$

$\frac{6}{11} < 0.55$

80. 0.832, 0.837, 0.839

81. $\frac{3}{7} \approx 0.4286$

$0.42, \frac{3}{7}, 0.43$

82. $\frac{18}{11} \approx 1.636, \frac{19}{12} \approx 1.583$

$\frac{19}{12}, 1.63, \frac{18}{11}$

83. $\frac{6}{7} \approx 0.8571, \frac{8}{9} = 0.\bar{8} \approx 0.8889, \frac{3}{4} = 0.75$

$\frac{3}{4}, \frac{6}{7}, \frac{8}{9}$

84.
$$\begin{aligned} \text{Area} &= \frac{1}{2} \cdot \text{base} \cdot \text{height} \\ &= 0.5 \cdot 4.6 \cdot 3 \\ &= 6.9 \end{aligned}$$

The area is 6.9 square feet.

85.
$$\begin{aligned} \text{Area} &= \frac{1}{2} \cdot \text{base} \cdot \text{height} \\ &= 0.5 \cdot 5.2 \cdot 2.1 \\ &= 5.46 \end{aligned}$$

The area is 5.46 square inches.

86.
$$\begin{aligned} x + 3.9 &= 4.2 \\ x + 3.9 - 3.9 &= 4.2 - 3.9 \\ x &= 0.3 \end{aligned}$$

87.
$$\begin{aligned} 70 &= y + 22.81 \\ 70 - 22.81 &= y + 22.81 - 22.81 \\ 47.19 &= y \end{aligned}$$

88.
$$\begin{aligned} 2x &= 17.2 \\ \frac{2x}{2} &= \frac{17.2}{2} \\ x &= 8.6 \end{aligned}$$

89.
$$\begin{aligned} 1.1y &= 88 \\ 10(1.1y) &= 10(88) \\ 11y &= 880 \\ \frac{11y}{11} &= \frac{880}{11} \\ y &= 80 \end{aligned}$$

90.
$$\begin{aligned} \frac{x}{4} &= 0.12 \\ 4\left(\frac{x}{4}\right) &= 4(0.12) \\ x &= 0.48 \end{aligned}$$

91.
$$\begin{aligned} 6.8 &= \frac{y}{5} \\ (6.8)5 &= \frac{y}{5} \cdot 5 \\ 34.0 &= y \end{aligned}$$

92.
$$\begin{aligned} x + 0.78 &= 1.2 \\ x + 0.78 - 0.78 &= 1.2 - 0.78 \\ x &= 0.42 \end{aligned}$$

93.
$$\begin{aligned} 0.56 &= 2x \\ \frac{0.56}{2} &= \frac{2x}{2} \\ 0.28 &= x \end{aligned}$$

94.
$$\begin{aligned} 1.3x - 9.4 &= 0.4x + 8.6 \\ 1.3x - 0.4x - 9.4 &= 0.4x - 0.4x + 8.6 \\ 0.9x - 9.4 + 9.4 &= 8.6 + 9.4 \\ 0.9x &= 18 \\ \frac{0.9x}{0.9} &= \frac{18}{0.9} \\ x &= 20 \end{aligned}$$

95.
$$\begin{aligned} 3(x - 1.1) &= 5x - 5.3 \\ 3x - 3.3 &= 5x - 5.3 \\ 3x - 3.3 + 3.3 &= 5x - 5.3 + 3.3 \\ 3x &= 5x - 2 \\ 3x - 5x &= 5x - 5x - 2 \\ -2x &= -2 \\ x &= 1 \end{aligned}$$

96. $\sqrt{64} = 8$ since $8^2 = 64$.

97. $\sqrt{144} = 12$ since $12^2 = 144$.

98. $\sqrt{36} = 6$ since $6^2 = 36$.

99. $\sqrt{1} = 1$ since $1^2 = 1$.

100. $\sqrt{\frac{4}{25}} = \frac{2}{5}$ since $\left(\frac{2}{5}\right)^2 = \frac{4}{25}$.

101. $\sqrt{\frac{1}{100}} = \frac{1}{10}$ since $\left(\frac{1}{10}\right)^2 = \frac{1}{100}$.

102.
$$\begin{aligned} a^2 + b^2 &= c^2 \\ 12^2 + 5^2 &= c^2 \\ 144 + 25 &= c^2 \\ 169 &= c^2 \\ \sqrt{169} &= c \\ 13 &= c \end{aligned}$$

103.
$$\begin{aligned} a^2 + b^2 &= c^2 \\ 20^2 + 21^2 &= c^2 \\ 400 + 441 &= c^2 \\ 841 &= c^2 \\ \sqrt{841} &= c \\ 29 &= c \end{aligned}$$

104.
$$\begin{aligned} a^2 + b^2 &= c^2 \\ 9^2 + b^2 &= 14^2 \\ 81 + b^2 &= 196 \\ b^2 &= 115 \\ b &= \sqrt{115} \\ b &\approx 10.7 \end{aligned}$$

105.
$$\begin{aligned} a^2 + b^2 &= c^2 \\ 124^2 + b^2 &= 155^2 \\ 15,376 + b^2 &= 24,025 \\ b^2 &= 8649 \\ b &= \sqrt{8649} \\ b &= 93 \end{aligned}$$

106.
$$\begin{aligned} a^2 + b^2 &= c^2 \\ 66^2 + 56^2 &= c^2 \\ 4356 + 3136 &= c^2 \\ 7492 &= c^2 \\ \sqrt{7492} &= c \\ 86.6 &\approx c \end{aligned}$$

107.
$$\begin{aligned} a^2 + b^2 &= c^2 \\ 20^2 + 20^2 &= c^2 \\ 400 + 400 &= c^2 \\ 800 &= c^2 \\ \sqrt{800} &= c \\ 28.28 &\approx c \end{aligned}$$
The diagonal is approximately 28.28 centimeters.

108.
$$\begin{aligned} a^2 + b^2 &= c^2 \\ 90^2 + b^2 &= 126^2 \\ 8100 + b^2 &= 15,876 \\ b^2 &= 7776 \\ b &= \sqrt{7776} \\ b &\approx 88.2 \end{aligned}$$
The height of the building is approximately 88.2 feet.

Chapter 5 Test

1. 45.092 is forty-five and ninety-two thousandths.

2. Three thousand and fifty-nine thousandths is 3000.059.

3.
$$\begin{array}{r} 2.893 \\ 4.210 \\ +\ 10.492 \\ \hline 17.595 \end{array}$$

4.
$$\begin{array}{r} -47.92 \\ -\ 3.28 \\ \hline -51.20 \end{array}$$

5. $9.83 - 30.25 = 9.83 + (-30.25)$
Subtract the absolute values.
$$\begin{array}{r} 30.25 \\ -9.83 \\ \hline 20.42 \end{array}$$
Since $30.25 > 9.83$, the answer is negative.
-20.42

6.
$$\begin{array}{r} 10.2 \\ \times\ 4.01 \\ \hline 102 \\ 0 \\ 40\ 800 \\ \hline 40.902 \end{array}$$

7. $0.23\overline{)0.00843}$
becomes
$$\begin{array}{r} 0.0366\ldots \approx 0.037 \\ 23\overline{)0.8430} \\ \underline{-69} \\ 153 \\ \underline{-138} \\ 150 \\ \underline{-138} \\ 15 \end{array}$$
$(-0.00843) \div (-0.23) \approx 0.037$

8. 34.8923 rounded to the nearest tenth is 34.9.

9. 0.8623 rounded to the nearest thousandth is 0.862.

10. $25.0909 < 25.9090$

11. $\frac{4}{9} = 0.444\ldots$
$\frac{4}{9} < 0.445$

12. $0.345 = \frac{345}{1000} = \frac{69}{200}$

13. $24.73 = 24\frac{73}{100}$

14. $\frac{13}{26} = \frac{1}{2} = 0.5$

15. $\frac{16}{17} \approx 0.941$

$$\begin{array}{r} 0.9411 \\ 17\overline{)\,16.0000} \\ \underline{-153} \\ 70 \\ \underline{-68} \\ 20 \\ \underline{-17} \\ 30 \\ \underline{-17} \\ 13 \end{array}$$

16. $(-0.6)^2 + 1.57 = 0.36 + 1.57 = 1.93$

17.
$$\begin{aligned} \frac{0.23+1.63}{-0.3} &= \frac{1.86}{-0.3} \\ &= -\frac{186}{30} \\ &= -\frac{62 \cdot 3}{10 \cdot 3} \\ &= -\frac{62}{10} \\ &= -6.2 \end{aligned}$$

18. $2.4x - 3.6 - 1.9x - 9.8$
$= (2.4 - 1.9)x - 3.6 - 9.8$
$= 0.5x - 13.4$

19. $\sqrt{49} = 7$

20. $\sqrt{157} = 12.530$

21. $\sqrt{\frac{64}{100}} = \frac{8}{10} = \frac{4}{5}$

22.
$$\begin{aligned} 0.2x + 1.3 &= 0.7 \\ 10(0.2x + 1.3) &= 10(0.7) \\ 2x + 13 &= 7 \\ 2x + 13 - 13 &= 7 - 13 \\ 2x &= -6 \\ x &= -3 \end{aligned}$$

23.
$$\begin{aligned} 2(x + 5.7) &= 6x - 3.4 \\ 2x + 11.4 &= 6x - 3.4 \\ 2x - 2x + 11.4 &= 6x - 2x - 3.4 \\ 11.4 &= 4x - 3.4 \\ 11.4 + 3.4 &= 4x - 3.4 + 3.4 \\ 14.8 &= 4x \\ 3.7 &= x \end{aligned}$$

24.
$$\begin{aligned} a^2 + b^2 &= c^2 \\ 4^2 + 4^2 &= c^2 \\ 16 + 16 &= c^2 \\ 32 &= c^2 \\ \sqrt{32} &= c \\ 5.66 &\approx c \end{aligned}$$
The length of the hypotenuse is approximately 5.66 centimeters.

25.
$$\begin{aligned} \text{Area} &= \frac{1}{2} \cdot \text{base} \cdot \text{height} \\ &= 0.5 \cdot 4.2 \cdot 1.1 \\ &= 2.31 \end{aligned}$$
The area is 2.31 square miles.

26. $A = 123.8(80) = 9904$
9904 square feet of lawn
$9904 \times 0.02 = 198.08$
She needs 198.08 ounces of insecticide.

27.
$$\begin{aligned} \text{Circumference} &= 2 \cdot \pi \cdot \text{radius} \\ &= 2 \cdot \pi \cdot 9 \\ &= 18\pi \\ &\approx 18(3.14) \\ &= 56.52 \end{aligned}$$
The circumference is 18π miles $\approx$ 56.52 miles.

28. $1.4\overline{)120}$

becomes

$$\begin{array}{r} 85.7 \approx 86 \\ 14\overline{)1200.0} \\ \underline{-112} \\ 80 \\ \underline{-70} \\ 10\;0 \\ \underline{-9\;8} \\ 2 \end{array}$$

It will take 86 high-density disks.

29.

$$\begin{array}{r} 14.2 \\ 16.1 \\ \underline{+\ 23.7} \end{array} \quad \text{Estimate} \quad \begin{array}{r} 14 \\ 16 \\ \underline{+\ 24} \\ 54 \end{array}$$

The total distance is about 54 miles.

30. Frozen broccoli consumption went from 2.4 pounds to 2.1 pounds, so it decreased.
$7.7 - 5.8 = 1.9$
The average American ate 1.9 pounds more in 1998 than in 1992.

Chapter 6

Pretest

1. The ratio of 15 to 19 is $\frac{15}{19}$.

2. The ratio of $3\frac{4}{5}$ to $9\frac{1}{8}$ is $\frac{3\frac{4}{5}}{9\frac{1}{8}}$.

3. $\frac{5.1}{7.9} = \frac{5.1 \cdot 10}{7.9 \cdot 10} = \frac{51}{79}$

4. $\frac{\$30}{\$20} = \frac{3 \cdot \$10}{2 \cdot \$10} = \frac{3}{2}$

5. $\frac{\$6}{3 \text{ pounds}} = \frac{3}{3 \cdot 1 \text{ pound}} = \frac{\$2}{1 \text{ pound}}$

6. $\frac{292.5 \text{ miles}}{15 \text{ gallons}} = \frac{19.5 \text{ miles}}{1 \text{ gallon}}$
   $= 19.5$ miles per gallon

7. $\frac{\$3.27}{3 \text{ boxes}} = \frac{\$1.09}{1 \text{ box}} = \$1.09$ per box

8. $\frac{\text{credits}}{\text{classes}} = \frac{\text{credits}}{\text{classes}}$
   $\frac{18}{6} = \frac{15}{5}$

9. $\frac{\text{days}}{\text{weeks}} = \frac{\text{days}}{\text{weeks}}$
   $\frac{49}{7} = \frac{35}{5}$

10. $\frac{3}{5} = \frac{21}{34}?$
    $5 \cdot 21 = 3 \cdot 34?$
    $105 \neq 102$
    Since the cross products are not equal, the proportion is not true.

11. $\frac{72}{9} = \frac{216}{27}?$
    $9 \cdot 216 = 72 \cdot 27?$
    $1944 = 1944$
    Since the cross products are equal, the proportion is true.

12. $\frac{22}{x} = \frac{66}{12}$
    $66x = 22 \cdot 12$
    $66x = 264$
    $\frac{66x}{66} = \frac{264}{66}$
    $x = 4$

13. $\frac{x}{8} = \frac{0.6}{2.4}$
    $8 \cdot 0.6 = 2.4x$
    $4.8 = 2.4x$
    $\frac{4.8}{2.4} = \frac{2.4x}{2.4}$
    $2 = x$

14. $\frac{\frac{1}{2}}{8} = \frac{3}{y}$
    $8 \cdot 3 = \frac{1}{2}y$
    $24 = \frac{1}{2}y$
    $2 \cdot 24 = 2 \cdot \frac{1}{2}y$
    $48 = y$

15. $\frac{1\frac{1}{3}}{\frac{6}{5}} = \frac{n}{9}$

$$\frac{6}{5}n = 9 \cdot 1\frac{1}{3}$$
$$\frac{6}{5}n = 9 \cdot \frac{4}{3}$$
$$\frac{6}{5}n = \frac{3 \cdot 3 \cdot 4}{3}$$
$$\frac{6}{5}n = 12$$
$$\frac{5}{6} \cdot \frac{6}{5}n = \frac{5}{6} \cdot 12$$
$$n = \frac{5 \cdot 6 \cdot 2}{6}$$
$$n = 10$$

16. $\frac{\text{miles}}{\text{inches}} = \frac{\text{miles}}{\text{inches}}$

$$\frac{20 \text{ miles}}{1 \text{ inch}} = \frac{x \text{ miles}}{5\frac{3}{8} \text{ inches}}$$
$$\frac{20}{1} = \frac{x}{5\frac{3}{8}}$$
$$1 \cdot x = 20 \cdot 5\frac{3}{8}$$
$$x = 20 \cdot \frac{43}{8}$$
$$x = \frac{4 \cdot 5 \cdot 43}{2 \cdot 4}$$
$$x = \frac{215}{2} = 107\frac{1}{2}$$

$5\frac{3}{8}$ inches represents $107\frac{1}{2}$ miles.

17. Since the triangles are similar, we can compare any of the corresponding sides to find the ratio.

$$\frac{5}{10} = \frac{1}{2}$$

18. $\frac{6}{18} = \frac{2}{n}$

$$18 \cdot 2 = 6n$$
$$36 = 6n$$
$$\frac{36}{6} = \frac{6n}{6}$$
$$6 = n$$

19. $\frac{3.5}{24.5} = \frac{n}{38.5}$

$$24.5n = 3.5 \cdot 38.5$$
$$24.5n = 134.75$$
$$\frac{24.5n}{24.5} = \frac{134.75}{24.5}$$
$$n = 5.5$$

20. $\frac{\text{tree height}}{\text{shadow length}} = \frac{\text{tree height}}{\text{shadow length}}$

$$\frac{40 \text{ feet}}{22 \text{ feet}} = \frac{36 \text{ feet}}{x \text{ feet}}$$
$$\frac{40}{22} = \frac{36}{x}$$
$$22 \cdot 36 = 40x$$
$$792 = 40x$$
$$\frac{792}{40} = \frac{40x}{40}$$
$$19.8 = x$$

A 36-foot tree casts a 19.8-foot shadow.

Section 6.1

Practice Problems

1. The ratio is $\frac{20}{23}$.

2. The ratio of 10.3 to 15.1 is $\frac{10.3}{15.1}$.

3. The ratio of $3\frac{1}{3}$ to $12\frac{1}{5}$ is $\frac{3\frac{1}{3}}{12\frac{1}{5}}$.

4. $\frac{\$8}{\$6} = \frac{8}{6} = \frac{4 \cdot 2}{3 \cdot 2} = \frac{4}{3}$

5. $\frac{1.71}{4.56} = \frac{1.71 \cdot 100}{4.56 \cdot 100} = \frac{171}{456} = \frac{3 \cdot 57}{8 \cdot 57} = \frac{3}{8}$

6. $\frac{\text{work miles}}{\text{total miles}} = \frac{4800}{15,000} = \frac{8 \cdot 600}{25 \cdot 600} = \frac{8}{25}$

7. a. $\frac{\text{shortest side}}{\text{longest side}} = \frac{6\text{ m}}{10\text{ m}} = \frac{3 \cdot 2}{5 \cdot 2} = \frac{3}{5}$

   b. $\frac{\text{longest side}}{\text{perimeter}} = \frac{10\text{ m}}{6\text{ m} + 8\text{ m} + 10\text{ m}} = \frac{10\text{m}}{24\text{ m}} = \frac{5 \cdot 2}{12 \cdot 2} = \frac{5}{12}$

Exercise Set 6.1

1. The ratio of 11 to 14 is $\frac{11}{14}$.

3. The ratio of 23 to 10 is $\frac{23}{10}$.

5. The ratio of 151 to 201 is $\frac{151}{201}$.

7. The ratio of 2.8 to 7.6 is $\frac{2.8}{7.6}$.

9. The ratio of 5 to $7\frac{1}{2}$ is $\frac{5}{7\frac{1}{2}}$.

11. The ratio of $3\frac{3}{4}$ to $1\frac{2}{3}$ is $\frac{3\frac{3}{4}}{1\frac{2}{3}}$.

13. $\frac{16}{24} = \frac{2 \cdot 8}{3 \cdot 8} = \frac{2}{3}$

15. $\frac{7.7}{10} = \frac{7.7 \cdot 10}{10 \cdot 10} = \frac{77}{100}$

17. $\frac{4.63}{8.21} = \frac{4.63 \cdot 100}{8.21 \cdot 100} = \frac{463}{821}$

19. $\frac{9\text{ inches}}{12\text{ inches}} = \frac{3 \cdot 3}{4 \cdot 3} = \frac{3}{4}$

21. $\frac{10\text{ hours}}{24\text{ hours}} = \frac{5 \cdot 2}{12 \cdot 2} = \frac{5}{12}$

23. $\frac{\$32}{\$100} = \frac{8 \cdot 4}{25 \cdot 4} = \frac{8}{25}$

25. $\frac{24\text{ days}}{14\text{ days}} = \frac{12 \cdot 2}{7 \cdot 2} = \frac{12}{7}$

27. $\frac{32,000\text{ bytes}}{46,000\text{ bytes}} = \frac{16 \cdot 2000}{23 \cdot 2000} = \frac{16}{23}$

29. $\frac{\text{vacation / other miles}}{\text{total miles}} = \frac{900\text{ miles}}{15,000\text{ miles}} = \frac{300 \cdot 3}{300 \cdot 50} = \frac{3}{50}$

31. $\frac{\text{length}}{\text{width}} = \frac{30\text{ feet}}{18\text{ feet}} = \frac{5 \cdot 6}{3 \cdot 6} = \frac{5}{3}$

33. perimeter = 8 + 15 + 17 = 40 feet
    $\frac{\text{longest side}}{\text{perimeter}} = \frac{17\text{ feet}}{40\text{ feet}} = \frac{17}{40}$

35. $\frac{\text{calories from fat}}{\text{total calories}} = \frac{200}{450} = \frac{4 \cdot 50}{9 \cdot 50} = \frac{4}{9}$

37. $\frac{\text{women}}{\text{men}} = \frac{125}{100} = \frac{25 \cdot 5}{25 \cdot 4} = \frac{5}{4}$

39. There are 6000 – 4500 = 1500 married students.
    $\frac{\text{single students}}{\text{married students}} = \frac{4500}{1500} = \frac{3 \cdot 1500}{1 \cdot 1500} = \frac{3}{1}$

41. $\frac{\text{red blood cells}}{\text{platelet cells}} = \frac{600}{40} = \frac{40 \cdot 15}{40 \cdot 1} = \frac{15}{1}$

43. $\frac{\$6.15}{\$0.25} = \frac{123 \cdot 0.05}{5 \cdot 0.05} = \frac{123}{5}$

45. $\frac{25\text{ medals}}{205\text{ medals}} = \frac{5 \cdot 5}{41 \cdot 5} = \frac{5}{41}$

47.
$$\begin{array}{r} 2.3 \\ 9\overline{)\,20.7} \\ \underline{-18} \\ 27 \\ \underline{-27} \\ 0 \end{array}$$

49. $3.7\overline{)0.555}$ Move the decimal points 1 place.
$$\begin{array}{r} 0.15 \\ 37\overline{)5.55} \\ \underline{-37} \\ 185 \\ \underline{-185} \\ 0 \end{array}$$

51. States without bicycle helmet laws
$= 50 - 15$
$= 35$
$$\frac{15 \text{ states}}{35 \text{ states}} = \frac{3 \cdot 5}{7 \cdot 5} = \frac{3}{7}$$

53. Answers may vary.

55. $\frac{\text{bruised}}{\text{total}} = \frac{3}{3+30} = \frac{3}{33} = \frac{1 \cdot 3}{11 \cdot 3} = \frac{1}{11}$
No, the shipment should not be refused.

57. Answers may vary.

Section 6.2

Practice Problems

1. $\frac{12 \text{ commercials}}{45 \text{ minutes}} = \frac{3 \cdot 4 \text{ commercials}}{3 \cdot 15 \text{ minutes}}$
$= \frac{4 \text{ commercials}}{15 \text{ minutes}}$

2. $\frac{1680 \text{ dollars}}{8 \text{ weeks}} = \frac{8 \cdot 210 \text{ dollars}}{8 \cdot 1 \text{ week}}$
$= \frac{210 \text{ dollars}}{1 \text{ week}}$

3. $\frac{236 \text{ miles}}{12 \text{ gallons}} = \frac{4 \cdot 59 \text{ miles}}{4 \cdot 3 \text{ gallons}}$
$= \frac{59 \text{ miles}}{3 \text{ gallons}}$

4. $\frac{3600 \text{ feet}}{12 \text{ seconds}}$
$= \frac{12 \cdot 300 \text{ feet}}{12 \cdot 1 \text{ second}}$
$= \frac{300 \text{ feet}}{1 \text{ second}}$ or 300 feet per second

5. $\frac{52 \text{ bushels}}{8 \text{ trees}} = \frac{6.5 \text{ bushels}}{1 \text{ tree}}$
or 65 bushels/tree
$$\begin{array}{r} 6.5 \\ 8\overline{)\,52.0} \\ \underline{-48} \\ 4\,0 \\ \underline{-\,4\,0} \\ 0 \end{array}$$

6. $\frac{\$170}{5 \text{ days}} = \frac{5 \cdot \$34}{5 \cdot 1 \text{ day}} = \frac{\$34}{1 \text{ day}}$
The rental agency charges \$34 per day.

7. 11-ounce size:
unit price $= \frac{\$2.32}{11 \text{ ounces}}$
$\approx \frac{\$0.21}{1 \text{ ounce}}$ or \$0.21 per ounce
16-ounce size:
unit price $= \frac{\$3.59}{16 \text{ ounces}}$
$\approx \frac{\$0.22}{1 \text{ ounce}}$ or \$0.22 per ounce
Thus, the 11-ounce bag is the better buy.

Exercise Set 6.2

1. $\frac{5 \text{ shrubs}}{15 \text{ feet}} = \frac{1 \text{ shrub}}{3 \text{ feet}}$

3. $\frac{15 \text{ returns}}{100 \text{ sales}} = \frac{3 \text{ returns}}{20 \text{ sales}}$

5. $\frac{8 \text{ phone lines}}{36 \text{ employees}} = \frac{2 \text{ phone lines}}{9 \text{ employees}}$

7. $\frac{18 \text{ gallons of pesticide}}{4 \text{ acres of crops}} = \frac{9 \text{ gallons}}{2 \text{ acres}}$

9. $\frac{6 \text{ flight attendants}}{200 \text{ passengers}} = \frac{3 \text{ flight attendants}}{100 \text{ passengers}}$

11. $\frac{355 \text{ calories}}{10 \text{ fluid ounces}} = \frac{71 \text{ calories}}{2 \text{ fluid ounces}}$

13. $\frac{375 \text{ riders}}{5 \text{ subway cars}} = \frac{75 \text{ riders}}{1 \text{ subway car}}$
$= 75$ riders per car

15. $\frac{330 \text{ calories}}{3 \text{ ounces}} = \frac{110 \text{ calories}}{1 \text{ ounce}}$
$= 110$ calories per ounce

17. $\frac{144 \text{ diapers}}{24 \text{ babies}} = \frac{6 \text{ diapers}}{1 \text{ baby}}$
$= 6$ diapers per baby

19. $\frac{\$1,000,000}{20 \text{ years}} = \frac{\$50,000}{1 \text{ year}} = \$50,000$ per year

21. $\frac{600 \text{ kilometers}}{90 \text{ minutes}}$
$= \frac{20 \text{ kilometers}}{3 \text{ minutes}}$
$= 6\frac{2}{3}$ kilometers per minute
≈ 6.67 kilometers per minute

23. $\frac{2,450,000 \text{ voters}}{2 \text{ senators}}$
$= \frac{1,225,000 \text{ voters}}{1 \text{ senator}}$
$= 1,225,000$ voters per senator

25. $\frac{12,000 \text{ good}}{40 \text{ defective}} = \frac{300 \text{ good}}{1 \text{ defective}}$
$= 300$ good per defective

27. $\frac{12 \text{ million tons dust and dirt}}{25 \text{ million acres}}$
$= \frac{0.48 \text{ ton dust and dirt}}{1 \text{ acre}}$
$= 0.48$ ton dust and dirt per acre

29. $\frac{\$20,000,000}{400 \text{ species}} = \frac{\$50,000}{1 \text{ species}}$
$= \$50,000$ per species

31. $\frac{\$123,219,000}{24 \text{ libraries}} = \frac{\$5,134,125}{1 \text{ library}}$
$= \$5,134,125$ per library

33. **a.** Greer:
$\frac{250 \text{ boards}}{8 \text{ hours}}$
$= 31.25$ computer boards per hour

b. Lamont:
$\frac{400 \text{ boards}}{12 \text{ hours}} \approx 33.3$ boards per hour

c. Lamont can assemble about 33.3 boards in the same time that Greer assembles 31.25 boards. Thus, Lamont assembles them faster.

35. $\frac{\$57.50}{5 \text{ compact disks}} = \frac{\$11.50}{1 \text{ compact disk}}$
$= \$11.50$ per compact disk

37. $\frac{\$1.19}{7 \text{ bananas}} = \frac{\$0.17}{1 \text{ banana}} = \$0.17$ per banana

39. 8-ounce size:
$\frac{\$1.19}{8 \text{ ounces}} \approx \$0.149$ per ounce
12-ounce size:
$\frac{\$1.59}{12 \text{ ounces}} \approx \$0.133$ per ounce
The 12-ounce size costs less per ounce, thus it is the better buy.

41. 16-ounce size:
$\frac{\$1.69}{16 \text{ ounces}} \approx \$0.106$ per ounce
6-ounce size:
$\frac{\$0.69}{6 \text{ ounces}} = \$0.115$ per ounce
The 16-ounce size costs less per ounce, thus it is the better buy.

43. 12-ounce size:
$\frac{\$2.29}{12 \text{ ounces}} \approx \$0.191$ per ounce
8-ounce size:
$\frac{\$1.49}{8 \text{ ounces}} \approx \$0.186$ per ounce
The 8-ounce size costs less per ounce, thus it is the better buy.

45. 100-pack size:
$\frac{\$0.59}{100 \text{ napkins}} \approx \$0.006$ per napkin
180-pack size:
$\frac{\$0.93}{180 \text{ napkins}} \approx \$0.005$ per napkin
The package of 180 napkins costs less per napkin thus, it is the better buy.

47.
$$\begin{array}{r} 1.7 \\ \times\ 6 \\ \hline 10.2 \end{array}$$

49.
$$\begin{array}{r} 3.7 \\ \times\ 1.2 \\ \hline 74 \\ 370 \\ \hline 4.44 \end{array}$$

51. $2.3\overline{)4.37}$ Move the decimal points 1 place
$$\begin{array}{r} 1.9 \\ 23.\overline{)43.7} \\ \underline{-23} \\ 207 \\ \underline{-207} \\ 0 \end{array}$$

53.

Miles Driven	Miles per Gallon (round to the nearest tenth)
79,543 − 79,286 = 257	$\frac{257}{13.4} = 19.2$
79,895 − 79,543 = 352	$\frac{352}{15.8} = 22.3$
80,242 − 79,895 = 347	$\frac{347}{16.1} = 21.6$

55. $\frac{11,674 \text{ steps}}{7759 \text{ feet}}$
$$\begin{array}{r} 1.50 \\ 7759\overline{)11,674.00} \\ \underline{-7759} \\ 39150 \\ \underline{-38795} \\ 3550 \\ \underline{-\quad 0} \\ 3550 \end{array}$$
Rounded to the nearest tenth, the unit rate is 1.5 steps per foot.

57. $\frac{477,121 \text{ students}}{20,039 \text{ teachers}}$
$$\begin{array}{r} 23.8 \\ 20,039\overline{)477,121.0} \\ \underline{-40078} \\ 76341 \\ \underline{-60117} \\ 162240 \\ \underline{-160312} \\ 1928 \end{array}$$
Rounded to the nearest whole student, the unit rate is 24 students per teacher.

59. Answers may vary.

Section 6.3

Practice Problems

1. a. $\frac{\text{right}}{\text{wrong}} = \frac{\text{right}}{\text{wrong}}$
$\frac{24}{6} = \frac{4}{1}$

b. $\frac{\text{Cubs fans}}{\text{Mets fans}} = \frac{\text{Cubs fans}}{\text{Met fans}}$

$\frac{32}{18} = \frac{16}{9}$

2. $\frac{3}{6} = \frac{4}{8}?$
$6 \cdot 4 = 3 \cdot 8?$
$24 = 24$
Since the cross products are equal, the proportion is true.

3. $\frac{3.6}{6} = \frac{5.4}{8}?$
$6 \cdot 5.4 = 3.6 \cdot 8?$
$32.4 \neq 28.8$
Since the cross products are not equal, the proportion is not true.

4. $\frac{4\frac{1}{5}}{2\frac{1}{3}} = \frac{3\frac{3}{10}}{1\frac{5}{6}}?$
$2\frac{1}{3} \cdot 3\frac{3}{10} = 4\frac{1}{5} \cdot 1\frac{5}{6}?$
$\frac{7}{3} \cdot \frac{33}{10} = \frac{21}{5} \cdot \frac{11}{6}?$
$\frac{77}{10} = \frac{77}{10}$
Since the cross products are equal, the proportion is true.

5. $\frac{2}{7} = \frac{x}{35}$
$7 \cdot x = 2 \cdot 35$
$7x = 70$
$\frac{7x}{7} = \frac{70}{7}$
$x = 10$

6. $\frac{2}{15} = \frac{y}{60}$
$15y = 2 \cdot 60$
$15y = 120$
$\frac{15y}{15} = \frac{120}{15}$
$y = 8$

Check: $\frac{2}{15} = \frac{y}{60}$
$\frac{2}{15} = \frac{8}{60}$
$\frac{2}{15} = \frac{2}{15}$
Since $\frac{2}{15} = \frac{2}{15}$ is a true statement, 8 is the solution.

7. $\frac{8}{z} = \frac{1}{9}$
$1z = 8 \cdot 9$
$z = 72$

8. $\frac{y}{6} = \frac{0.7}{1.2}$
$6(0.7) = 1.2y$
$4.2 = 1.2y$
$\frac{4.2}{1.2} = \frac{1.2y}{1.2}$
$3.5 = y$

9. $\frac{15}{z} = \frac{8}{10}$
$8z = 150$
$\frac{8z}{8} = \frac{150}{8}$
$z = \frac{75}{4}$ or 18.75

10. $\frac{3.4}{1.8} = \frac{y}{3}$
$1.8y = 10.2$
$\frac{1.8y}{1.8} = \frac{10.2}{1.8}$
$y \approx 5.7$

Mental Math

1. Is $1 \cdot 6 = 2 \cdot 3$? Yes
True

2. Is $1 \cdot 15 = 3 \cdot 5$? Yes
True

3. Is $2 \cdot 3 = 1 \cdot 5$? No
False

4. Is 11 · 1 = 2 · 5? No
False

5. Is 3 · 4 = 2 · 6? Yes
True

6. Is 4 · 6 = 3 · 8? Yes
True

Exercise Set 6.3

1. $\frac{10 \text{ diamonds}}{6 \text{ opals}} = \frac{5 \text{ diamonds}}{3 \text{ opals}}$

3. $\frac{3 \text{ printers}}{12 \text{ computers}} = \frac{1 \text{ printer}}{4 \text{ computers}}$

5. $\frac{6 \text{ eagles}}{58 \text{ sparrows}} = \frac{3 \text{ eagles}}{29 \text{ sparrows}}$

7. $\frac{2\frac{1}{4} \text{ cups flour}}{24 \text{ cookies}} = \frac{6\frac{3}{4} \text{ cups flour}}{72 \text{ cookies}}$

9. $\frac{22 \text{ vanilla wafers}}{1 \text{ cup cookie crumbs}}$
$= \frac{55 \text{ vanilla wafers}}{2.5 \text{ cups cookie crumbs}}$

11. $\frac{15}{9} = \frac{5}{3}?$
$9 \cdot 5 = 15 \cdot 3?$
$45 = 45$
Since the cross products are equal, the proportion is true.

13. $\frac{8}{6} = \frac{9}{7}?$
$6 \cdot 9 = 8 \cdot 7?$
$54 \neq 56$
Since the cross products are not equal, the proportion is false.

15. $\frac{9}{36} = \frac{2}{8}?$
$362 = 9 \cdot 8?$
$72 = 72$
Since the cross products are equal, the proportion is true.

17. $\frac{5}{8} = \frac{625}{1000}?$
$8 \cdot 625 = 5 \cdot 1000?$
$5000 = 5000$
Since the cross products are equal, the proportion is true.

19. $\frac{0.8}{0.3} = \frac{0.2}{0.6}?$
$0.3 \cdot 0.2 = 0.8 \cdot 0.6?$
$0.06 \neq 0.48$
Since the cross products are not equal, the proportional is false.

21. $\frac{4.2}{8.4} = \frac{5}{10}?$
$8.4 \cdot 5 = 4.2 \cdot 10?$
$42 = 42$
Since the cross products are equal, the proportion is true.

23. $\frac{\frac{3}{4}}{\frac{4}{3}} = \frac{\frac{1}{2}}{\frac{8}{9}}?$
$\frac{4}{3} \cdot \frac{1}{2} = \frac{3}{4} \cdot \frac{8}{9}?$
$\frac{2 \cdot 2 \cdot 1}{3 \cdot 2} = \frac{3 \cdot 4 \cdot 2}{4 \cdot 3 \cdot 3}?$
$\frac{2}{3} = \frac{2}{3}$
Since the cross products are equal, the proportion is true.

25.
$$\frac{2\frac{2}{5}}{\frac{2}{3}} = \frac{\frac{10}{9}}{\frac{1}{4}}?$$
$$\frac{2}{3} \cdot \frac{10}{9} = 2\frac{2}{5} \cdot \frac{1}{4}?$$
$$\frac{2}{3} \cdot \frac{10}{9} = \frac{12}{5} \cdot \frac{1}{4}?$$
$$\frac{2 \cdot 10}{3 \cdot 9} = \frac{3 \cdot 4 \cdot 1}{5 \cdot 4}?$$
$$\frac{20}{27} = \frac{3}{5}?$$
$$27 \cdot 3 = 20 \cdot 5?$$
$$81 \neq 100$$

Since the cross products are not equal, the proportion is false.

27.
$$\frac{x}{5} = \frac{6}{10}$$
$$30 = 10x$$
$$\frac{30}{10} = \frac{10x}{10}$$
$$3 = x$$

29.
$$\frac{30}{10} = \frac{15}{y}$$
$$10 \cdot 15 = 30y$$
$$150 = 30y$$
$$\frac{150}{30} = \frac{30y}{30}$$
$$5 = y$$

31.
$$\frac{z}{8} = \frac{50}{100}$$
$$400 = 100z$$
$$\frac{400}{100} = \frac{100z}{400}$$
$$4 = z$$

33.
$$\frac{n}{6} = \frac{8}{15}$$
$$6 \cdot 8 = 15n$$
$$48 = 15n$$
$$\frac{48}{15} = \frac{15n}{15}$$
$$3.2 = n$$

35.
$$\frac{12}{10} = \frac{x}{16}$$
$$10x = 12 \cdot 16$$
$$10x = 192$$
$$\frac{10x}{10} = \frac{192}{10}$$
$$x = 19.2$$

37.
$$\frac{n}{\frac{6}{5}} = \frac{4\frac{1}{6}}{6\frac{2}{3}}$$
$$\frac{6}{5} \cdot 4\frac{1}{6} = 6\frac{2}{3} \cdot n$$
$$\frac{6}{5} \cdot \frac{25}{6} = \frac{20}{3}n$$
$$\frac{150}{30} = \frac{20}{3}n$$
$$5 = \frac{20}{3}n$$
$$\frac{3}{20} \cdot \frac{5}{1} = \frac{3}{20} \cdot \frac{20}{3}n$$
$$\frac{15}{20} = n$$
$$\frac{3}{4} = n$$

39.
$$\frac{\frac{3}{4}}{12} = \frac{y}{48}$$
$$12y = \frac{3}{4} \cdot 48$$
$$12y = 36$$
$$\frac{12y}{12} = \frac{36}{12}$$
$$y = 3$$

41.
$$\frac{\frac{2}{3}}{\frac{6}{9}} = \frac{12}{z}$$
$$\frac{6}{9} \cdot 12 = \frac{2}{3} z$$
$$\frac{3}{2} \cdot \frac{6}{9} \cdot \frac{12}{1} = \frac{3}{2} \cdot \frac{2}{3} z$$
$$\frac{3 \cdot 2 \cdot 3 \cdot 12}{2 \cdot 3 \cdot 3 \cdot 1} = z$$
$$12 = z$$

43
$$\frac{n}{0.6} = \frac{0.05}{12}$$
$$0.6 \cdot 0.05 = 12n$$
$$0.03 = 12n$$
$$\frac{0.03}{12} = \frac{12n}{12}$$
$$0.0025 = n$$

45.
$$\frac{3.5}{12.5} = \frac{7}{z}$$
$$12.5 \cdot 7 = 3.5z$$
$$87.5 = 3.5z$$
$$\frac{87.5}{3.5} = \frac{3.5z}{3.5}$$
$$25 = z$$

47.
$$\frac{3.2}{0.3} = \frac{x}{1.4}$$
$$0.3x = 3.2 \cdot 1.4$$
$$0.3x = 4.48$$
$$\frac{0.3x}{0.3} = \frac{4.48}{0.3}$$
$$x \approx 14.9$$

49.
$$\frac{z}{5.2} = \frac{0.08}{6}$$
$$5.2 \cdot 0.08 = 6z$$
$$0.416 = 6z$$
$$\frac{0.416}{6} = \frac{6z}{6}$$
$$0.07 \approx z$$

51.
$$\frac{9}{11} = \frac{x}{4}$$
$$11x = 36$$
$$\frac{11x}{11} = \frac{36}{11}$$
$$x \approx 3.3$$

53.
$$\frac{43}{17} = \frac{8}{z}$$
$$17 \cdot 8 = 43z$$
$$136 = 43z$$
$$\frac{136}{43} = \frac{43z}{43}$$
$$3.163 \approx z$$

55. $-8 < 8$

57. $-2 > -3$

59. $-5 < 0$

61. $-1\frac{1}{2} > -2\frac{1}{2}$

63.
$$\frac{n}{7} = \frac{0}{8}$$
$$7 \cdot 0 = 8n$$
$$0 = 8n$$
$$\frac{0}{8} = \frac{8n}{8}$$
$$0 = n$$

65.
$$\frac{n}{1150} = \frac{588}{483}$$
$$1150 \cdot 588 = 483n$$
$$676,200 = 483n$$
$$\frac{676,200}{483} = \frac{483n}{483}$$
$$1400 = n$$

67. $\frac{222}{1515} = \frac{37}{n}$

$1515 \cdot 37 = 222n$

$56,055 = 222n$

$\frac{52,055}{222} = \frac{222n}{222}$

$252.5 = n$

69. Answers may vary.

Integrated Review

1. $\frac{18}{20} = \frac{9 \cdot 2}{10 \cdot 2} = \frac{9}{10}$

2. $\frac{36}{100} = \frac{9 \cdot 4}{25 \cdot 4} = \frac{9}{25}$

3. $\frac{8.6}{10} = \frac{86}{100} = \frac{43 \cdot 2}{50 \cdot 2} = \frac{43}{50}$

4. $\frac{1.6}{4.6} = \frac{16}{46} = \frac{8 \cdot 2}{23 \cdot 2} = \frac{8}{23}$

5. $\frac{8.65}{6.95} = \frac{865}{695} = \frac{173 \cdot 5}{139 \cdot 5} = \frac{173}{139}$

6. $\frac{7.2}{8.4} = \frac{72}{84} = \frac{6 \cdot 12}{7 \cdot 12} = \frac{6}{7}$

7. $\frac{3\frac{1}{2}}{13} = \frac{3.5}{13} = \frac{35}{130} = \frac{7 \cdot 5}{26 \cdot 5} = \frac{7}{26}$

8. $\frac{1\frac{2}{3}}{2\frac{3}{4}} = \frac{5}{3} \div \frac{11}{4} = \frac{5}{3} \cdot \frac{4}{11} = \frac{5 \cdot 4}{3 \cdot 11} = \frac{20}{33}$

9. $\frac{8 \text{ inches}}{12 \text{ inches}} = \frac{8}{12} = \frac{2 \cdot 4}{3 \cdot 4} = \frac{2}{3}$

10. $\frac{3 \text{ hours}}{24 \text{ hours}} = \frac{3}{24} = \frac{1 \cdot 3}{8 \cdot 3} = \frac{1}{8}$

11. $\frac{3}{2\frac{7}{20}} = \frac{3}{1} \div \frac{47}{20} = \frac{3}{1} \cdot \frac{20}{47} = \frac{3 \cdot 20}{1 \cdot 47} = \frac{60}{47}$

12. $\frac{\$6232 \text{ million}}{\$944 \text{ million}} = \frac{6232}{944} = \frac{779 \cdot 8}{118 \cdot 8} = \frac{779}{118}$

13. $\frac{5 \text{ offices}}{20 \text{ graduate assistants}}$
$= \frac{1 \text{ office}}{4 \text{ graduate assistants}}$

14. $\frac{6 \text{ lights}}{15 \text{ feet}} = \frac{2 \text{ lights}}{5 \text{ feet}}$

15. $\frac{100 \text{ U.S. Senators}}{50 \text{ states}} = \frac{2 \text{ U.S. Senators}}{1 \text{ state}}$

16. $\frac{5 \text{ teachers}}{140 \text{ students}} = \frac{1 \text{ teacher}}{28 \text{ students}}$

17. $\frac{165 \text{ miles}}{3 \text{ hours}} = \frac{55 \text{ miles}}{1 \text{ hour}}$

$$\begin{array}{r} 55 \\ 3\overline{)165} \\ \underline{-15} \\ 15 \\ \underline{15} \\ 0 \end{array}$$

55 miles per hour or 55 mph

18. $\frac{560 \text{ feet}}{4 \text{ seconds}} = \frac{140 \text{ feet}}{1 \text{ second}}$

$$\begin{array}{r} 140 \\ 4\overline{)560} \\ \underline{-4} \\ 16 \\ \underline{-16} \\ 00 \\ \underline{-0} \\ 0 \end{array}$$

140 feet per second

19. $\frac{63 \text{ employees}}{3 \text{ fax lines}} = \frac{21 \text{ employees}}{1 \text{ fax line}}$

$$\begin{array}{r} 21 \\ 3\overline{)63} \\ \underline{-6} \\ 03 \\ \underline{-3} \\ 0 \end{array}$$

21 employees per fax line

20. $\frac{85 \text{ phone calls}}{5 \text{ teenagers}} = \frac{17 \text{ phone calls}}{1 \text{ teenager}}$

$$\begin{array}{r} 17 \\ 5\overline{)85} \\ \underline{-5} \\ 35 \\ \underline{-35} \\ 0 \end{array}$$

17 phone calls per teenager

21. 8-pound unit price $= \frac{\$2.16}{8 \text{ pounds}}$
$= \$0.27$ per pound

18-pound unit price $= \frac{\$4.99}{18 \text{ pounds}}$
$\approx \$0.28$ per pound

The 8-pound size is the better buy.

22. 100-plate unit price $= \frac{\$1.98}{100 \text{ plates}}$
$\approx \$0.020$ per plate

500-plate unit price $= \frac{\$8.99}{500 \text{ plates}}$
$\approx \$0.018$ per plate

The 500-plate size is the better buy.

23. 3-pack unit price $= \frac{\$2.39}{3 \text{ packs}}$
$\approx \$0.80$ per pack

8-pack unit price $= \frac{\$5.99}{8 \text{ packs}}$
$\approx \$0.75$ per pack

The 8-pack size is the better buy.

24. 4-battery unit price $= \frac{\$3.69}{4 \text{ batteries}}$
$\approx \$0.92$ per battery

10-battery unit price $= \frac{\$9.89}{10 \text{ batteries}}$
$\approx \$0.99$ per battery

The 4 batteries are the better buy.

Section 6.4

Practice Problems

1. $\frac{12 \text{ feet}}{1 \text{ inch}} = \frac{n \text{ feet}}{3\frac{1}{2} \text{ inches}}$

$$\frac{12}{1} = \frac{n}{3\frac{1}{2}}$$

$$1 \cdot n = 12 \cdot 3\frac{1}{2}$$

$$n = \frac{12}{1} \cdot \frac{7}{2}$$

$$n = 42$$

$3\frac{1}{2}$ inches corresponds to 42 feet.

2. $\frac{14 \text{ gallons}}{3 \text{ ounces}} = \frac{n \text{ gallons}}{16 \text{ ounces}}$

$$\frac{14}{3} = \frac{n}{16}$$
$$3 \cdot n = 14 \cdot 16$$
$$3n = 224$$
$$\frac{3n}{3} = \frac{224}{3}$$
$$n = 74\frac{2}{3}$$

$74\frac{2}{3}$ gallons of gas can be treated with a 16-ounce bottle of alcohol.

3. area = 260 feet · 4 feet = 1040 square feet

$$\frac{1 \text{ gallon}}{400 \text{ square feet}} = \frac{n \text{ gallons}}{1040 \text{ square feet}}$$
$$\frac{1}{400} = \frac{n}{1040}$$
$$400 \cdot n = 1 \cdot 1040$$
$$400n = 1040$$
$$\frac{400n}{400} = \frac{1040}{400}$$
$$n = 2.6$$

Since we must buy full gallons, 3 gallons is needed.

Exercise Set 6.4

1. Let x = the number of completed passes.

$$\frac{4 \text{ completed}}{9 \text{ attempted}} = \frac{x \text{ completed}}{27 \text{ attempted}}$$
$$9x = 27 \cdot 4$$
$$9x = 108$$
$$\frac{9x}{9} = \frac{108}{9}$$
$$x = 12$$

He completed 12 passes.

3. Let x = the amount of time.

$$\frac{30 \text{ min}}{4 \text{ pages}} = \frac{x \text{ min}}{22 \text{ pages}}$$
$$4x = 660$$
$$\frac{4x}{4} = \frac{660}{4}$$
$$x = 165$$

It takes 165 minutes to do 22 pages.

5. Let x = the number of students accepted.

$$\frac{2 \text{ accepts}}{7 \text{ applicants}} = \frac{x \text{ accepts}}{630 \text{ applicants}}$$
$$7x = 1260$$
$$\frac{7x}{7} = \frac{1260}{7}$$
$$x = 180$$

180 students were accepted.

7. Let x = the length of the wall.

$$\frac{1 \text{ inch}}{8 \text{ feet}} = \frac{2\frac{7}{8} \text{ inches}}{x \text{ feet}}$$
$$x = 8 \cdot 2\frac{7}{8}$$
$$x = 8 \cdot \frac{23}{8}$$
$$x = 23$$

The wall is 23 feet long.

9. Let x = the amount of floor space.

$$\frac{9 \text{ sq. ft}}{1 \text{ student}} = \frac{x \text{ sq. ft}}{30 \text{ students}}$$
$$x = 270$$

270 square feet is required.

11. Let x = the number of miles.

$$\frac{450\text{ mi}}{12\text{ gal}} = \frac{x}{1\frac{1}{2}\text{ gal}}$$
$$450 \cdot 1\frac{1}{2} = 12x$$
$$450 \cdot \frac{3}{2} = 12x$$
$$675 = 12x$$
$$\frac{675}{12} = \frac{12x}{12}$$
$$56.25 = x$$

To the nearest mile, Dave can go 56 miles.

13. Let x = the distance from Milan to Rome.

$$\frac{1\text{ cm}}{30\text{ km}} = \frac{15\text{ cm}}{x\text{ km}}$$
$$450 = x$$

It is 450 km from Milan to Rome.

15. Let x = the number of ounces of grapefruit juice.

$$\frac{3\text{ grapefruit}}{4\text{ cranberry}} = \frac{x\text{ ounces grapefruit}}{32\text{ ounces cranberry}}$$
$$4x = 96$$
$$\frac{4x}{4} = \frac{96}{4}$$
$$x = 24$$

24 ounces of grapefruit juice should be used.

17. Let x = the number of bags of fertilizer.

$$\text{Area of lawn} = 260\text{ feet} \cdot 180\text{ feet}$$
$$= 46{,}800\text{ square feet}$$
$$\frac{1\text{ bag}}{3000\text{ sq. ft}} = \frac{x}{46{,}800\text{ sq. ft}}$$
$$3000x = 46{,}800$$
$$\frac{3000x}{3000} = \frac{46{,}800}{3000}$$
$$x = 15.6$$

Since whole bags must be purchased, 16 bags are needed.

19. Let x = the value of Janet's home.

$$\frac{\$1.45\text{ taxes}}{\$100\text{ value}} = \frac{\$2349\text{ taxes}}{x\text{ value}}$$
$$1.45x = 234{,}900$$
$$\frac{1.45x}{1.45} = \frac{234{,}900}{1.45}$$
$$x = 162{,}000$$

The value of her house if $162,000.

21. Let x = the number of hits expected.

$$\frac{3\text{ hits}}{8\text{ at bats}} = \frac{x\text{ hits}}{40\text{ at bats}}$$
$$8x = 120$$
$$\frac{8x}{8} = \frac{120}{8}$$
$$x = 15$$

He would be expected to make 15 hits.

23. Let x = the number of people who prefer Coke.

$$\frac{2\text{ Coke}}{3\text{ people}} = \frac{x\text{ Coke}}{40\text{ people}}$$
$$3x = 80$$
$$\frac{3x}{3} = \frac{80}{3}$$
$$x = 26\frac{2}{3}$$

Rounded to the nearest person, 27 people are likely to prefer Coke.

25. Let x = the number of weeks it will last.

$$\frac{5\text{ boxes}}{3\text{ weeks}} = \frac{144\text{ boxes}}{x\text{ weeks}}$$
$$432 = 5x$$
$$\frac{432}{5} = \frac{5x}{5}$$
$$86.4 = x$$
$$x = 86.4$$

The envelopes will last 86 weeks.

27. Let n = the number of people.

$$\frac{1 \text{ person}}{625 \text{ square feet of lawn}} = \frac{n \text{ people}}{3750 \text{ square feet of lawn}}$$
$$625n = 3750$$
$$\frac{625n}{625} = \frac{3750}{625}$$
$$n = 6$$

The lawn supplies enough oxygen for 6 people.

29. Let n = the estimated height of Statue of Liberty.

$$\frac{42 \text{ feet}}{2 \text{ feet}} = \frac{n \text{ feet}}{5\frac{1}{3} \text{ feet}}$$
$$2n = 42 \cdot 5\frac{1}{3}$$
$$2n = \frac{42}{1} \cdot \frac{16}{3}$$
$$2n = 224$$
$$\frac{2n}{2} = \frac{224}{2}$$
$$n = 112$$

The estimate height is 112 feet. The actual height is 111 feet 1 inch. The difference in the estimated height and the actual height is $112 - 111\frac{1}{2} = \frac{11}{12}$ of a foot or 11 inches.

31. Let n = the number of pounds of sugar.

$$\frac{1 \text{ pound of sugar}}{2\frac{1}{4} \text{ cups packed sugar}} = \frac{n \text{ pounds sugar}}{6 \text{ cups packed sugar}}$$
$$2\frac{1}{4} \cdot n = 1 \cdot 6$$
$$\frac{9}{4} \cdot n = 6$$
$$\frac{4}{9} \cdot \frac{9}{4} \cdot n = \frac{4}{9} \cdot \frac{6}{1}$$
$$n = \frac{8}{3}$$
$$n = 2\frac{2}{3}$$

$2\frac{2}{3}$ pounds of sugar will be required.

33. Let n = the number of visits including a prescription.

$$\frac{7 \text{ prescriptions}}{10 \text{ visits}} = \frac{n \text{ prescriptions}}{620 \text{ visits}}$$
$$10n = 7 \cdot 620$$
$$10n = 4340$$
$$\frac{10n}{10} = \frac{4340}{10}$$
$$n = 434$$

434 emergency room visits included a prescription.

35. Let n = the number of calories.

$$\frac{4}{190} = \frac{9}{n}$$
$$9 \cdot 190 = 4n$$
$$1710 = 4n$$
$$\frac{1710}{4} = \frac{4n}{4}$$
$$427.5 = n$$

There are 427.5 calories in the 9-piece.

37. Let n = the number of cups of rock salt.

$$\frac{5}{1} = \frac{12}{n}$$
$$1 \cdot 12 = 5n$$
$$12 = 5n$$
$$\frac{12}{5} = \frac{5n}{5}$$
$$2.4 = n$$

Mix 2.4 cups of rock salt with the ice.

39. $15 = 3 \cdot 5$

41. $20 = 4 \cdot 5$
$= 2 \cdot 2 \cdot 5$
$= 2^2 \cdot 5$

43. $200 = 2 \cdot 100$
$= 2 \cdot 4 \cdot 25$
$= 2 \cdot 2 \cdot 2 \cdot 5 \cdot 5$
$= 2^3 \cdot 5^2$

45. $32 = 4 \cdot 8$
$= 2 \cdot 2 \cdot 2 \cdot 4$
$= 2 \cdot 2 \cdot 2 \cdot 2 \cdot 2$
$= 2^5$

47.
$$\frac{40 \text{ pounds}}{n \text{ feet}} = \frac{60 \text{ pounds}}{7 \text{ feet}}$$
$$60n = 40 \cdot 7$$
$$60n = 280$$
$$\frac{60n}{60} = \frac{280}{60}$$
$$n = \frac{14}{3}$$
$$n = 4\frac{2}{3}$$

The board will balance when the distance is $4\frac{2}{3}$ feet.

49. Answers may vary.

Section 6.5

Practice Problems

1. We are given the lengths of two corresponding sides. Their ratio is $\frac{9 \text{ meters}}{13 \text{ meters}} = \frac{9}{13}$.

2. Since the triangles are similar, corresponding sides are in proportion.

$$\frac{10}{5} = \frac{n}{4}$$
$$5n = 40$$
$$\frac{5n}{5} = \frac{40}{5}$$
$$n = 8$$

The missing length is 8 units.

3. The triangles formed are similar triangles, so

$\frac{5}{n} = \frac{8}{200}$

$8n = 1000$

$n = \frac{1000}{8}$

$n = 125$ feet

The height of the building is 125 feet.

Exercise Set 6.5

1. Since the triangles are similar, we can compare any of the corresponding sides to find the ratio.

$\frac{22}{11} = \frac{2}{1}$

3. Since the triangles are similar, we can compare any of the corresponding sides to find the ratio.

$\frac{12}{8} = \frac{3}{2}$

5. $\frac{3}{n} = \frac{6}{9}$

$6n = 27$

$\frac{6n}{6} = \frac{27}{6}$

$n = 4.5$

7. $\frac{12}{4} = \frac{18}{n}$

$72 = 12n$

$\frac{72}{12} = \frac{12n}{12}$

$6 = n$

9. $\frac{n}{3.75} = \frac{12}{9}$

$45 = 9n$

$\frac{45}{9} = \frac{9n}{9}$

$5 = n$

11. $\frac{18}{n} = \frac{40}{30}$

$40n = 540$

$\frac{40n}{40} = \frac{540}{40}$

$n = 13.5$

13. $\frac{n}{3.25} = \frac{17.5}{3.25}$

$56.875 = 3.25n$

$\frac{56.875}{3.25} = \frac{3.25n}{3.25}$

$17.5 = n$

15. $\frac{n}{2} = \frac{8\frac{1}{2}}{2\frac{1}{8}}$

$\left(2\frac{1}{8}\right)n = (2)8\frac{1}{2}$

$\frac{17}{8}n = 2 \cdot \frac{17}{2}$

$\frac{8}{17} \cdot \frac{17}{8}n = \frac{8}{17} \cdot \frac{17}{1}$

$n = 8$

17. $\frac{34}{n} = \frac{16}{10}$

$16n = 340$

$\frac{16n}{16} = \frac{340}{16}$

$n = 21.25$

19. $\dfrac{\text{height of building}}{\text{height of man}} = \dfrac{\text{length of building's shadow}}{\text{length of man's shadow}}$

$$\frac{x}{6} = \frac{75}{9}$$
$$9x = 450$$
$$x = 50$$

The building is 50 feet high.

21. $\dfrac{11}{x} = \dfrac{5}{2}$

$$5x = 22$$
$$\frac{5x}{5} = \frac{22}{5}$$
$$x = 4.4$$
$$\frac{14}{y} = \frac{5}{2}$$
$$5y = 28$$
$$\frac{5y}{5} = \frac{28}{5}$$
$$y = 5.6$$

The lengths are $x = 4.4$ feet and $y = 5.6$ feet.

23. $\dfrac{30}{24} = \dfrac{18}{x}$

$$30x = 432$$
$$x = 14.4$$

The shadow is 14.4 feet long.

25. $\dfrac{1}{4} = \dfrac{1 \cdot 25}{4 \cdot 25} = \dfrac{25}{100} = 0.25$

27. $\dfrac{13}{20} = \dfrac{13 \cdot 5}{20 \cdot 5} = \dfrac{65}{100} = 0.65$

29. $\dfrac{9}{10} = 0.9$

31. $\dfrac{5.2}{n} = \dfrac{7.8}{12.6}$

$$7.8n = 65.52$$
$$n = 8.4$$

33. Answers may vary.

Chapter 6 Review

1. $\frac{6000 \text{ people}}{4800 \text{ people}} = \frac{6000}{4800} = \frac{60}{48} = \frac{5}{4}$

2. $\frac{121 \text{ births}}{143 \text{ births}} = \frac{121}{143} = \frac{11}{13}$

3. $\frac{2\frac{1}{4} \text{ days}}{10 \text{ days}} = \frac{2\frac{1}{4}}{10}$
$= \frac{\frac{9}{4}}{10}$
$= \frac{9}{4} \div 10$
$= \frac{9}{4} \cdot \frac{1}{10}$
$= \frac{9}{40}$

4. $\frac{14 \text{ quarters}}{5 \text{ quarters}} = \frac{14}{5}$

5. $\frac{4 \text{ weeks}}{15 \text{ weeks}} = \frac{4}{15}$

6. $\frac{4 \text{ yards}}{8 \text{ yards}} = \frac{1}{2}$

7. $\frac{3\frac{1}{2} \text{ dollars}}{7 \text{ dollars}} = \frac{3\frac{1}{2}}{7}$
$= \frac{\frac{7}{2}}{7}$
$= \frac{7}{2} \div 7$
$= \frac{7}{2} \cdot \frac{1}{7}$
$= \frac{1}{2}$

8. $\frac{3.5 \text{ cm}}{75 \text{ cm}} = \frac{3.5}{75} = \frac{7}{150}$

9. $\frac{8 \text{ stillborn births}}{1000 \text{ live births}} = \frac{1 \text{ stillborn birth}}{125 \text{ live births}}$

10. $\frac{6 \text{ professors}}{20 \text{ assistants}} = \frac{3 \text{ professors}}{10 \text{ assistants}}$

11. $\frac{15 \text{ pages}}{6 \text{ minutes}} = \frac{5 \text{ pages}}{2 \text{ minutes}}$

12. $\frac{8 \text{ computers}}{6 \text{ hours}} = \frac{4 \text{ computers}}{3 \text{ hours}}$

13. $\frac{468 \text{ miles}}{9 \text{ hours}} = \frac{52 \text{ miles}}{1 \text{ hour}} = 52$ miles per hour
or 52 mph

14. $\frac{180 \text{ ft}}{12 \text{ sec}} = \frac{15 \text{ ft}}{1 \text{ sec}} = 15$ ft per sec

15. $\frac{\$0.93}{3 \text{ pears}} = \frac{\$0.31}{1 \text{ pear}}$
$= \$0.31$ per pear

16. $\frac{\$6.96}{4 \text{ diskettes}} = \frac{\$1.74}{1 \text{ diskette}}$
$= \$1.74$ per diskette

17. $\frac{260 \text{ kilometers}}{4 \text{ hours}} = \frac{65 \text{ kilometer}}{1 \text{ hour}}$
$= 65$ km per hour

18. $\frac{8 \text{ gal}}{6 \text{ acres}} = \frac{1\frac{1}{3} \text{ gal}}{1 \text{ acre}} = 1\frac{1}{3}$ gal per acre

19. $\frac{\$184 \text{ for books}}{5 \text{ college courses}}$
$= \frac{\$36.80}{1 \text{ college course}}$
$= \$36.80$ per college course

20. $\frac{52 \text{ bushels}}{4 \text{ trees}} = \frac{13 \text{ bushels}}{1 \text{ tree}}$
$= 13$ bushels per tree

21. 8-ounce size:
$\frac{\$0.99}{8 \text{ oz}} \approx \frac{\$0.124}{1 \text{ oz}} = \$0.124$ per ounce
12-ounce size:
$\frac{\$1.69}{12 \text{ oz}} \approx \frac{\$0.141}{1 \text{ oz}} = \$0.141$ per ounce
The 8-ounce size is the better buy.

22. 18-ounce size:
$\frac{\$1.49}{18 \text{ oz}} \approx \frac{\$0.083}{1 \text{ oz}} = \$0.083$ per ounce
28-ounce size:
$\frac{\$2.39}{28 \text{ oz}} \approx \frac{\$0.085}{1 \text{ oz}} = \$0.085$ per ounce
The 18-ounce size is the better buy.

23. 16-ounce size:
$\frac{\$0.59}{16 \text{ oz}} \approx \frac{\$0.037}{1 \text{ oz}} = \$0.037$ per ounce
$\frac{1}{2}$-gallon size:
$\frac{1}{2}$ gallon = 64 ounces
$$\frac{\$1.69}{\frac{1}{2} \text{ gallon}} = \frac{\$1.69}{64 \text{ oz}}$$
$$\approx \frac{\$0.026}{1 \text{ oz}}$$
$= \$0.026$ per ounce
1-gallon size:
1-gallon = 128 ounces
$$\frac{\$2.26}{1 \text{ gallon}} = \frac{\$2.26}{128 \text{ oz}}$$
$$\approx \frac{\$0.018}{1 \text{ oz}}$$
$= \$0.018$ per ounce
The 1-gallon size is the better buy.

24. 12-ounce size:
$\frac{\$0.59}{12 \text{ oz}} \approx \frac{\$0.0492}{1 \text{ oz}} = \$0.0492$ per ounce
16-ounce size:
$\frac{\$0.79}{16 \text{ oz}} \approx \frac{\$0.0494}{1 \text{ oz}} = \$0.0494$ per ounce
32-ounce size:
$\frac{\$1.19}{32 \text{ oz}} \approx \frac{\$0.0372}{1 \text{ oz}} = \$0.0372$ per ounce
The 32-oz size is the better buy.

25. $\frac{20 \text{ men}}{14 \text{ women}} = \frac{10 \text{ men}}{7 \text{ women}}$

26. $\frac{50 \text{ tries}}{4 \text{ successes}} = \frac{25 \text{ tries}}{2 \text{ successes}}$

27. $\frac{16 \text{ sandwiches}}{8 \text{ players}} = \frac{2 \text{ sandwiches}}{1 \text{ player}}$

28. $\frac{12 \text{ tires}}{3 \text{ cars}} = \frac{4 \text{ tires}}{1 \text{ car}}$

29. $\frac{21}{8} = \frac{16}{4}?$
$8 \cdot 16 = 21 \cdot 4?$
$128 \neq 84$
Since the cross products are not equal, the proportion is false.

30. $\frac{3}{5} = \frac{60}{100}?$
$5 \cdot 60 = 3 \cdot 100?$
$300 = 300$
Since the cross products are equal, the proportion is true.

31. $\frac{3.1}{6.2} = \frac{0.8}{0.16}?$
$62 \cdot 0.8 = 3.1 \cdot 0.16?$
$4.96 \neq 0.496$
Since the cross products are not equal, the proportion is false.

32. $\frac{3.75}{3} = \frac{7.5}{6}?$
$3 \cdot 7.5 = 3.75 \cdot 6?$
$22.5 = 22.5$
Since the cross products are equal, the proportion is true.

33. $\frac{x}{6} = \frac{15}{18}$
$15 \cdot 6 = 18x$
$90 = 18x$
$\frac{90}{18} = \frac{18x}{18}$
$5 = x$

34.
$$\frac{y}{9}=\frac{5}{3}$$
$$9\cdot 5=3y$$
$$45=3y$$
$$\frac{45}{3}=\frac{3y}{3}$$
$$15=y$$

35.
$$\frac{4}{13}=\frac{10}{x}$$
$$10\cdot 13=4x$$
$$130=4x$$
$$\frac{130}{4}=\frac{4x}{4}$$
$$\frac{65}{2}=x$$
$$32.5=x$$

36.
$$\frac{8}{5}=\frac{9}{z}$$
$$5\cdot 9=8z$$
$$45=8z$$
$$\frac{45}{8}=\frac{8z}{8}$$
$$5.625=z$$

37.
$$\frac{16}{3}=\frac{y}{6}$$
$$3y=16\cdot 6$$
$$3y=96$$
$$\frac{3y}{3}=\frac{96}{3}$$
$$y=32$$

38.
$$\frac{x}{3}=\frac{9}{2}$$
$$3\cdot 9=2x$$
$$27=2x$$
$$\frac{27}{2}=\frac{2x}{2}$$
$$13.5=x$$

39.
$$\frac{x}{5}=\frac{27}{2\frac{1}{4}}$$
$$5\cdot 27=\left(2\frac{1}{4}\right)x$$
$$135=\frac{9}{4}x$$
$$\frac{4}{9}\cdot 135=\frac{4}{9}\cdot\frac{9}{4}x$$
$$\frac{4\cdot 9\cdot 15}{9}=x$$
$$60=x$$

40.
$$\frac{2\frac{1}{2}}{6}=\frac{3}{z}$$
$$6\cdot 3=\left(2\frac{1}{2}\right)z$$
$$18=\frac{5}{2}z$$
$$\frac{2}{5}\cdot 18=\frac{2}{5}\cdot\frac{5}{2}z$$
$$\frac{36}{5}=z$$
$$7\frac{1}{5}=z$$

41.
$$\frac{x}{0.4}=\frac{4.7}{3}$$
$$3x=(4.7)(0.4)$$
$$3x=1.88$$
$$\frac{3x}{3}=\frac{1.88}{3}$$
$$x=\frac{1.88}{3}$$
$$x\approx 0.63$$

42.
$$\frac{0.07}{0.3}=\frac{7.2}{n}$$
$$(0.3)(7.2)=0.07n$$
$$2.16=0.07n$$
$$\frac{2.16}{0.07}=\frac{0.07n}{0.07}$$
$$30.9\approx n$$
rounded to the nearest tenth

43. Let x = the number of completed passes.

$$\frac{3 \text{ completed passes}}{7 \text{ attempted passes}} = \frac{x \text{ completed passes}}{32 \text{ attempted passes}}$$
$$\frac{3}{7} = \frac{x}{32}$$
$$7x = 3 \cdot 32$$
$$7x = 96$$
$$\frac{7x}{7} = \frac{96}{7}$$
$$x \approx 13.7$$
$$x \approx 14$$

44. Let x = the number of attempts.

$$\frac{3 \text{ passes}}{7 \text{ attempts}} = \frac{15 \text{ passes}}{x \text{ attempts}}$$
$$\frac{3}{7} = \frac{15}{x}$$
$$7 \cdot 15 = 3x$$
$$105 = 3x$$
$$\frac{105}{3} = \frac{3x}{3}$$
$$35 = x$$

45. Let x = the number of bags of pesticide.
Garden area = 180 · 175 = 31,500 square feet

$$\frac{1 \text{ bag}}{4000 \text{ sq. feet}} = \frac{x \text{ bags}}{31{,}500 \text{ sq. feet}}$$
$$\frac{1}{4000} = \frac{x}{31{,}500}$$
$$4000x = 31{,}500$$
$$\frac{4000x}{4000} = \frac{31{,}500}{4000}$$
$$x = 7.875$$

Purchase 8 bags of pesticide.

46. Let x = the number of bags of pesticide.
Garden area = 250 · 250 = 62,500 square feet

$$\frac{1 \text{ bag}}{4000 \text{ sq. feet}} = \frac{x \text{ bags}}{62{,}500 \text{ sq. feet}}$$
$$4000x = 62{,}500$$
$$\frac{4000x}{4000} = \frac{62{,}500}{4000}$$
$$x = 15.625$$

Purchase 16 bags of pesticide.

47. Let x = the number of miles.

$$\frac{420 \text{ miles}}{11 \text{ gallons}} = \frac{x \text{ miles}}{1\frac{1}{2} \text{ gallons}}$$
$$\frac{420}{11} = \frac{x}{1\frac{1}{2}}$$
$$11x = 420 \cdot 1\frac{1}{2}$$
$$11x = 420 \cdot \frac{3}{2}$$
$$11x = 630$$
$$\frac{11x}{11} = \frac{630}{11}$$
$$x = 57\frac{3}{11}$$

Tom can only drive $57\frac{3}{11}$ miles on $1\frac{1}{2}$ gallons of gas so he will not be able to drive to a gas station 65 miles away.

48. Let x = the number of gallons of gas.

$$\frac{11 \text{ gal}}{420 \text{ miles}} = \frac{x \text{ gal}}{3000 \text{ miles}}$$
$$\frac{11}{420} = \frac{x}{3000}$$
$$33{,}000 = 420x$$
$$\frac{33{,}000}{420} = \frac{420x}{420}$$
$$78.5 \approx x$$

Tom will burn about 79 gallons.

49. Let x = the house value.

$$\frac{\$100 \text{ house value}}{\$1.15 \text{ tax}} = \frac{x \text{ house value}}{\$627.90 \text{ tax}}$$
$$\frac{100}{1.15} = \frac{x}{627.90}$$
$$(100) \cdot (627.90) = 1.15x$$
$$62{,}790 = 1.15x$$
$$\frac{62{,}790}{1.15} = \frac{1.15x}{1.15}$$
$$54{,}600 = x$$

The house value is \$54,600.

50. $\frac{1.15}{\$100} = \frac{\$x}{\$89,000}$

$\frac{1.15}{100} = \frac{x}{89,000}$

$100x = 1.15 \cdot 89,000$

$100x = 102,350$

$\frac{100x}{100} = \frac{102,350}{100}$

$x = 1023.50$

The property tax is \$1023.50.

51. Let x = the length in feet.

$\frac{1 \text{ inch}}{12 \text{ feet}} = \frac{3\frac{3}{8} \text{ inch}}{x \text{ feet}}$

$\frac{1}{12} = \frac{3\frac{3}{8}}{x}$

$\left(3\frac{3}{8}\right)(12) = x$

$\left(\frac{27}{8}\right) \cdot 12 = x$

$\frac{27 \cdot 4 \cdot 3}{4 \cdot 2} = x$

$\frac{81}{2} = x$

$40\frac{1}{2} = x$

The wall is $40\frac{1}{2}$ feet.

52. Let x = the blueprint measurement in inches.

$\frac{1 \text{ inch}}{12 \text{ feet}} = \frac{x \text{ inches}}{99 \text{ feet}}$

$\frac{1}{12} = \frac{x}{99}$

$12x = 99 \cdot 1$

$\frac{12x}{12} = \frac{99}{12}$

$x = \frac{33}{4}$

$x = 8\frac{1}{4}$

The blueprint measurement should be $8\frac{1}{4}$ inches.

53. $\frac{20}{8} = \frac{x}{15}$

$20 \cdot 15 = 8x$

$300 = 8x$

$\frac{300}{8} = \frac{8x}{4}$

$\frac{75}{2} = x$

$37.5 = x$

54. $\frac{x}{20} = \frac{16}{24}$

$20 \cdot 16 = 24x$

$320 = 24x$

$\frac{320}{24} = \frac{24x}{24}$

$\frac{40}{3} = x$

$13\frac{1}{3} = x$

55. $\frac{x}{5.8}=\frac{24}{8}$

$8x=(5.8)(24)$

$8x=139.2$

$\frac{8x}{8}=\frac{139.2}{8}$

$x=17.4$

56. $\frac{8\frac{2}{3}}{n}=\frac{12\frac{1}{2}}{9\frac{3}{8}}$

$\frac{\frac{26}{3}}{n}=\frac{\frac{25}{2}}{\frac{75}{8}}$

$\left(\frac{26}{3}\right)\left(\frac{75}{8}\right)=\left(\frac{25}{2}\right)n$

$\frac{325}{4}=\frac{25}{2}x$

$\frac{2}{25}\cdot\frac{325}{4}=\frac{2}{25}\cdot\frac{25}{2}x$

$\frac{13}{2}=x$

$6\frac{1}{2}=x$

57. Let x = building height.

$\frac{\text{man shadow}}{\text{building shadow}}=\frac{\text{man height}}{\text{building height}}$

$\frac{7\text{ feet}}{42\text{ feet}}=\frac{5\frac{1}{2}\text{ feet}}{x\text{ feet}}$

$\frac{7}{42}=\frac{5\frac{1}{2}}{x}$

$7x=\left(5\frac{1}{2}\right)(42)$

$7x=\frac{11}{2}\cdot 42$

$7x=231$

$\frac{7x}{7}=\frac{231}{7}$

$x=33$

The building is 33 feet tall.

58. $\frac{2\text{ inches}}{24\text{ feet}}=\frac{x\text{ inches}}{10\text{ feet}}=\frac{y\text{ inches}}{26\text{ feet}}$

$\frac{2}{24}=\frac{x}{10}$

$2\cdot 10=24x$

$20=24x$

$\frac{20}{24}=\frac{24x}{24}$

$\frac{5}{6}=x$

$\frac{2}{24}=\frac{y}{26}$

$2\cdot 26=24y$

$52=24y$

$\frac{52}{24}=\frac{24y}{24}$

$\frac{13}{6}=y$

$2\frac{1}{6}=y$

x is $\frac{5}{6}$ inches and y is $2\frac{1}{6}$ inches.

Chapter 6 Test

1. $\frac{4500\text{ trees}}{6500\text{ trees}}=\frac{9\cdot 500}{13\cdot 500}=\frac{9}{13}$

2. $\frac{\$75}{\$10}=\frac{15\cdot 5}{2\cdot 5}=\frac{15}{2}$

3. $\frac{28\text{ men}}{4\text{ women}}=\frac{7\text{ men}}{1\text{ woman}}$

4. $\frac{9\text{ inches of rain}}{30\text{ days}}=\frac{3\text{ inches of rain}}{10\text{ days}}$

5. $\dfrac{650 \text{ kilometers}}{8 \text{ hours}} = 81.25$ kilometers per hour

$$\begin{array}{r} 81.25 \\ 8\overline{)650.00} \\ \underline{-64} \\ 10 \\ \underline{-8} \\ 20 \\ \underline{-16} \\ 40 \\ \underline{-40} \\ 0 \end{array}$$

6. $\dfrac{8 \text{ inches of rain}}{12 \text{ hours}} = \dfrac{2}{3}$ inches rain per hour

$$\begin{array}{r} 0.66\ldots \\ 12\overline{)\ 8.00} \\ \underline{-7\,2} \\ 80 \\ \underline{-72} \\ 8 \end{array}$$

7. $\dfrac{140 \text{ students}}{5 \text{ teachers}} = \dfrac{28 \text{ students}}{1 \text{ teacher}}$
$= 28$ students per teacher

8. 8-ounce size:
$\dfrac{\$1.19}{8 \text{ oz}} \approx \$0.149$ per ounce
12-ounce size:
$\dfrac{\$1.89}{12 \text{ oz}} \approx \$0.158$ per ounce
The 8-ounce size costs less per ounce, thus, it is a better buy.

9. 16-ounce size:
$\dfrac{\$1.49}{16 \text{ oz}} \approx \$0.093$ per ounce
24-ounce size:
$\dfrac{\$2.39}{24 \text{ oz}} \approx \$0.100$ per ounce
The 16-ounce size costs less per ounce, thus it is the better buy.

10. $\dfrac{28}{16} = \dfrac{14}{8}?$

$16 \cdot 14 = 28 \cdot 8?$

$224 = 224$

Since the cross products are equal, the proportion is true.

11. $\dfrac{3.6}{2.2} = \dfrac{1.9}{1.2}?$

$2.2 \cdot 1.9 = 3.6 \cdot 1.2?$

$4.18 \neq 4.32$

Since the cross products are not equal, the proportion is false.

12.
$$\begin{aligned} \frac{n}{3} &= \frac{15}{9} \\ 3 \cdot 15 &= 9n \\ 45 &= 9n \\ \frac{45}{9} &= \frac{9n}{9} \\ 5 &= n \end{aligned}$$

13.
$$\begin{aligned} \frac{8}{x} &= \frac{11}{6} \\ 11 \cdot x &= 8 \cdot 6 \\ 11x &= 48 \\ \frac{11x}{11} &= \frac{48}{11} \\ x &= \frac{48}{11} = 4\frac{4}{11} \end{aligned}$$

14.
$$\frac{\frac{4}{3}}{7} = \frac{y}{\frac{1}{4}}$$
$$\frac{3}{7}y = 4 \cdot \frac{1}{4}$$
$$\frac{3}{7}y = 1$$
$$\frac{7}{3} \cdot \frac{3}{7}y = \frac{7}{3} \cdot 1$$
$$y = \frac{7}{3}$$

15.
$$\frac{1.5}{5} = \frac{2.4}{n}$$
$$1.5n = 2.4 \cdot 5$$
$$1.5n = 12$$
$$\frac{1.5n}{1.5} = \frac{12}{1.5}$$
$$n = 8$$

16. Let n = the number of feet.
$$\frac{2}{9} = \frac{11}{n}$$
$$2n = 99$$
$$\frac{2n}{2} = \frac{99}{2}$$
$$n = 49\frac{1}{2}$$
The home is $49\frac{1}{2}$ feet long.

17. Let x = the length of time.
$$\frac{3}{80} = \frac{x}{100}$$
$$80x = 300$$
$$\frac{80x}{80} = \frac{300}{80}$$
$$x = \frac{15}{4} = 3\frac{3}{4}$$
It will take $3\frac{3}{4}$ hours.

18. Let n = the number of grams of medicine.
$$\frac{10}{15} = \frac{n}{80}$$
$$15n = 800$$
$$\frac{15n}{15} = \frac{800}{15}$$
$$n = \frac{160}{3} = 53\frac{1}{3}$$
The standard dose is $53\frac{1}{3}$ grams.

19. Let n = the number of cartons.
$$\frac{6}{86} = \frac{8}{n}$$
$$6n = 688$$
$$\frac{6n}{6} = \frac{688}{6}$$
$$n = \frac{344}{3}$$
$$n = 114\frac{2}{3}$$
He can pack $114\frac{2}{3}$ cartons.

20. Let n = number that drink coffee
$$\frac{12}{25} = \frac{n}{31,000}$$
$$25n = 372,000$$
$$n = 14,880$$
The expected number is 14,880 adults.

21.
$$\frac{5}{8} = \frac{n}{12}$$
$$8n = 5 \cdot 12$$
$$8n = 60$$
$$\frac{8n}{8} = \frac{60}{8}$$
$$n = 7.5$$

Chapter 7

Pretest

1. $\frac{12}{100} = 12\%$

2. $57\% = 57.\% = 0.57$

3. $2.75 = 275\%$

4. $7.5\% = \frac{7.5}{100} - \frac{75}{1000} = \frac{3}{40}$

5. $\frac{3}{20} = \frac{3 \cdot 5}{20 \cdot 5} = \frac{15}{100} = 15\%$

6. $18\% \cdot 50 = x$

7. $4\% \cdot x = 89$

8. $\frac{82}{b} = \frac{90}{100}$

9. $\frac{48}{112} = \frac{p}{100}$

10. $x \cdot 80 = 16$
$\frac{x \cdot 80}{80} = \frac{16}{80}$
$x = 0.2$
$x = 20\%$
16 is 20% of 80.

11. $x = 1.5\% \cdot 220$
$x = 0.015 \cdot 220$
$x = 3.3$
3.3 is 1.5% of 220.

12. $32 = 16\% \cdot x$
$32 = 0.16 \cdot x$
$\frac{32}{0.16} = \frac{0.16 \cdot x}{0.16}$
$200 = x$
32 is 16% of 200.

13. What number is 1.2% of 250?
$x = 1.2\% \cdot 250$
$x = 0.012 \cdot 250$
$x = 3$
3 lightbulbs were defective.

14. What number is 2% of 4200?
$x = 2\% \cdot 4200$
$x = 0.02 \cdot 4200$
$x = 84$
The enrollment decreased by 84 students.
The current enrollment is 4200 – 84 = 4316 students.

15. $\$34.93 = r \cdot \$499$
$\frac{\$34.93}{\$499} = \frac{r \cdot \$499}{\$499}$
$0.07 = r$
$7\% = r$
The sales tax rate is 7%.

16. $\$448 = 2\% \cdot x$
$\$448 = 0.02 \cdot x$
$\frac{\$448}{0.02} = \frac{0.02 \cdot x}{0.02}$
$\$22,400 = x$
His sales were $22,400.

17. discount $= 12\% \cdot \$650$
$= 0.12 \cdot \$650$
$= \$78$
sales price = $650 – 78 = $572

18. simple interest = principal · rate · time
$= \$600 \cdot 8\% \cdot 3$
$= \$600 \cdot (0.08) \cdot 3$
$= \$144$

19. The compound interest factor for 6 years at 6% compounded quarterly is 1.42950.
total amount = $5000 · 1.42950 = $7147.50

20. $\text{monthly payment} = \dfrac{\text{principal interest}}{\text{number of payments}}$
$= \dfrac{\$700 + \$196}{2 \cdot 12}$
$= \dfrac{\$896}{24}$
$\approx \$37.33$

Section 7.1

Practice Problems

1. $\frac{23}{100} = 23\%$

2. $\frac{29}{100} = 29\%$

3. $89\% = 89.\% = 0.89$

4. $2.7\% = 0.027$

5. $150\% = 150.\% = 1.5$

6. $0.69\% = 0.0069$

7. $0.19 = 19.\% = 19\%$

8. $1.75 = 175.\% = 175\%$

9. $0.044 = 004.4\% = 4.4\%$

10. $0.7 = 070.\% = 70\%$

11. $25\% = \frac{25}{100} = \frac{1 \cdot 25}{4 \cdot 25} = \frac{1}{4}$

12. $2.3\% = \dfrac{2.3}{100} = \dfrac{2.3 \cdot 10}{100 \cdot 10} = \dfrac{23}{1000}$

13. $150\% = \frac{150}{100} = \frac{3 \cdot 50}{2 \cdot 50} = \frac{3}{2}$

14. $66\frac{2}{3}\% = \dfrac{66\frac{2}{3}}{100}$
$= \dfrac{\frac{200}{3}}{100}$
$= \dfrac{200}{3} \div 100$
$= \dfrac{200}{3} \cdot \dfrac{1}{100}$
$= \dfrac{2}{3}$

15. $8\% = \frac{8}{100} = \frac{2 \cdot 4}{25 \cdot 4} = \frac{2}{25}$

16. $\frac{1}{2} = \frac{1}{2} \cdot 100\% = \frac{100}{2}\% = 50\%$

17. $\dfrac{7}{40} = \dfrac{7}{40} \cdot 100\% = \dfrac{700}{40}\% = \dfrac{35}{2}\% = 17\dfrac{1}{2}\%$

18. $2\frac{1}{4} = \frac{9}{4} \cdot 100\% = \frac{900}{4}\% = 225\%$

19. $\frac{3}{17} = \frac{3}{17} \cdot 100\% = \frac{300}{17}\% \approx 17.65\%$

```
       17.647 ≈ 17.65
17) 300.000
   -17
    130
   -119
     11 0
    -10 2
        80
      - 68
       120
      -119
         1
```

20. $25\% = 25.\% = 0.25$

21. $1\frac{1}{4} = \frac{5}{4} \cdot 100\% = \frac{500}{4}\% = 125\%$

Mental Math

1. $\dfrac{13}{100} = 13\%$

2. $\frac{92}{100} = 92\%$

3. $\frac{87}{100} = 87\%$

4. $\frac{71}{100} = 71\%$

5. $\frac{1}{100} = 1\%$

6. $\frac{2}{100} = 2\%$

Exercise Set 7.1

1. $\frac{81}{100} = 81\%$

3. $\frac{9}{100} = 9\%$

5. The largest section of the circle graph is chocolate chip. Therefore, chocolate chip was the most preferred cookie.
$\frac{52}{100} = 52\%$

7. $\frac{12}{100} = 12\%$

9. $48\% = 48.\% = 0.48$

11. $6\% = 6.\% = 0.06$

13. $100\% = 100.\% = 1.00 = 1$

15. $61.3\% = 0.613$

17. $2.8\% = 0.028$

19. $0.6\% = 0.006$

21. $300\% = 300.\% = 3.00 = 3$

23. $32.58\% = 0.3258$

25. $73.7\% = 0.737$

27. $25\% = 25.\% = 0.25$

29. $80.2\% = 0.802$

31. $46.2\% = 0.462$

33. $3.1 = 3.10 = 310.\% = 310\%$

35. $29 = 29.00 = 2900.\% = 2900\%$

37. $0.003 = 000.3\% = 0.3\%$

39. $0.22 = 022.\% = 22\%$

41. $0.056 = = 005.6\% = 5.6\%$

43. $0.3328 = 033.28\% = 33.28\%$

45. $3.00 = 300.\% = 300\%$

47. $0.7 = 0.70 = 070.\% = 70\%$

49. $0.10 = 010.\% = 10\%$

51. $0.753 = 075.3\% = 75.3\%$

53. $0.38 = 038.\% = 38\%$

55. $4\% = \frac{4}{100} = \frac{4 \cdot 1}{4 \cdot 25} = \frac{1}{25}$

57. $4.5\% = \frac{4.5}{100} = \frac{45}{1000} = \frac{5 \cdot 9}{5 \cdot 200} = \frac{9}{200}$

59. $175\% = \frac{175}{100} = \frac{25 \cdot 7}{25 \cdot 4} = \frac{7}{4} = 1\frac{3}{4}$

61. $73\% = \frac{73}{100}$

63. $12.5\% = \frac{12.5}{100} = \frac{125}{1000} = \frac{125 \cdot 1}{125 \cdot 8} = \frac{1}{8}$

65. $6.25\% = \frac{6.25}{100} = \frac{625}{10{,}000} = \frac{625}{625 \cdot 16} = \frac{1}{16}$

67. $10\frac{1}{3}\% = \frac{10\frac{1}{3}}{100}$
$= \frac{\frac{31}{3}}{100}$
$= \frac{31}{3} \div 100$
$= \frac{31}{3} \cdot \frac{1}{100}$
$= \frac{31}{300}$

69. $22\frac{3}{8}\% = \frac{22\frac{3}{8}}{100}$
$= \frac{\frac{179}{8}}{100}$
$= \frac{179}{8} \div 100$
$= \frac{179}{8} \cdot \frac{1}{100}$
$= \frac{179}{800}$

71. $\frac{3}{4} = \frac{3}{4} \cdot 100\% = \frac{300}{4}\% = 75\%$

73. $\frac{7}{10} = \frac{7}{10} \cdot 100\% = \frac{700}{10}\% = 70\%$

75. $\frac{2}{5} = \frac{2}{5} \cdot 100\% = \frac{200}{5}\% = 40\%$

77. $\frac{59}{100} = \frac{59}{100} \cdot 100\% = \frac{59 \cdot 100}{100}\% = 59\%$

79. $\frac{17}{50} = \frac{17}{50} \cdot 100\% = \frac{1700}{50}\% = 34\%$

81. $\frac{3}{8} = \frac{3}{8} \cdot 100\%$
$= \frac{300}{8}\%$
$= \frac{4 \cdot 75}{4 \cdot 2}\%$
$= \frac{75}{2}\%$
$= 37\frac{1}{2}\%$

83. $\frac{5}{16} = \frac{5}{16} \cdot 100\% = \frac{500}{16}\% = 31\frac{1}{4}\%$

85. $\frac{2}{3} = \frac{2}{3} \cdot 100\% = \frac{200}{3}\% = 66\frac{2}{3}\%$

87. $2\frac{1}{2} = \frac{5}{2} \cdot 100\% = \frac{500}{2}\% = 250\%$

89. $1\frac{9}{10} = \frac{19}{10} \cdot 100\% = \frac{1900}{10}\% = 190\%$

91. $\frac{7}{11} = \frac{7}{11} \cdot 100\% = \frac{700}{11}\% \approx 63.64\%$

$$\begin{array}{r} 63.636 \approx 63.64 \\ 11\overline{)\ 700.000} \\ \underline{-66} \\ 40 \\ \underline{-33} \\ 70 \\ \underline{-66} \\ 40 \\ \underline{-33} \\ 70 \\ \underline{-66} \\ 4 \end{array}$$

93. $\frac{4}{15} = \frac{4}{15} \cdot 100\% = \frac{400}{15}\% \approx 26.67\%$

$$\begin{array}{r} 26.666 \approx 26.67 \\ 15\overline{)400.00} \\ \underline{-30} \\ 100 \\ \underline{-90} \\ 100 \\ \underline{-90} \\ 100 \\ \underline{-90} \\ 10 \end{array}$$

95. $\frac{1}{7} = \frac{1}{7} \cdot 100\% = \frac{100}{7}\% \approx 14.29\%$

$$\begin{array}{r} 14.285 \approx 14.29 \\ 7\overline{)100.000} \\ \underline{-7} \\ 30 \\ \underline{-28} \\ 20 \\ \underline{-14} \\ 60 \\ \underline{-56} \\ 40 \\ \underline{-35} \\ 5 \end{array}$$

97. $\frac{11}{12} = \frac{11}{12} \cdot 100\% = \frac{1100}{12}\% \approx 91.67\%$

$$\begin{array}{r} 91.666 \approx 91.67 \\ 12\overline{)1100.000} \\ \underline{-108} \\ 20 \\ \underline{-12} \\ 80 \\ \underline{-72} \\ 80 \\ \underline{-72} \\ 80 \\ \underline{-72} \\ 8 \end{array}$$

99.

Percent	Decimal	Fraction
35%	0.35	$\frac{7}{20}$
20%	0.2	$\frac{1}{5}$
50%	0.5	$\frac{1}{2}$
70%	0.7	$\frac{7}{10}$
37.5%	0.375	$\frac{3}{8}$

101.

Percent	Decimal	Fraction
40%	0.4	$\frac{2}{5}$
23.5%	0.235	$\frac{47}{200}$
80%	0.8	$\frac{4}{5}$
$33\frac{1}{3}\%$	$0.333\overline{3}$	$\frac{1}{3}$
87.5%	0.875	$\frac{7}{8}$
7.5%	0.075	$\frac{3}{40}$

103. $18.8\% = 0.188 = \frac{188}{1000} = \frac{47}{250}$

105. $\frac{17}{200} = \frac{17}{200} \cdot 100\% = \frac{1700}{200}\% = \frac{17}{2}\% = 8.5\%$

$$\begin{array}{r} 8.5 \\ 2\overline{)17.0} \\ \underline{-16} \\ 10 \\ \underline{-10} \\ 0 \end{array}$$

107. $67.4\% = 0.674$

109. $\frac{3}{10} = \frac{3 \cdot 10}{10 \cdot 10} = \frac{30}{100} = 30\%$

111. $3n = 45$

$\frac{3n}{3} = \frac{45}{3}$

$n = 15$

113. $-8n = 80$

$\frac{-8n}{-8} = \frac{80}{-8}$

$n = -10$

115. $6n = -72$

$\frac{6n}{6} = \frac{-72}{6}$

$n = -12$

117. $\frac{850}{736} \approx 1.155 = 115.5\%$

119. A fraction written as a percent is greater than 100% when the numerator is greater than the denominator.

121. Answers may vary.

123. Database administrators, computer support specialists, and other computer scientists is predicted to be the fastest growing occupation.

125. $85\% = 85.\% = 0.85$

127. Answers may vary.

Section 7.2

Practice Problems

1. $6 = x \cdot 24$

2. $1.8 = 20\% \cdot x$

3. $x = 40\% \cdot 3.6$

4. $42\% \cdot 50 = x$

5. $15\% \cdot x = 9$

6. $x \cdot 150 = 90$

7. $x = 20\% \cdot 85$
$x = 0.2 \cdot 85$
$x = 17$
The number is 17.

8. $90\% \cdot 150 = x$
$0.9 \cdot 150 = x$
$135 = x$
The number is 135.

9. $15\% \cdot x = 1.2$
$0.15 \cdot x = 1.2$
$\frac{0.15 \cdot x}{0.15} = \frac{1.2}{0.15}$
$x = 8$
The number is 8.

10. $27 = 4\frac{1}{2}\% \cdot x$
$27 = 0.045 \cdot x$
$\frac{27}{0.045} = \frac{0.045 \cdot x}{0.045}$
$600 = x$
The number is 600.

11. $x \cdot 80 = 8$
$\frac{x \cdot 80}{80} = \frac{8}{80}$
$x = 0.10$
$x = 10\%$
The percent is 10%.

12. $35 = x \cdot 25$
$\frac{35}{25} = \frac{x \cdot 25}{25}$
$1.4 = x$
$140\% = x$
The percent is 140%.

Mental Math

1. percent: 42
base: 50
amount: 21

2. percent: 30
base: 65
amount: 19.5

3. percent: 125
base: 86
amount: 107.5

4. percent: 110
base: 90
amount: 99

Exercise Set 7.2

1. $15\% \cdot 72 = x$

3. $30\% \cdot x = 80$

5. $x \cdot 90 = 20$

7. $1.9 = 40\% \cdot x$

9. $x = 9\% \cdot 43$

11. $10\% \cdot 35 = x$
$0.10 \cdot 35 = x$
$3.5 = x$

13. $x = 14\% \cdot 52$
$x = 0.14 \cdot 52$
$x = 7.28$

15. $3000 = 5\% \cdot x$
$30 = 0.05x$
$\frac{30}{0.05} = \frac{0.05x}{0.05}$
$600 = x$

17. $1.2 = 12\% \cdot x$
$1.2 = 0.12x$
$\frac{1.2}{0.12} = \frac{0.12x}{0.12}$
$10 = x$

19. $66 = x \cdot 60$
$\frac{66}{60} = \frac{x \cdot 60}{60}$
$1.1 = x$
$110\% = x$

21. $16 = x \cdot 50$
$\frac{16}{50} = \frac{x \cdot 50}{50}$
$0.32 = x$
$32\% = x$

23. $0.1 = 10\% \cdot x$
$0.1 = 0.10x$
$\frac{0.1}{0.10} = \frac{0.10x}{0.10}$
$1 = x$

25. $125\% \cdot 36 = x$
$1.25 \cdot 36 = x$
$45 = x$

27. $82.5 = 16\frac{1}{2}\% \cdot x$
$82.5 = 0.165 \cdot x$
$\frac{82.5}{0.165} = \frac{0.165 \cdot x}{0.165}$
$500 = x$

29. $2.58 = x \cdot 50$
$\frac{2.58}{50} = \frac{x \cdot 50}{50}$
$0.0516 = x$
$5.16\% = x$

31. $x = 42\% \cdot 60$
$x = 0.42 \cdot 60$
$x = 25.2$

33. $x \cdot 150 = 67.5$
$\frac{x \cdot 150}{150} = \frac{67.5}{150}$
$x = 0.45$
$x = 45\%$

35. $120\% \cdot x = 42$
$1.2 \cdot x = 42$
$\frac{1.2 \cdot x}{1.2} = \frac{42}{1.2}$
$x = 35$

37. $\frac{27}{n} = \frac{9}{10}$
$9 \cdot n = 27 \cdot 10$
$\frac{9 \cdot n}{9} = \frac{270}{9}$
$n = 30$

39. $\frac{n}{5} = \frac{8}{11}$
$5 \cdot 8 = 11 \cdot n$
$\frac{40}{11} = \frac{11 \cdot n}{11}$
$3\frac{7}{11} = n$

41. $\frac{17}{12} = \frac{n}{20}$

43. $\frac{8}{9} = \frac{14}{n}$

45. $1.5\% \cdot 45{,}775 = x$
$0.015 \cdot 45{,}775 = x$
$686.625 = x$

47. $22{,}113 = 180\% \cdot x$
$22{,}113 = 1.8 \cdot x$
$\frac{22{,}113}{1.8} = \frac{1.8 \cdot x}{1.8}$
$12{,}285 = x$

Section 7.3

Practice Problems

1. $\frac{55}{b} = \frac{15}{100}$

2. $\frac{35}{70} = \frac{p}{100}$

3. $\frac{a}{68} = \frac{25}{100}$

4. $\frac{520}{b} = \frac{65}{100}$

5. $\frac{65}{50} = \frac{p}{100}$

6. $\frac{a}{80} = \frac{36}{100}$

7. $\frac{a}{120} = \frac{8}{100}$
$a \cdot 100 = 120 \cdot 8$
$a \cdot 100 = 960$
$\frac{a \cdot 100}{100} = \frac{960}{100}$
$a = 9.6$
Therefore, 9.6 is 8% of 120.

8. $\frac{60}{b} = \frac{75}{100}$
$b \cdot 75 = 6000$
$\frac{b \cdot 75}{75} = \frac{6000}{75}$
$b = 80$
Therefore, 75% of 80 is 60.

9. $\frac{15}{b} = \frac{5}{100}$
$5 \cdot b = 1500$
$\frac{5 \cdot b}{5} = \frac{1500}{5}$
$b = 300$
Therefore, 15 is 5% of 300.

10. $\frac{5}{40} = \frac{p}{100}$
$40 \cdot p = 500$
$\frac{40 \cdot p}{40} = \frac{500}{40}$
$p = 12.5$
Therefore, 12.5% of 40 is 5.

11. $\frac{336}{160} = \frac{p}{100}$
$160 \cdot p = 33,600$
$\frac{160 \cdot p}{160} = \frac{33,600}{160}$
$p = 210$
Therefore, 210% of 160 is 336.

Mental Math

1. amount: 12.6
base: 42
percent: 30

2. amount: 201
base: 300
percent: 67

3. amount: 102
base: 510
percent: 20

4. amount: 248
base: 620
percent: 40

Exercise Set 7.3

1. $\frac{a}{65} = \frac{32}{100}$

3. $\frac{75}{b} = \frac{40}{100}$

5. $\frac{70}{200} = \frac{p}{100}$

7. $\frac{2.3}{b} = \frac{58}{100}$

9. $\frac{a}{130} = \frac{19}{100}$

11.
$$\frac{a}{55}=\frac{10}{100}$$
$$\frac{a}{55}=\frac{1}{10}$$
$$a\cdot 10=55\cdot 1$$
$$a\cdot 10=55$$
$$\frac{a\cdot 10}{10}=\frac{55}{10}$$
$$a=5.5$$
Therefore, 10% of 55 is 5.5.

13.
$$\frac{a}{105}=\frac{18}{100}$$
$$\frac{a}{105}=\frac{9}{50}$$
$$a\cdot 50=105\cdot 9$$
$$\frac{a\cdot 50}{50}=\frac{945}{50}$$
$$a=18.9$$
Therefore, 18% of 105 is 18.9.

15.
$$\frac{60}{b}=\frac{15}{100}$$
$$\frac{60}{b}=\frac{3}{20}$$
$$60\cdot 20=b\cdot 3$$
$$1200=b\cdot 3$$
$$\frac{1200}{3}=\frac{b\cdot 3}{3}$$
$$400=b$$
Therefore, 15% of 400 is 60.

17.
$$\frac{7.8}{b}=\frac{78}{100}$$
$$\frac{7.8}{b}=\frac{39}{50}$$
$$7.8\cdot 50=b\cdot 39$$
$$390=b\cdot 39$$
$$\frac{390}{39}=\frac{b\cdot 39}{39}$$
$$10=b$$
Therefore, 78% of 10 is 7.8.

19.
$$\frac{105}{84}=\frac{p}{100}$$
$$\frac{5}{4}=\frac{p}{100}$$
$$5\cdot 100=4\cdot p$$
$$500=4\cdot p$$
$$\frac{500}{4}=\frac{4\cdot p}{4}$$
$$125=p$$
Therefore, 125% of 84 is 105.

21.
$$\frac{14}{50}=\frac{p}{100}$$
$$\frac{7}{25}=\frac{p}{100}$$
$$7\cdot 100=25\cdot p$$
$$700=25\cdot p$$
$$\frac{700}{25}=\frac{25\cdot p}{25}$$
$$28=p$$
Therefore, 28% of 50 is 14.

23.
$$\frac{2.9}{b}=\frac{10}{100}$$
$$\frac{2.9}{b}=\frac{1}{10}$$
$$2.9\cdot 10=b\cdot 1$$
$$29=b$$
Therefore, 10% of 29 is 2.9.

25.
$$\frac{a}{80}=\frac{2.4}{100}$$
$$a\cdot 100=80\cdot 2.4$$
$$a\cdot 100=192$$
$$\frac{a\cdot 100}{100}=\frac{192}{100}$$
$$a=1.92$$
Therefore, 2.4% of 80 is 1.92.

27.
$$\frac{160}{b}=\frac{16}{100}$$
$$\frac{160}{b}=\frac{4}{25}$$
$$160\cdot 25=b\cdot 4$$
$$4000=b\cdot 4$$
$$\frac{4000}{4}=\frac{b\cdot 4}{4}$$
$$1000=b$$
Therefore, 16% of 1000 is 160.

29. $$\frac{348.6}{166} = \frac{p}{100}$$
$$348.6 \cdot 100 = 166 \cdot p$$
$$34,860 = 166 \cdot p$$
$$\frac{34,860}{166} = \frac{166 \cdot p}{166}$$
$$210 = p$$
Therefore, 210% of 166 is 348.6.

31. $$\frac{a}{62} = \frac{89}{100}$$
$$a \cdot 100 = 62 \cdot 89$$
$$a \cdot 100 = 5518$$
$$\frac{a \cdot 100}{100} = \frac{5518}{100}$$
$$a = 55.18$$
Therefore, 89% of 62 is 55.18.

33. $$\frac{3.6}{8} = \frac{p}{100}$$
$$3.6 \cdot 100 = 8 \cdot p$$
$$360 = 8 \cdot p$$
$$\frac{360}{8} = \frac{8 \cdot p}{8}$$
$$45 = p$$
Therefore, 45% of 8 is 3.6.

35. $$\frac{119}{b} = \frac{140}{100}$$
$$\frac{119}{b} = \frac{7}{5}$$
$$119 \cdot 5 = b \cdot 7$$
$$595 = b \cdot 7$$
$$\frac{595}{7} = \frac{b \cdot 7}{7}$$
$$85 = b$$
Therefore, 140% of 85 is 119.

37. $\frac{11}{16} + \frac{3}{16} = \frac{11+3}{16} = \frac{14}{16} = \frac{7 \cdot 2}{8 \cdot 2} = \frac{7}{8}$

39. $$\begin{array}{rcr} 3\frac{1}{2} & = & 3\frac{15}{30} \\ -\frac{11}{30} & = & -\frac{11}{30} \\ \hline & & 3\frac{4}{30} = 3\frac{2}{15} \end{array}$$

41. $$\begin{array}{r} \overset{1}{0.41} \\ +\,0.29 \\ \hline 0.70 \end{array}$$ or 0.7

43. $$\begin{array}{r} 2.38 \\ -\,0.19 \\ \hline 2.19 \end{array}$$
Check:
$$\begin{array}{r} \overset{1}{2.19} \\ +\,0.19 \\ \hline 2.38 \end{array}$$

45. $$\frac{a}{53,862} = \frac{22.3}{100}$$
$$a \cdot 100 = 53,862 \cdot 22.3$$
$$a \cdot 100 = 1,201,122.6$$
$$\frac{a \cdot 100}{100} = \frac{1,201,122.6}{100}$$
$$a = 12,011.226$$
Therefore, 22.3% of 53,862 rounded to the nearest tenth is 12,011.2.

47. $$\frac{8652}{b} = \frac{119}{100}$$
$$8652 \cdot 100 = b \cdot 119$$
$$865,200 = b \cdot 119$$
$$\frac{865,200}{119} = \frac{b \cdot 119}{119}$$
$$7270.6 \approx b$$
Therefore, to the nearest tenth 8625 is 119% of 7270.6.

Integrated Review

1. $0.12 = 12\%$
2. $0.68 = 68\%$
3. $\frac{1}{4} = 0.25 = 25\%$
4. $\frac{1}{2} = 0.50 = 50\%$
5. $5.2 = 520\%$
6. $7.8 = 780\%$
7. $\frac{3}{50} = \frac{3}{50} \cdot \frac{2}{2} = \frac{6}{100} = 6\%$

8. $\frac{11}{25}=\frac{11}{25}\cdot\frac{4}{4}=\frac{44}{100}=44\%$

9. $2\frac{1}{2}=\frac{5}{2}=\frac{5}{2}\cdot\frac{50}{50}=\frac{250}{100}=250\%$

10. $3\frac{1}{4}=\frac{13}{4}=\frac{3}{14}\cdot\frac{25}{25}=\frac{325}{100}=325\%$

11. $0.03 = 3\%$

12. $0.05 = 5\%$

13. $65\% = 65.\% = 0.65$

14. $31\% = 31.\% = 0.31$

15. $8\% = 8.\% = 0.08$

16. $7\% = 7.\% = 0.07$

17. $142\% = 142.\% = 1.42$

18. $538\% = 538.\% = 5.38$

19. $2.9\% = 0.029$

20. $6.6\% = 0.066$

21. $3\% = \frac{3}{100}$

22. $8\% = \frac{8}{100} = \frac{2}{25}$

23. $5.25\% = \frac{5.25}{100} = \frac{525}{10{,}000} = \frac{21}{400}$

24. $12.75\% = \frac{12.75}{100} = \frac{1275}{10{,}000} = \frac{51}{400}$

25. $38\% = \frac{38}{100} = \frac{19}{50}$

26. $45\% = \frac{45}{100} = \frac{9}{20}$

27. $12\frac{1}{3}\% = \frac{37}{3}\% = \frac{37}{3}\div 100 = \frac{37}{3}\cdot\frac{1}{100} = \frac{37}{300}$

28.
$$\begin{aligned}16\frac{2}{3} &= \frac{50}{3}\% \\ &= \frac{50}{3}\div 100 \\ &= \frac{50}{3}\cdot\frac{1}{100} \\ &= \frac{50}{300} \\ &= \frac{1}{6}\end{aligned}$$

29.
$$\begin{aligned}\frac{a}{70} &= \frac{12}{100} \\ \frac{a}{70} &= \frac{3}{25} \\ a\cdot 25 &= 70\cdot 3 \\ a\cdot 25 &= 210 \\ \frac{a\cdot 25}{25} &= \frac{210}{25} \\ a &= 8.4\end{aligned}$$
12% of 70 is 8.4.

30.
$$\begin{aligned}\frac{36}{b} &= \frac{36}{100} \\ \frac{36}{b} &= \frac{9}{25} \\ 36\cdot 25 &= b\cdot 9 \\ 900 &= b\cdot 9 \\ \frac{900}{9} &= \frac{b\cdot 9}{9} \\ 100 &= b\end{aligned}$$
36 is 36% of 100.

31.
$$\begin{aligned}\frac{212.5}{b} &= \frac{85}{100} \\ \frac{212.5}{b} &= \frac{17}{20} \\ 212.5\cdot 20 &= b\cdot 17 \\ 4250 &= b\cdot 17 \\ \frac{4250}{17} &= \frac{b\cdot 17}{17} \\ 250 &= b\end{aligned}$$
212.5 is 85% of 250.

32.
$$\frac{66}{55}=\frac{p}{100}$$
$$66\cdot 100=55\cdot p$$
$$6600=55\cdot p$$
$$\frac{6600}{55}=\frac{55\cdot p}{55}$$
$$120=p$$
66 is 120% of 55.

33.
$$\frac{23.8}{85}=\frac{p}{100}$$
$$23.8\cdot 100=85\cdot p$$
$$2380=85\cdot p$$
$$\frac{2380}{85}=\frac{85\cdot p}{85}$$
$$28=p$$
23.8 is 28% of 85.

34.
$$\frac{a}{200}=\frac{38}{100}$$
$$\frac{a}{200}=\frac{19}{50}$$
$$a\cdot 50=200\cdot 19$$
$$a\cdot 50=3800$$
$$\frac{a\cdot 50}{50}=\frac{3800}{50}$$
$$a=76$$
38% of 200 is 76.

35.
$$\frac{a}{44}=\frac{25}{100}$$
$$\frac{a}{44}=\frac{1}{4}$$
$$a\cdot 4=44\cdot 1$$
$$a\cdot 4=44$$
$$\frac{a\cdot 4}{4}=\frac{44}{4}$$
$$a=11$$
11 is 25% of 44.

36.
$$\frac{128.7}{99}=\frac{p}{100}$$
$$128.7\cdot 100=99\cdot p$$
$$12,870=99\cdot p$$
$$\frac{12,870}{99}=\frac{99\cdot p}{99}$$
$$130=p$$
130% of 99 is 128.7.

37.
$$\frac{215}{250}=\frac{p}{100}$$
$$\frac{43}{50}=\frac{p}{100}$$
$$43\cdot 100=50\cdot p$$
$$4300=50\cdot p$$
$$\frac{4300}{50}=\frac{50\cdot p}{50}$$
$$86=p$$
86% of 250 is 215.

38.
$$\frac{a}{84}=\frac{45}{100}$$
$$\frac{a}{84}=\frac{9}{20}$$
$$a\cdot 20=84\cdot 9$$
$$a\cdot 20=756$$
$$\frac{a\cdot 20}{20}=\frac{756}{2}$$
$$a=37.8$$
37.8 is 45% of 84.

39.
$$\frac{63}{b}=\frac{42}{100}$$
$$\frac{63}{b}=\frac{21}{50}$$
$$63\cdot 50=b\cdot 21$$
$$3150=b\cdot 21$$
$$\frac{3150}{21}=\frac{b\cdot 21}{21}$$
$$150=b$$
63 is 42% of 150.

40. $\frac{58.9}{b} = \frac{95}{100}$

$\frac{58.9}{b} = \frac{19}{20}$

$58.9 \cdot 20 = b \cdot 19$

$1178 = b \cdot 19$

$\frac{1178}{19} = \frac{b \cdot 19}{19}$

$62 = b$

58.9 is 95% of 62.

Section 7.4

Practice Problems

1. 775 is 31% of what number?

$775 = 31\% \cdot x$

$775 = 0.31 \cdot x$

$\frac{775}{0.31} = \frac{0.31 \cdot x}{0.31}$

$2500 = x$

There are 2500 students that go to Euclid University.

2. 130 is what percent of 190?

$130 = x \cdot 190$

$\frac{130}{190} = \frac{x \cdot 190}{190}$

$0.684 \approx x$

$68.4\% \approx x$

About 68.4% of the calories are from fat.

3. What number is 94% of 109 million?

$x = 94\% \cdot 109$

$x = 0.94 \cdot 109$

$x = 102.46$

The increase in vehicles is 102.46 million. The number of vehicles in 1998 is 109 million + 102.46 million = 211.46 million.

4. $\text{percent increase} = \frac{\text{amount of increase}}{\text{original amount}}$

$= \frac{333 - 285}{285}$

$= \frac{48}{285}$

≈ 0.168

$= 16.8\%$

The attendance increased by about 16.8%.

5. $\text{percent decrease} = \frac{\text{amount of decrease}}{\text{original amount}}$

$= \frac{20,145 - 18,430}{20,145}$

$= \frac{1715}{20,145}$

≈ 0.085

$= 8.5\%$

The population decreased by about about 8.5%.

Exercise Set 7.4

1. Let n = the number of bolts inspected.

$1.5\% \cdot n = 24$

$0.015 \cdot n = 24$

$\frac{0.015 \cdot n}{0.015} = \frac{24}{0.015}$

$n = 1600$

1600 bolts were inspected.

3. Let x = the hours billed per 40 hour week.

$x = 75\% \cdot 40$

$x = (0.75)(40)$

$x = 30$

The owner can bill 30 hours.

5. Let x = the percent of income spent on food.

$\$300 = x \cdot \$2000$

$\frac{300}{2000} = \frac{x \cdot 2000}{2000}$

$0.15 = x$

$0.15 \cdot 100\% = x$

$15\% = x$

She spends 15% of her monthly income on food.

7. Let x = the number of defective components.

$x = 1.04\% \cdot 28,350$
$x = 0.0104 \cdot 28,350$
$x = 294.84$
$x \approx 295$

295 defective components are expected.

9. Let x = the percent of members with some direct connection.

$82 = x \cdot 535$
$\frac{82}{535} = \frac{x \cdot 535}{535}$
$0.153 = x$
$0.153 \cdot 100\% = x$
$15.3\% = x$

15.3% of the members had some direct connection.

11. Let n = the number of registered dental hygienists in the U.S.

$98.3\% \cdot n = 98,400$
$0.983 \cdot n = 98,400$
$\frac{0.983n}{0.983} = \frac{98,400}{0.983}$
$n \approx 100,102$

There are 100,102 dental hygienists in the U.S.

13. Let x = the percent of calories from fat.

$35 = x \cdot 120$
$\frac{35}{120} = \frac{x \cdot 120}{120}$
$0.292 = x$
$0.292 \cdot 100\% = x$
$29.2\% = x$

29.2% of the food's total calories is from fat.

15. What is 19% of 3,340,000?

$19\% \cdot 3,340,000 = x$
$0.19 \cdot 3,340,000 = x$
$634,600 = x$

In 2000, the sales are 3,340,000 + 634,600 = 3,974,600.
What is 19% of 3,974,600?

$19\% \cdot 3,974,600 = x$
$0.19 \cdot 3,974,600 = x$
$755,174 = x$

In 2001, the sales are 3,974,600 + 755,174 = 4,729,774.

17. 20% of \$170 is what number?

$20\% \cdot \$170 = n$
$0.2 \cdot \$170 = n$
$\$34.00 = n$

Their new bill is \$170 – \$34 = \$136.

19. 4.5% of \$19,286 is what number?

$4.5\% \cdot \$19,286 = n$
$0.045 \cdot \$19,286 = n$
$\$867.87 = n$

The increase is \$867.87. The new price is \$19,286 + \$867.87 = \$20,153.87.

21. 33% of 26.8 million is what number?

$33\% \cdot 26.8 \text{ million} = n$
$0.33 \cdot 26,800,000 = n$
$8,844,000 = n$

projected population:
26.8 million + 8.844 million
= 35.644 million
The increase is 8.844 million. The projected population is 35.644 million.

	Original Amount	New Amount	Amount of Increase	Percent Increase
23.	40	50	$50 - 40 = 10$	$\frac{10}{40} = 0.25 = 25\%$
25.	85	187	$187 - 85 = 102$	$\frac{102}{85} = 1.2 = 120\%$

	Original Amount	New Amount	Amount of Decrease	Percent Decrease
27.	8	6	$8 - 6 = 2$	$\frac{2}{8} = 0.25 = 25\%$
29.	160	40	$160 - 40 = 120$	$\frac{120}{160} = 0.75 = 75\%$

31. percent decrease
$$\begin{aligned} &= \frac{150-84}{150} \\ &= \frac{66}{150} \\ &= 0.44 \\ &= 44\% \end{aligned}$$

33. percent increase
$$\begin{aligned} &= \frac{23.7-19.5}{19.5} \\ &= \frac{4.2}{19.5} \\ &\approx 0.215 \\ &= 21.5\% \end{aligned}$$

35. percent decrease
$$\begin{aligned} &= \frac{10,845-10,700}{10,845} \\ &= \frac{145}{10,845} \\ &\approx 0.013 \\ &= 1.3\% \end{aligned}$$

37. percent decrease
$$\begin{aligned} &= \frac{52.1-7.2}{52.1} \\ &= \frac{44.9}{52.1} \\ &\approx 0.862 \\ &= 86.2\% \end{aligned}$$

39. $\text{percent increase} = \frac{300{,}000 - 75{,}000}{75{,}000}$
$= \frac{225{,}000}{75{,}000}$
$= 3$
$= 300\%$

41. $\text{percent increase} = \frac{29{,}000 - 16{,}000}{16{,}000}$
$= \frac{13{,}000}{16{,}000}$
$= 0.8125$
$= 81.25\%$ or 81.3% when rounded to the nearest tenth

43.
$$\begin{array}{r} 0.12 \\ \times\ 38 \\ \hline 96 \\ 360 \\ \hline 4.56 \end{array}$$

45.
$$\begin{array}{r} 9.20 \\ +\ 1.98 \\ \hline 11.18 \end{array}$$

47.
$$\begin{array}{r} 78.00 \\ -\ 19.46 \\ \hline 58.54 \end{array}$$

49. $\text{percent increase} = \frac{28{,}700 - 26{,}518}{26{,}518}$
$= \frac{2182}{26{,}518}$
≈ 0.082
$= 8.2\%$

51. The increased number is double the original number.

53. Answers may vary.

Section 7.5

Practice Problems

1. $\text{tax} = 6\% \cdot \$29.90$
$= 0.06 \cdot \$29.9$
$= \$1.79$

$\text{total} = \$29.90 + \$1.79$
$= \$31.69$

The tax is \$1.79 and the total is \$31.69.

2. $\$1080 = r \cdot \$13{,}500$
$\frac{\$1080}{\$13{,}500} = \frac{r \cdot 13{,}500}{13{,}500}$
$0.08 = r$
$8\% = r$

The sales tax rate is 8%.

3. $\text{commission} = 6.6\% \cdot \$37{,}632$
$= 0.066 \cdot \$37{,}632$
$= \$2483.712$

His commission is \$2483.71.

4. $\$1290 = r \cdot \$8600$
$\frac{\$1290}{\$8600} = \frac{r \cdot 8600}{8600}$
$0.15 = r$
$15\% = r$

The commission rate is 15%.

5. $\text{discount} = 15\% \cdot \$700$
$= 0.15 \cdot \$700$
$= \$105$

$\text{sale price} = \$700 - \$105$
$= \$595$

The discount is \$105 and the sale price is \$595.

Exercise Set 7.5

1. $\text{tax} = 5\% \cdot \$150.00$
$= 0.05 \cdot \$150.00$
$= \$7.50$

The sales tax is \$7.50.

3. $\text{tax} = 7.5\% \cdot \$799$
$= 0.075 \cdot \$799$
$= \$59.925$
$\text{total} = \$799 + \$59.925$
$= \$858.925$
The total price is $858.93.

5. $\$54 = r \cdot \$600$
$\frac{\$54}{\$600} = r$
$0.09 = r$
$9\% = r$
The sales tax rate is 9%.

7. $\text{tax} = 8.5\% \cdot \$220$
$= 0.085 \cdot \$220$
$= \$18.70$
$\text{total} = \$220 + \$18.70$
$= \$238.70$
The total price is $238.70.

9. $\text{tax} = 6.5\% \cdot \$1800$
$= 0.065 \cdot \$1800$
$= \$117$
$\text{total} = \$1800 + \$117 = \$1917$
The total price is $1917.

11. $\$920 = 8\% \cdot n$
$\$920 = 0.08 \cdot n$
$\frac{\$920}{0.08} = n$
$\$11,500 = n$
The purchase price is $11,500.

13. $\text{total purchase} = \$90 + \$15 = \$105$
$\text{tax} = 7\% \cdot \$105$
$= 0.07 \cdot \$105$
$= \$7.35$
$\text{total} = \$105 + \$7.35 = \$112.35$
The total price is $112.35.

15. $\$98.70 = r \cdot \$1645$
$\frac{\$98.70}{\$1645} = r$
$0.06 = r$
$6\% = r$
The sales tax rate is 6%.

17. $\text{commission} = 4\% \cdot \$1,236,856$
$= 0.04 \cdot \$1,236,856$
$= \$49,474.24$
Her commission was $49,474.24.

19. $\$1380.40 = r \cdot \$9860.00$
$\frac{\$1380.40}{\$9860.00} = r$
$0.14 = r$
$14\% = r$
The commission rate is 14%.

21. $\text{commission} = 1.5\% \cdot \$125,900$
$= 0.015 \cdot \$125,900$
$= \$1888.50$
His commission is $1888.50.

23. $\$2565 = r \cdot \$85,500$
$\frac{\$2565}{\$85,500} = r$
$0.03 = r$
$3\% = r$
The commission rate is 3%.

	Original Price	Discount Rate	Amount of Discount	Sale Price
25.	\$68.00	10%	\$68.00 · 10% = \$68.00 · 0.10 = \$6.80	\$68.00 – \$6.80 = \$61.20
27.	\$96.50	50%	\$96.50 · 50% = \$9650 · 0.5 = \$48.25	\$96.50 – \$48.25 = \$48.25
29.	\$215.00	35%	\$215.00 · 35% = \$215.00 · 0.35 = \$75.25	\$215.00 – \$75.25 = \$139.75
31.	\$21,700.00	15%	\$21,700.00 · 15% = \$21,700.00 · 0.15 = \$3255.00	\$21,700.00 – \$3255.00 = \$18,445.00

33.
$$\begin{aligned}\text{discount} &= 15\% \cdot \$300\\ &= 0.15 \cdot \$300\\ &= \$45\\ \text{sale price} &= \$300 - \$45\\ &= \$255\end{aligned}$$

35. $2000 \cdot 0.3 \cdot 2 = 600 \cdot 2 = 1200$

37. $400 \cdot 0.03 \cdot 11 = 12 \cdot 11 = 132$

39.
$$\begin{aligned}600 \cdot 0.04 \cdot \frac{2}{3} &= 24 \cdot \frac{2}{3}\\ &= \frac{24}{1} \cdot \frac{2}{3}\\ &= \frac{24 \cdot 2}{3}\\ &= \frac{48}{3}\\ &= 16\end{aligned}$$

41.
$$\begin{aligned}\text{tax} &= 7.5\% \cdot \$24,966\\ &= 0.075 \cdot \$24,966\\ &= \$1872.45\\ \text{total price} &= \$24,966 + \$1872.45\\ &= \$26,838.45\end{aligned}$$

The total price is \$26,838.45.

43. 60% discount:
$\$50 \cdot 60\% = \$50 \cdot 0.6 = \$30$
discount price = $\$50 - \$30 = \$20$
30% discount followed by a 35% discount
$\$50 \cdot 30\% = \$50 \cdot 0.3 = \$15$
$\$50 - \$15 = \$35$
$\$35 \cdot 35\% = \$35 \cdot 0.35 = \$12.25$
discount price = $\$35 - \$12.25 = \$22.75$
A discount of 60% is better.

Section 7.6

Practice Problems

1. $I = P \cdot R \cdot T$
$= \$750 \cdot 8\% \cdot 3$
$= \$750 \cdot 0.08 \cdot 3$
$= \$180$

2. time = 9 months = $\frac{9}{12}$ year = $\frac{3}{4}$ year
$I = P \cdot R \cdot T$
$= \$800 \cdot 20\% \cdot \frac{3}{4}$
$= \$800 \cdot 0.2 \cdot \frac{3}{4}$
$= \$120$
She paid \$120 in interest.

3. time = 6 months = $\frac{6}{12}$ year = $\frac{1}{2}$ year
$I = P \cdot R \cdot T$
$= \$500 \cdot 12\% \cdot \frac{1}{2}$
$= \$500 \cdot 0.12 \cdot \frac{1}{2}$
$= \$30$
total amount $= \$500 + \$30$
$= \$530$

4. total amount
= original principal · compound interest factor
$= \$5500 \cdot 1.41902$
$= \$7804.61$

5. interest earned
= total amount – original principal
$= \$9933.14 - \$5500$
$= \$4433.14$

6. monthly payment
$= \dfrac{\text{principal} + \text{interest}}{\text{total number of payments}}$
$= \dfrac{\$3000 + \$1123.58}{3 \cdot 12}$
$\approx \$114.54$

The monthly payment is \$114.54

Calculator Explorations

1. $\left(1 + \frac{0.09}{4}\right)^{(4 \cdot 5)} \approx 1.56051$

2. $\left(1 + \frac{0.14}{365}\right)^{(365 \cdot 15)} \approx 8.16288$

3. $\left(1 + \frac{0.11}{1}\right)^{(1 \cdot 20)} \approx 8.06231$

4. $\left(1 + \frac{0.07}{2}\right)^{(2 \cdot 1)} \approx 1.07123$

5. Compound interest factor $= \left(1 + \frac{0.06}{4}\right)^{(4 \cdot 4)}$
≈ 1.26899
$\$500 \cdot 1.26899 \approx \$634.50$

6. Compound interest factor
$= \left(1 + \frac{0.05}{365}\right)^{(365 \cdot 19)}$
≈ 2.58554
$\$2500 \cdot 2.58554 = \$6463.85$

Exercise Set 7.6

1. simple interest = principal · rate · time
$= \$200 \cdot 8\% \cdot 2$
$= \$200 \cdot (0.08) \cdot 2$
$= \$32$

3. simple interest = principal · rate · time
$= \$160 \cdot 11.5\% \cdot 4$
$= \$160 \cdot (0.115) \cdot 4$
$= \$73.60$

5. $$\begin{aligned}\text{simple interest} &= \text{principal} : \text{rate} \cdot \text{time}\\ &= \$5000 \cdot 10\% \cdot 1\frac{1}{2}\\ &= \$5000 \cdot (0.10) \cdot 1.5\\ &= \$750\end{aligned}$$

7. $$\begin{aligned}\text{simple interest} &= \text{principle} \cdot \text{rate} \cdot \text{time}\\ &= \$375 \cdot 18\% \cdot \frac{6}{12}\\ &= \$375 \cdot (0.18) \cdot (0.5)\\ &= \$33.75\end{aligned}$$

9. $$\begin{aligned}\text{simple interest} &= \text{principal} \cdot \text{rate} \cdot \text{time}\\ &= \$2500 \cdot 16\% \cdot \frac{21}{12}\\ &= \$2500 \cdot (0.16) \cdot 1.75\\ &= \$700\end{aligned}$$

11. $$\begin{aligned}\text{simple interest} &= \text{principal} \cdot \text{rate} \cdot \text{time}\\ &= \$62,500 \cdot 12.5\% \cdot 2\\ &= \$62,500 \cdot (0.125) \cdot 2\\ &= \$15,625\end{aligned}$$

total = \$62,500 + \$15,625 = \$78,125.
The total amount paid is \$78,125.

13. $$\begin{aligned}\text{simple interest} &= \text{principal} \cdot \text{rate} \cdot \text{time}\\ &= \$5000 \cdot 9\% \cdot \frac{15}{12}\\ &= \$5000 \cdot (0.09) \cdot 1.25\\ &= \$562.50\end{aligned}$$

total = \$5000 + \$562.50 = \$5562.50
The total amount received is \$5562.50

15. $$\begin{aligned}\text{simple interest} &= \text{principal} \cdot \text{rate} \cdot \text{time}\\ &= \$8500 \cdot 12\% \cdot 4\\ &= \$8500 \cdot (0.12) \cdot 4\\ &= \$4080\end{aligned}$$

total = \$8500 + \$4080 = \$12,580
She pays back a total of \$12,580.

17. $$\begin{aligned}\text{total} &= \text{principal} \cdot \text{compound interest factor}\\ &= \$6150(7.61226)\\ &= \$46,815.399\end{aligned}$$

The total amount is \$46,815.40.

19. $$\begin{aligned}\text{total} &= \text{principal} \cdot \text{compound interest factor}\\ &= \$1560(1.49176)\\ &= \$2327.1456\end{aligned}$$

The total amount is \$2327.15.

21. total = principal · compound interest factor
$= \$10,000(5.81636)$
$= \$58,163.60$
The total amount is \$58,163.60.

23. total = principal · compound interest factor
$= \$2675(1.09000)$
$= \$2915.75$
compound interest
= total amount − original principal
$= \$2915.75 - \$2675$
$= \$240.75$
The interest earned is \$240.75.

25. total = original principal · compound interest factor
$= \$2000(1.46933)$
$= \$2938.66$
compound interest = total amount − original principal
$= \$2938.66 - \$2000$
$= \$938.66$
The interest earned is \$938.66.

27. total = original principal · compound interest factor
$= \$2000(1.48595)$
$= \$2971.90$
compound interest = total amount − original principal
$= \$2971.90 - \$2000$
$= \$971.90$
The interest earned is \$971.90.

29.
$$\text{monthly payment} = \frac{\text{principal} + \text{interest}}{\text{number of payments}} = \frac{\$1500 + \$61.88}{6} = \frac{\$1561.88}{6} \approx \$260.31$$
The monthly payment is \$260.31.

31.
$$\text{monthly payment} = \frac{\text{principal} + \text{interest}}{\text{number of payments}} = \frac{\$20,000 + \$10,588.70}{48} = \frac{\$30,588.70}{48} \approx \$637.26$$
The monthly payment is \$637.26.

33. $-5 + (-24) = -29$

35. $(-5)(-10) = 50$

37. $\dfrac{7-10}{3} = \dfrac{-3}{3} = -1$

39. Answers may vary.

41. Answers may vary.

Chapter 7 Review

1. $\frac{37}{100} = 37\%$

2. $\frac{77}{100} = 77\%$

3. $83\% = 0.83$

4. $75\% = 0.75$

5. $73.5\% = 0.735$

6. $1.5\% = 0.015$

7. $125\% = 1.25$

8. $145\% = 1.45$

9. $0.5\% = 0.005$

10. $0.7\% = 0.007$

11. $200\% = 2.00$ or 2

12. $400\% = 4.00$ or 4

13. $26.25\% = 0.2625$

14. $85.34\% = 0.8534$

15. $2.6 = 260\%$

16. $0.055 = 5.5\%$

17. $0.35 = 35\%$

18. $1.02 = 102\%$

19. $0.725 = 72.5\%$

20. $0.25 = 25\%$

21. $0.076 = 7.6\%$

22. $0.085 = 8.5\%$

23. $0.75 = 75\%$

24. $0.65 = 65\%$

25. $4.00 = 400\%$

26. $9.00 = 900\%$

27. $1\% = \frac{1}{100}$

28. $10\% = \frac{10}{100} = \frac{1}{10}$

29. $25\% = \frac{25}{100} = \frac{1}{4}$

30. $8.5\% = \frac{8.5}{100} = \frac{85}{1000} = \frac{17}{200}$

31. $10.2\% = \frac{10.2}{100} = \frac{102}{1000} = \frac{51}{500}$

32.
$$\begin{aligned} 16\frac{2}{3}\% &= \frac{16\frac{2}{3}}{100} \\ &= 16\frac{2}{3} \div 100 \\ &= \frac{50}{3} \cdot \frac{1}{100} \\ &= \frac{50}{300} \\ &= \frac{1}{6} \end{aligned}$$

33. $33\frac{1}{3}\% = \dfrac{33\frac{1}{3}}{100} = 33\frac{1}{3} \div 100 = \dfrac{100}{3} \cdot \dfrac{1}{100} = \dfrac{1}{3}$

34. $110\% = \frac{110}{100} = \frac{11}{10}$ or $1\frac{1}{10}$

35. $\dfrac{1}{5} = \dfrac{1}{5} \cdot 100\% = \dfrac{100\%}{5} = 20\%$

36. $\frac{7}{10} = \frac{7}{10} \cdot 100\% = 70\%$

37. $\frac{5}{6} \cdot 100\% = \frac{500\%}{6} = \frac{250\%}{3} = 83\frac{1}{3}\%$

38. $\frac{5}{8} = \frac{5}{8} \cdot 100\%$
$= \frac{500\%}{8}$
$= \frac{125}{2}\%$
$= 62\frac{1}{2}\%$ or 62.5%

39. $1\frac{2}{3} = \frac{5}{3} = \frac{5}{3} \cdot 100\% = \frac{500\%}{3} = 166\frac{2}{3}\%$

40. $1\frac{1}{4} = \frac{5}{4} = \frac{5}{4} \cdot 100\% = 125\%$

41. $\frac{3}{5} = \frac{3}{5} \cdot 100\% = \frac{300\%}{5} = 60\%$

42. $\frac{1}{16} = \frac{1}{16} \cdot 100\%$
$= \frac{100\%}{16}$
$= \frac{25}{4}\%$
$= 6\frac{1}{4}\%$ or 6.25%

43. $90\% = \frac{90}{100} = \frac{9}{10}$; $10\% = \frac{10}{100} = \frac{1}{10}$

44. $96\% = \frac{96}{100} = \frac{24}{25}$

45. $\frac{21}{25} = \frac{21}{25} \cdot 100\% = 84\%$

46. $150\% = 1.50$ or 1.5

47. $1250 = 1.25\% \cdot n$
$1250 = 0.0125n$
$\frac{1250}{0.0125} = n$
$100{,}000 = n$

48. $n = 33\frac{1}{3}\% \cdot 24{,}000$
$n = \frac{33\frac{1}{3}}{100} \cdot 24{,}000$
$n = 33\frac{1}{3} \cdot 240$
$n = \frac{100}{3} \cdot 240$
$n = 8000$

49. $124.2 = n \cdot 540$
$\frac{124.2}{540} = n$
$0.23 = n$
$23\% = n$

50. $22.9 = 20\% \cdot n$
$22.9 = 0.2n$
$\frac{22.9}{0.2} = n$
$114.5 = n$

51. $n = 40\% \cdot 7500$
$n = (0.40)(7500)$
$n = 3000$

52. $693 = n \cdot 462$
$\frac{693}{462} = n$
$1.5 = n$
$150\% = n$

53. $\frac{104.5}{b} = \frac{25}{100}$
$b \cdot 25 = 104.5 \cdot 100$
$b \cdot 25 = 10{,}450$
$b = \frac{10{,}450}{25}$
$b = 418$
Therefore, 25% of 418 is 104.5.

54. $\frac{16.5}{b} = \frac{5.5}{100}$
$b \cdot 5.5 = 16.5 \cdot 100$
$b \cdot 5.5 = 1650$
$b = \frac{1650}{5.5}$
$b = 300$
Therefore, 5.5% of 300 is 16.5.

55. $\frac{a}{180} = \frac{36}{100}$
$a \cdot 100 = 36 \cdot 180$
$a \cdot 100 = 6480$
$a = \frac{6480}{100}$
$a = 64.8$
Therefore, 36% of 180 is 64.8.

56. $\frac{63}{35} = \frac{p}{100}$
$63 \cdot 100 = 35 \cdot p$
$6300 = 35 \cdot p$
$\frac{6300}{35} = p$
$180 = p$
Therefore, 180% of 35 is 63.

57. $\frac{93.5}{85} = \frac{p}{100}$
$93.5 \cdot 100 = 85 \cdot p$
$9350 = 85 \cdot p$
$\frac{9350}{85} = p$
$110 = p$
Therefore, 110% of 85 is 93.5.

58. $\frac{a}{500} = \frac{33}{100}$
$a \cdot 100 = 500 \cdot 33$
$a \cdot 100 = 16,500$
$a = \frac{16,500}{100}$
$a = 165$
Therefore, 33% of 500 is 165.

59. What percent of 2000 is 1320?
$n \cdot 2000 = 1320$
$n = \frac{1320}{2000}$
$n = 0.66$
$n = 66\%$

60. What percent of 12,360 is 2000?
$n \cdot 12,360 = 2000$
$n = \frac{2000}{12,360} \approx 0.16$
About 16% of entering freshmen are enrolled in Basic College Mathematics.

61. $\text{percent increase} = \frac{\text{amount of increase}}{\text{original amount}}$
$= \frac{33-16}{16}$
$= \frac{17}{16}$
$= 1.0625$
$= 106.25\%$

62. $\text{percent decrease} = \frac{\text{amount of decrease}}{\text{original amount}}$
$= \frac{675-534}{675}$
$= \frac{141}{675}$
$= 0.208\overline{8}$
$\approx 20.9\%$

63. $\text{decrease} = 4\% \cdot \$215,000$
$= 0.04 \cdot 215,000 = \$8600$
$\text{Next year} = \text{this year} - \text{decrease}$
$= \$215,000 - \$8600 = \$206,400$

64. 15% of \$11.50 is what number?
$(0.15)(\$11.50) = n$
$\$1.725 = n$
New hourly rate is
$\$11.50 + \$1.725 = \$13.225 \approx \$13.23$

65. Sales tax is
$(5.5\%)(\$250) = (0.055)(\$250) = \$13.75.$
Total price of the coat is
$\$250 + \$13.75 = \$263.75.$

66. Sales tax is
$(4.5\%)(\$25.50) = (0.045)(\$25.50)$
$= \$1.1475$
$\approx \$1.15$

67. $\text{Commission} = 5\% \cdot \$100,000$
$= 0.05 \cdot \$100,000$
$= \$5000$

68. $\text{Commission} = (7.5\%)(\$4005)$
$= (0.075)(\$4005)$
$= \$300.375$
$\approx \$300.38$

69. $$\begin{aligned}\text{discount} &= 30\% \cdot \$3000\\ &= 0.3 \cdot \$3000\\ &= \$900\\ \text{sale price} &= \$3000 - \$900\\ &= \$2100\end{aligned}$$

The discount is \$900 and the sale price is \$2100.

70. $$\begin{aligned}\text{discount} &= 10\% \cdot \$90\\ &= 0.1 \cdot \$90\\ &= \$9\\ \text{sale price} &= \$90 - \$9\\ &= \$81\end{aligned}$$

The discount is \$9 and the sale price is \$81.

71. $$\begin{aligned}\text{simple interest} &= \text{principal} \cdot \text{rate} \cdot \text{time}\\ &= \$4000 \cdot 12\% \cdot \frac{3}{12}\\ &= \$4000 \cdot 0.12 \cdot 0.25\\ &= \$120\end{aligned}$$

72. $$\begin{aligned}\text{simple interest} &= \text{principle} \cdot \text{rate} \cdot \text{time}\\ &= \$1200 \cdot 15\% \cdot \frac{8}{12}\\ &= \$1200 \cdot 0.15 \cdot \frac{2}{3}\\ &= \$120\end{aligned}$$

total due = \$1200 + \$120 = \$1320

73. The compound interest factor for 15 years at 12% compounded annually is 5.47357.

$$\begin{aligned}\text{total amount} &= \$5500 \cdot 5.47357\\ &= \$30{,}104.635\end{aligned}$$

The total is \$30,104.64.

74. The compound interest factor for 10 years at 11% compounded semiannually is 2.91776.

$$\begin{aligned}\text{total amount} &= \$6000 \cdot 2.91776\\ &= \$17{,}506.56\end{aligned}$$

The total is \$17,506.56.

75. The compound interest factor for 5 years at 12% compounded quarterly is 1.80611.

total amount = \$100 · 1.80611 ≈ \$180.61

Interest earned

= total amount – original principal

= \$180.61 – \$100 = \$80.61

76. The compound interest factor for 20 years at 18% compounded quarterly is 33.83010.

$$\begin{aligned}\text{Total amount} &= \$1000 \cdot 33.83010\\ &= \$33{,}830.10\\ \text{Interest earned} &= \$33{,}830.10 - \$1000\\ &= \$32{,}830.10\end{aligned}$$

Chapter 7 Test

1. $85\% = 85.\% = 0.85$

2. $500\% = 500.\% = 5$

3. $0.6\% = 0.006$

4. $0.056 = 5.6\%$

5. $6.1 = 6.10 = 610.\% = 610\%$

6. $0.35 = 35\%$

7. $120\% = \frac{120}{100} = \frac{6}{5}$

8. $38.5\% = \frac{38.5}{100} = \frac{385}{1000} = \frac{77}{200}$

9. $0.2\% = \frac{0.2}{100} = \frac{2}{1000} = \frac{1}{500}$

10. $\frac{11}{20} = \frac{11}{20} \cdot \frac{5}{5} = \frac{55}{100} = 55\%$

11. $\frac{3}{8} = 0.375 = 37.5\%$

12. $1\frac{3}{4} = \frac{7}{4} = \frac{7}{4} \cdot \frac{25}{25} = \frac{175}{100} = 175\%$

13. $\frac{1}{5} = \frac{1}{5} \cdot 100\% = \frac{100}{5}\% = 20\%$

14. $64\% = \frac{64}{100} = \frac{16}{25}$

15. $n = 42\% \cdot 80$

$n = (0.42)(80)$

$n = 33.6$

Therefore, 42% of 80 is 33.6.

16. $0.6\% \cdot n = 7.5$
$0.006 \cdot n = 7.5$
$n = \frac{7.5}{0.006}$
$n = 1250$
Therefore, 0.6% of 1250 is 7.5.

17. $567 = x \cdot 756$
$\frac{567}{756} = \frac{x \cdot 756}{x}$
$0.75 = x$
$75\% = x$
Therefore, 75% of 756 is 567.

18. 12% of 320 is the amount of copper.
$12\% \cdot 320 = n$
$0.12 \cdot 320 = n$
$38.4 = n$
The alloy contains 38.4 pounds of copper.

19. 20% of what is $11,350?
$0.20n = \$11,350$
$n = \frac{\$11,350}{0.20}$
$n = \$56,750$
The value is $56,750.

20. $\text{tax} = 1.25\% \cdot \$354$
$= (0.0125)(\$354)$
$= \$4.425$
$\approx \$4.43$
Total price is $354 + $4.43 = $358.43.

21. $\text{percent increase} = \frac{26,460 - 25,200}{25,200}$
$= \frac{1260}{25,200}$
$= 0.05$
$= 5\%$
The population increased 5%.

22. $\text{discount} = 15\% \cdot \$120$
$= 0.15 \cdot \$120$
$= \$18$
$\text{sale price} = \$120 - \$18$
$= \$102$

23. $\text{commission} = 4\% \cdot \$9875$
$= 0.04 \cdot \$9875$
$= \$395$
His commission is $395.

24. $\$1.53 = r \cdot \$152.99$
$\frac{\$1.53}{\$152.99} = r$
$0.01 \approx r$
$1\% = r$

25. $\text{simple interest} = \text{principal} \cdot \text{rate} \cdot \text{time}$
$= \$2000 \cdot 9.25\% \cdot 3\frac{1}{2}$
$= \$2000 \cdot (0.0925) \cdot 3.5$
$= \$647.50$

26. The compound interest factor for 5 years at 8% compounded annually is 1.46933.
$\text{total amount} = \$1365 \cdot 1.46933$
$= \$2005.63545$
The total amount is $2005.64.

27. $\text{interest} = \text{principal} \cdot \text{rate} \cdot \text{time}$
$\text{interest} = \$400 \cdot 13.5\% \cdot \frac{6}{12}$
$= \$400 \cdot (0.135) \cdot 0.5$
$= \$27.00$
Total amount due the bank = $400 + $27
= $427

28. $\text{percent decrease} = \frac{21,212 - 17,637}{21,212}$
$= \frac{3575}{21,212}$
≈ 0.169
$= 16.9\%$
The percent decrease is about 16.9%.

Chapter 8

Pretest

1.

Activity	Hours/Day	% of Day	Degrees of Circle
Attending colege classes	4 hours	$16\frac{2}{3}\%$	$16\frac{2}{3}\% \cdot 360° = 60°$
Studying	3 hours	12.5%	$12.5\% \cdot 360° = 45°$
Working	5 hours	$20\frac{5}{6}\%$	$20\frac{5}{6}\% \cdot 360° = 75°$
Sleeping	8 hours	$33\frac{1}{3}\%$	$33\frac{1}{3}\% \cdot 360° = 120°$
Driving	1 hour	$4\frac{1}{6}\%$	$4\frac{1}{6}\% \cdot 360° = 15°$
Other	3 hours	12.5%	$12.5\% \cdot 360° = 45°$

2. The lowest point occurs at April. The least number of burglaries occurred in April.

3. July had 400 burglaries.

4. September had 700 burglaries.

5.
$$3x + y = -1$$
$$3(-2) + 7 = -1$$
$$-6 + 7 = -1$$
$$1 = -1 \quad \text{False}$$
No, (–2, 7) is not a solution.

6. a.
$$y = 2x - 6$$
$$y = 2(0) - 6$$
$$y = 0 - 6$$
$$y = -6$$
The ordered pair is (0, –6).

b.
$$y = 2x - 6$$
$$y = 2(-3) - 6$$
$$y = -6 - 6$$
$$y = -12$$
The ordered pairs is (–3, –12).

c. $$\begin{aligned} y &= 2x - 6 \\ 0 &= 2x - 6 \\ 0 + 6 &= 2x - 6 + 6 \\ 6 &= 2x \\ \frac{6}{2} &= \frac{2x}{2} \\ 3 &= x \end{aligned}$$

The ordered pair is (3, 0).

7. Find 3 ordered pairs that are solutions to $y = 4x - 1$.

Let $x = 0$.

$y = 4x - 1$
$y = 4(0) - 1$
$y = 0 - 1$
$y = -1$

The ordered pair is (0, –1).

Let $x = 1$.

$y = 4x - 1$
$y = 4(1) - 1$
$y = 4 - 1$
$y = 3$

The ordered pair is (1, 3).

Let $x = -1$.

$y = 4x - 1$
$y = 4(-1) - 1$
$y = -4 - 1$
$y = -5$

The ordered pair is (–1, –5).

Plot the 3 ordered pairs. Then draw the line that contains them.

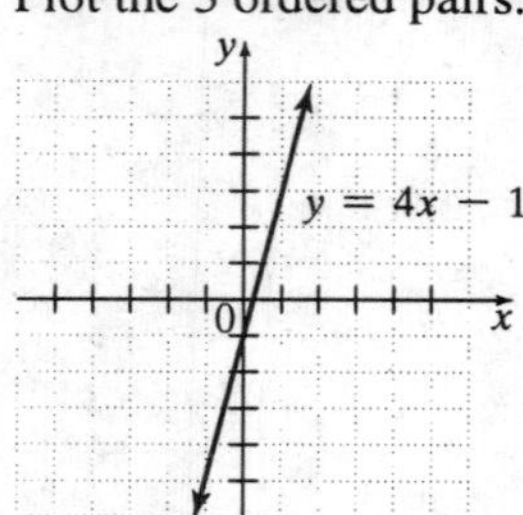

8. $$\text{mean} = \frac{28+36+81+73+28+74+31+74+64+25+74}{11}$$
$$= \frac{588}{11}$$
$$\approx 53.5$$

median: write the numbers in order to identify the middle number.

25, 28, 28, 31, 36, 64, 73, 74, 74, 74, 81

median = 64

mode: 74 is the mode because it is the number that occurs most often.

9.

Grade	Point Value	Credit Hours	(Point Value) · (Credit Hours)
B	3	4	12
B	3	3	9
A	4	3	12
C	2	5	10
D	1	2	2
Totals		17	45

Grade Point Average = $\frac{45}{17} \approx 2.65$

10. possible outcomes: 1, 2, 3, 4, 5, 6

probability of a 4 = $\frac{1}{6}$

11. possible outcomes: 1, 2, 3, 4, 5, 6

probability of a number greater than 3 = $\frac{3}{6} = \frac{1}{2}$

12. possible outcomes: 1, 2, 3, 4, 5, 6

probability of a 3 or a 5 = $\frac{2}{6} = \frac{1}{3}$

Section 8.1

Practice Problems

1. $\frac{\text{10 people preferring oatmeal raisin cookies}}{\text{100 people}} = \frac{10}{100} = \frac{1}{10}$

2. Either elementary or junior high schools
= 59% + 17%
= 76%

3. $\text{amount} = 17\% \cdot 65{,}800$
$= 0.17 \cdot 65{,}800$
$= 11{,}186$

Thus, 11,186 junior high schools had CD-ROM technology in 1999.

4. Freshman $30\% \cdot 360° = 108°$
Sophomores $27\% \cdot 360° = 97.2° \approx 97°$
Juniors $25\% \cdot 360° = 90°$
Seniors $18\% \cdot 360° = 64.8° \approx 65°$

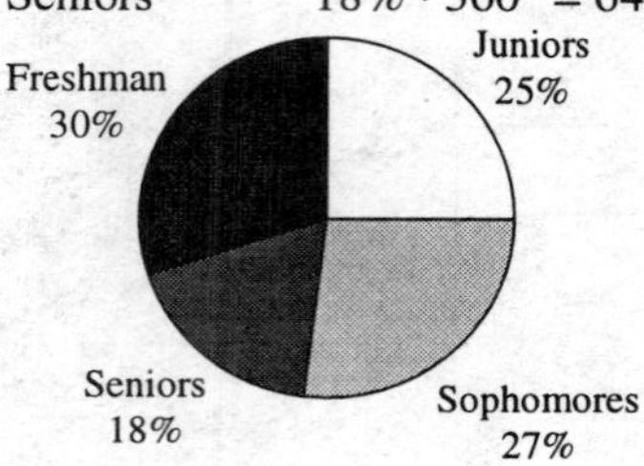

Exercise Set 8.1

1. The largest sector, or 320 students, corresponds to where most college students live. Most college students live at parent or guardian's home.

3. $\frac{\text{Students living in campus housing}}{\text{total students}}$
$= \frac{180}{700} = \frac{9}{35}$

5. $\frac{\text{Students living in campus housing}}{\text{students living at home}}$
$= \frac{180}{320} = \frac{9}{16}$

7. The largest sector is 30% which corresponds to "less than \$20." Less than \$20 is the price range that is purchased most often.

9. The sectors which represent "less than \$30" are "less than \$20" and "\$20 – \$29."
percent = 30% + 23% = 53%

11. amount = $30\% \cdot 4700 = 0.30 \cdot 4700 = 1410$
1410 Central High teenagers spent less than \$20.

13. amount = $28\% \cdot 4700 = 0.28 \cdot 4700 = 1316$
1316 Central High teenagers spent \$30–\$49.

15. The sectors which represent fiction are "Adult Fiction" and Children's Fiction."
percent = 33% + 22% = 55%

17. The second-largest sector is 25% which corresponds to "Nonfiction." Nonfiction is the second-largest category of books.

19.
$$\begin{aligned}\text{amount} &= 25\% \cdot 125,600\\ &= 0.25 \cdot 125,600\\ &= 31,400 \text{ books}\end{aligned}$$
This library has 31,400 nonfiction books.

21.
$$\begin{aligned}\text{amount} &= 22\% \cdot 125,600\\ &= 0.25 \cdot 125,600\\ &= 31,400 \text{ books}\end{aligned}$$
This library has 27,632 books in children's fiction.

23.
$$\begin{aligned}\text{amount} &= (17\% + 3\%) \cdot 125,600\\ &= 20\% \cdot 125,600\\ &= 0.2 \cdot 125,600\\ &= 25,120\end{aligned}$$
This library has 25,120 books in reference or other category.

25. Under 3 days $7\% \cdot 360° = 25.2°$
3 – 10 days $18\% \cdot 360° = 64.8°$
11 – 20 days $14\% \cdot 360° = 50.4°$
21 or more days $61\% \cdot 360° = 219.6°$

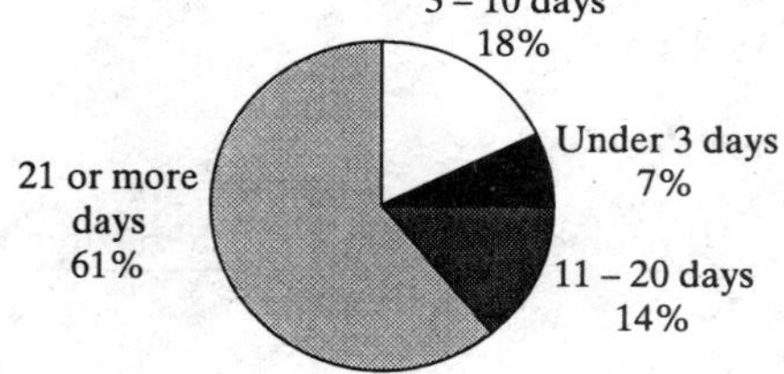

27. $20 = 4 \cdot 5 = 2 \cdot 2 \cdot 5 = 2^2 \cdot 5$

29. $40 = 8 \cdot 5 = 2 \cdot 2 \cdot 2 \cdot 5 = 2^3 \cdot 5$

31. $85 = 5 \cdot 17$

33. Pacific Ocean:
$$\begin{aligned}49\% \cdot 264,489,800 &= 0.49 \cdot 264,489,800\\ &= 129,600,002\end{aligned}$$
The Pacific Ocean is 129,600,002 square kilometers.

35. Indian Ocean:

$$21\% \cdot 264,489,800 = 0.21 \cdot 264,489,800 = 55,542,858$$

The Indian Ocean is 55,542,858 square kilometers.

37. Answers may vary.

Section 8.2

Practice Problems

1. a. United States corresponds to 14 symbols, and each symbol represents 50 billion kilowatt hours of energy. This means that the United States generates approximately $14 \cdot 50$ or 700 billion kilowatt hours.

b. Japan corresponds to 6.5 symbols and Ukraine corresponds to 1.5 symbols. Thus, Japan and Ukraine combined generate approximately $(6.5 + 1.5) \cdot 50$ or 400 billion kilowatt hours.

2. a. First locate the bar that represents insects. Go to the top of the bar and move horizontally left until the scale is reached. The height of the bar on the scale is 20. There are approximately 20 species of insects that are endangered.

b. The most endangered species is represented by the tallest bar. The tallest bar corresponds to birds.

3.

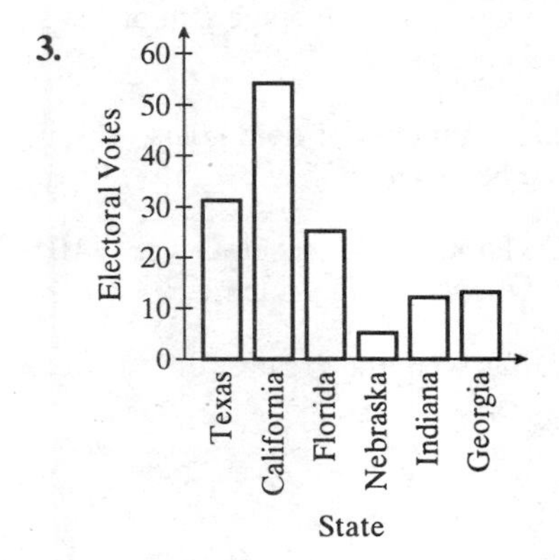

4. a. The month with the lowest temperature corresponds to the lowest point. Follow the lowest point down to the horizontal month scale, which reads January.

b. Locate 25°F on the vertical scale and move across horizontally until you reach a darkened point. The month for that point is December.

c. The points above the temperature of 70°F represent the months June, July, and August.

Exercise Set 8.2

1. The year 1998 has the most cars, so the greatest number of automobiles was manufactured in 1998.

3. There are eight cars for 1996, and each car represents 500 automobiles. Approximately $8 \cdot 500 = 4000$ automobiles were manufactured in 1996.

5. Compare each year with the previous year to see that the production of automobiles decreased from the previous year in 1994, 1995, and 1999.

7. The pictograph must show 8 cars for 4000 manufactured automobiles $\left(\frac{4000}{500} = 8\right)$. The years 1993 and 1996 both show 8 cars.

9. The year 1992 has seven chickens and each chicken represents 3 ounces of chicken, so approximately $7 \cdot 3 = 21$ ounces of chicken were consumed per person per week in 1992.

11. To be greater than 21 ounces, the pictograph must show more than seven chickens $\left(\frac{21}{3} = 7\right)$. The years 1994, 1996, and 1988 have more than seven chickens.

13. There is one more chicken for 1992 than for 1988, so there was an increase of approximately $1 \cdot 3 = 3$ ounces per person per week.

15. The pictograph shows more chickens each year. This indicates a trend of increasing consumption of chicken per person per week in the United States.

17. The tallest bar corresponds to the month of April, so April has the most tornado-related deaths.

19. Draw a horizontal line from the top of the bar above May to the scale on the left. The approximate number of tornado-related deaths is 14.

21. Look for bars that extend above the horizontal line at 5. The months of February, March, April, May, and June have over 5 tornado-related deaths.

23. The longest bar corresponds to Tokyo with an estimated population of 26.5 million or 26,500,000 people.

25. The two U.S. cities are New York City and Los Angeles. New York City is the largest, with an estimated population of 16.2 million or 16,200,000 people.

27. Tokyo's population is an estimated 26.5 million and Bombay's population is an estimated 14.5 million. Tokyo's population is 265 million – 14.5 million = 12 million or 12,000,000 larger than Bombay's.

29.

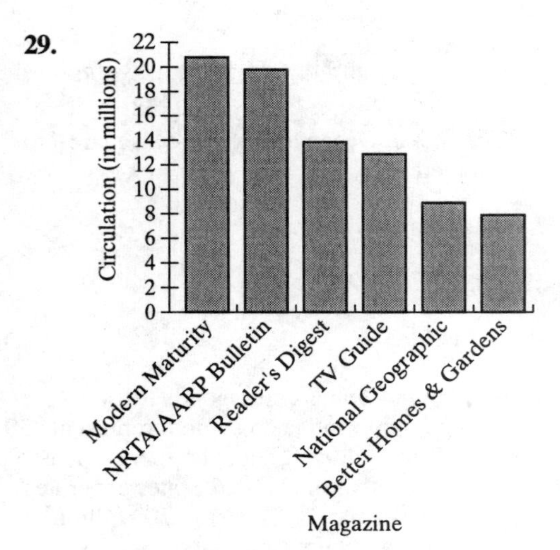

31.

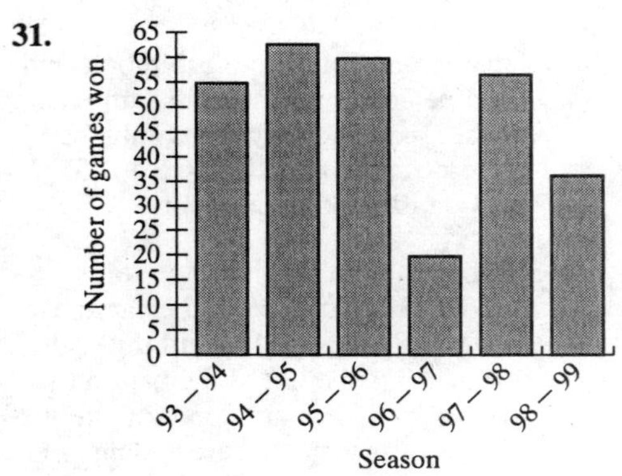

33. In 1985, there was an average of approximately 54.5 field goals per game.

35. 1975 had the highest average number of field goals per game.

37. The average number of field goals per game increased between 1955 and 1965.

39. The weekly box office receipts were $101 million two weeks after release.

41. The receipts were the greatest the 1st week after release.

43. Week 1 receipts were \$106 million. Week 8 receipts were \$12 million.
Decrease in receipts
= \$106 million – \$12 million
= \$94 million

45. During week 8, *The Phantom Menace* grossed \$12 million at the box office.

47. $30\% \cdot 12 = 0.3 \cdot 12 = 3.6$

49. $10\% \cdot 62 = 0.1 \cdot 62 = 6.2$

51. $\frac{1}{4} = 0.25 = 25\%$

53. $\frac{17}{50} = 0.34 = 34\%$

55. 83°F was the high temperature reading on Thursday.

57. The lowest temperature occurred on Sunday wit 68°F.

59. The greatest difference between high and low temperatures occurred on Tuesday with a difference of 86 – 73 = 13°F.

61. Look at the two bars for each continent. Since the bars representing North America have the largest difference in height, North America had the greatest change in percent of U.S. immigrants.

63. The highest bar representing 1998 data is for North America. North America accounts for the greatest percent of U.S. immigrants in 1998.

65. Answers may vary.

Section 8.3

Practice Problems

1.

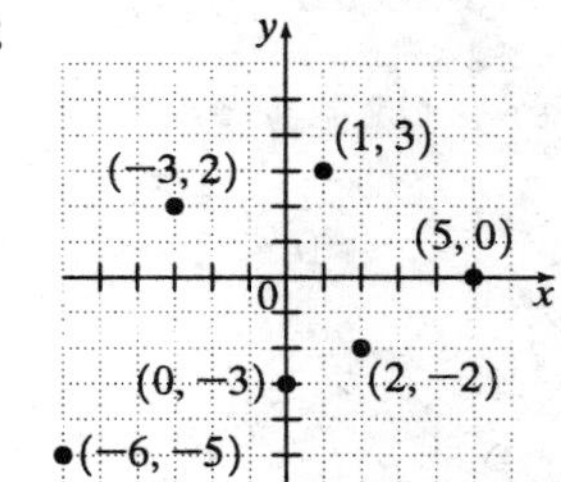

2. Point A has coordinates (5, 0).
Point B has coordinates (–2, 0).
Point C has coordinates (–3, –4).
Point D has coordinates (0, –1).
Point E has coordinates (2, 1).

3.
$$x + 3y = -12$$
$$0 + 3(-4) = -12$$
$$-12 = -12 \text{ True}$$
Yes (0, – 4) is a solution.

4.

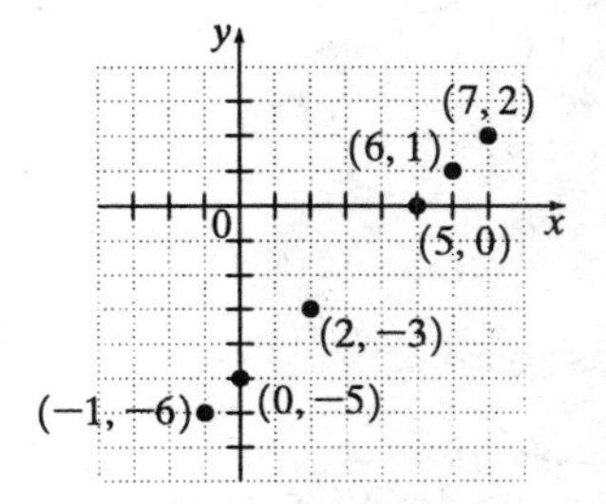

5. a.
$$x + y = 10$$
$$5 + y = 10$$
$$5 + y - 5 = 10 - 5$$
$$y = 5$$
The solution is (5, 5).

b.
$$x + y = 10$$
$$0 + y = 10$$
$$y = 10$$
The solution is (0, 10).

c.
$$\begin{aligned} x+y&=10\\ x+(-2)&=10\\ x-2&=10\\ x-2+2&=10+2\\ x&=12 \end{aligned}$$
The solution is (12, –2).

Exercise Set 8.3

1.

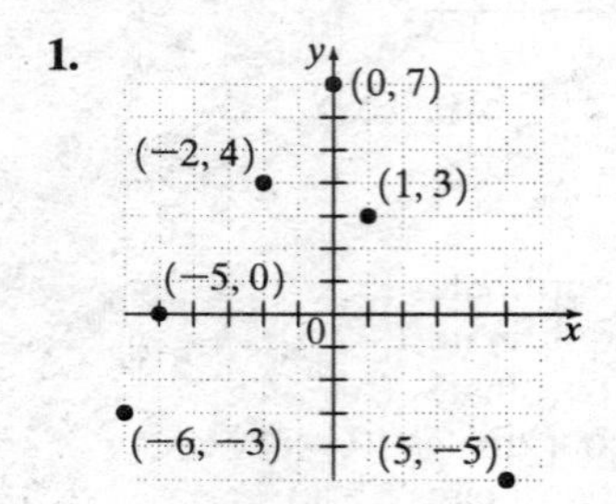

3.

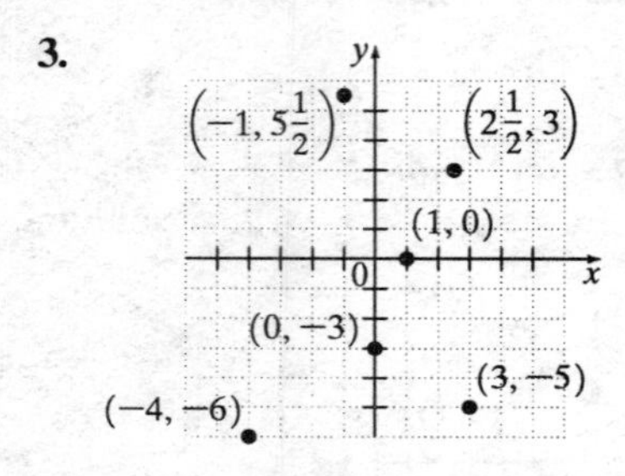

5. Point A has coordinates (0, 0).

Point B has coordinates $\left(3\frac{1}{2},\ 0\right)$.

Point C has coordinates (3, 2).
Point D has coordinates (–1, 3).
Point E has coordinates (–2, –2).
Point F has coordinates (0, –1).
Point G has coordinates (2, –1).

7. $y=-10x$

$0=-10(0)$

$0=0$ True

Yes, (0, 0) is a solution.

9. $x-y=3$

$1-2=3$

$-1=3$ False

No, (1, 2) is not a solution.

11. $y=2x+1$

$-3=2(-2)+1$

$-3=-4+1$

$-3=-3$ True

Yes, (–2, –3) is a solution.

13. $y=-4x$

$-8=-4\cdot 2$

$-8=-8$ True

Yes, (2, –8) is a solution.

15. $2x+3y=10$

$2(5)+3(0)=10$

$10+0=10$

$10=10$ True

Yes, (5, 0) is a solution.

17. $x-5y=-1$

$3-5(1)=-1$

$3-5=-1$

$-2=-1$ False

No, (3, 1) is not a solution. of $x-5y=-1$.

19. $2x+y=5$; (1, 3), (0, 5), (3, –1)

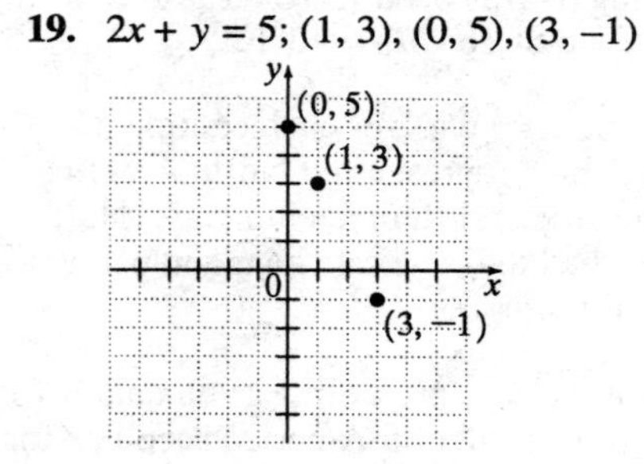

21. $x=5y$; (5, 1), (0, 0), (–5, –1)

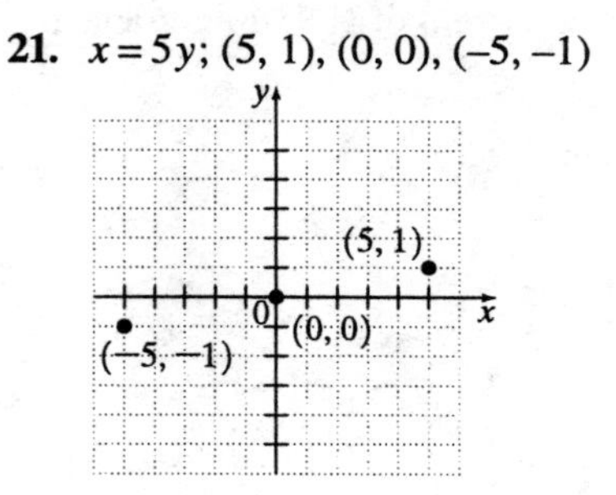

23. $x - y = 7$; (8, 1), (2, –5), (0, –7)

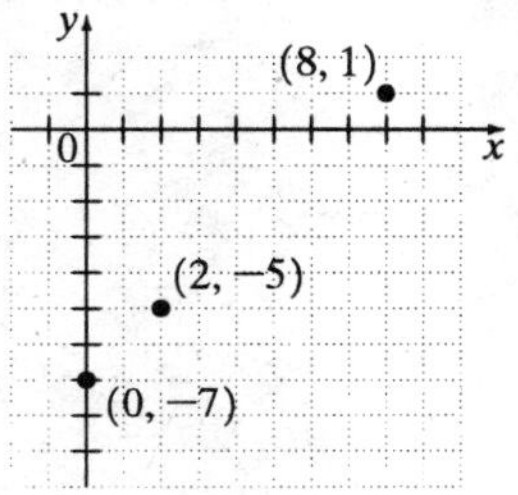

25. $y = 8x$
$y = 8(1)$
$y = 8$
The solution is (1, 8).
$y = 8x$
$y = 8(0)$
$y = 0$
The solution is (0, 0), the origin.

$$y = 8x$$
$$-16 = 8x$$
$$\frac{-16}{8} = \frac{8x}{8}$$
$$-2 = x$$

The solution is (–2, –16).

27.
$$x + y = 14$$
$$2 + y = 14$$
$$2 + y - 2 = 14 - 2$$
$$y = 12$$

The solution is (2, 12).

$$x + y = 14$$
$$x + (-8) = 14$$
$$x - 8 + 8 = 14 + 8$$
$$x = 22$$

The solution is (22, –8).

$$x + y = 14$$
$$0 + y = 14$$
$$y = 14$$

The solution is (0, 14).

29. $y = x + 5$
$y = 1 + 5$
$y = 6$
The solution is (1, 6).

$$y = x + 5$$
$$7 = x + 5$$
$$7 - 5 = x + 5 - 5$$
$$2 = x$$

The solution is (2, 7).
$y = x + 5$
$y = 3 + 5$
$y = 8$
The solution is (3, 8).

31. $y = 3x - 5$
$y = 3(1) - 5$
$y = 3 - 5$
$y = -2$
The solution is (1, –2).
$y = 3x - 5$
$y = 3(2) - 5$
$y = 6 - 5$
$y = 1$
The solution is (2, 1).
$y = 3x - 5$
$y = 3(3) - 5$
$y = 9 - 5$
$y = 4$
The solution is (3, 4).

33.
$$y = -x$$
$$0 = -x$$
$$\frac{0}{-1} = \frac{-x}{-1}$$
$$0 = x$$

The solution is (0, 0).
$y = -x$
$y = -(2)$
$y = -2$
The solution is (2, –2).

$$y = -x$$
$$2 = -x$$
$$\frac{2}{-1} = \frac{-x}{-1}$$
$$-2 = x$$

The solution is (–2, 2).

35.
$$x + y = -2$$
$$-2 + y = -2$$
$$-2 + y + 2 = -2 + 2$$
$$y = 0$$
The solution is (–2, 0).
$$x + y = -2$$
$$1 + y = -2$$
$$1 + y - 1 = -2 - 1$$
$$y = -3$$
The solution is (1, –3).
$$x + y = -2$$
$$x + 5 = -2$$
$$x + 5 - 5 = -2 - 5$$
$$x = -7$$
The solution is (–7, 5).

37. $5.6 - 3.9 = 1.7$

39.
```
    5.6
  × 3.9
   5 04
  16 80
  21.84
```

41. $(0.236)(-100) = -23.6$

43.
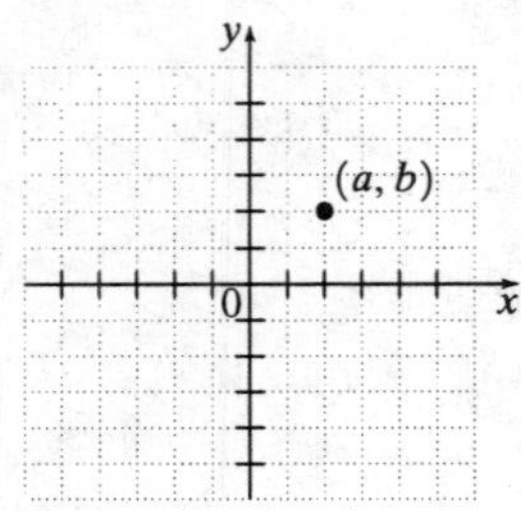

True

45.
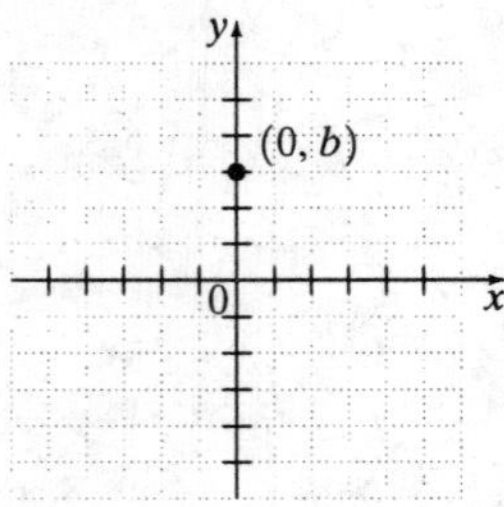

True

47.
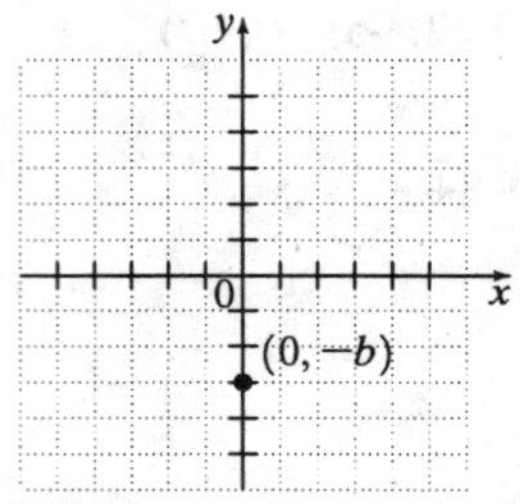

False

49.
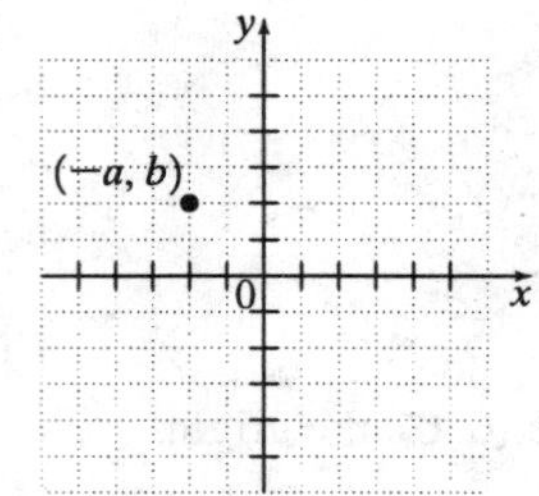

False

51. Answers may vary.

Integrated Review

1. For the year 2000, the graph has 10 pictures. Each picture represents 5 pounds. The approximate number of pounds of pork consumed per person is $10 \cdot 5 = 50$ pounds.

2. For the year 1998, the graph has $10\frac{1}{2}$ pictures. Each picture represents 5 pounds. The approximate number of pounds of pork consumed per person is $10.5 \cdot 5 = 52.5$ pounds.

3. The number of pounds of pork consumed was the greatest in 1999.

4. The number of pounds of pork consumed was the least in 1997.

5. The dam with the greatest height is represented by the tallest bar. The height of this bar is approximately 755 feet. Therefore, the Oroville Dam has the greatest height of approximately 755 feet.

6. The bar with a height between 625 feet and 650 feet is the one representing New Bullards Bar Dam. The New Bullards Bar Dam has a height of approximately 635 feet.

7. Find the difference in the heights of the bars representing these dams.
$725 - 710 = 15$
The Hoover Dam is approximately 15 feet taller than the Glen Canyon Dam.

8. There are 4 bars that have heights over 700 feet. Therefore, there are 4 United States dams that are over 700 feet.

9. Locate the highest points. Thursday and Saturday with temperatures of 100°F are the days with the highest temperature.

10. Locate the lowest point. Monday with a temperature of 82°F is the day with the lowest temperature.

11. There are three points below 90°F. The days corresponding to these points are Sunday, Monday, and Tuesday.

12. There are four points above 90°F. The days corresponding to these points are Wednesday, Thursday, Friday, and Saturday.

13. Whole milk: $35\% \cdot 200 = 70$ quart containers

14. Skim milk: $26\% \cdot 200 = 52$ quart containers

15. Buttermilk: $1\% \cdot 200 = 2$ quart containers

16. Flavored reduced fat and skim milk:
$3\% \cdot 200 = 6$ quart containers

17.

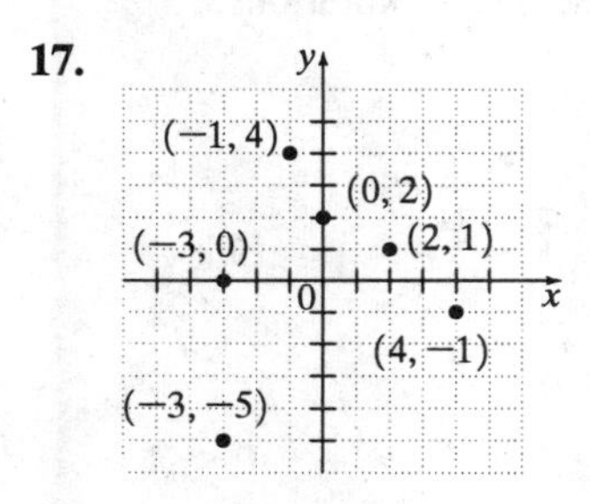

18. $x = 3y$
$1 = 3(3)$
$1 = 9$ False
No, it is not a solution.

19. $x + y = -6$
$-2 + (-4) = -6$
$-6 = -6$ True
Yes, it is a solution.

20. $x - y = 6$
$0 - y = 6$
$-y = 6$
$\frac{-y}{-1} = \frac{6}{-1}$
$y = -6$
The solution is (0, −6).
$x - y = 6$
$x - 0 = 6$
$x = 6$
The solution is (6, 0).
$x - y = 6$
$2 - y = 6$
$2 - y - 2 = 6 - 2$
$-y = 4$
$\frac{-y}{-1} = \frac{4}{-1}$
$y = -4$
The solution is (2, −4).

Section 8.4

Practice Problems

1.

y
$x - y = 1$
(3, 2)
(0, −1)
0
(1, 0)
x

2. Let $x = 0$.
$y - x = 6$
$y - 0 = 6$
$y = 6$
The ordered pair is (0, 6).
Let $y = 0$.
$y - x = 6$
$0 - x = 6$
$-x = 6$
$\frac{-x}{-1} = \frac{6}{-1}$
$x = -6$
The ordered pair is (–6, 0).
Let $x = 4$.
$y - x = 6$
$y - 4 = 6$
$y - 4 + 4 = 6 + 4$
$y = 10$
The ordered pair is (4, 10). Plot the three ordered pairs. Then draw the line through them.

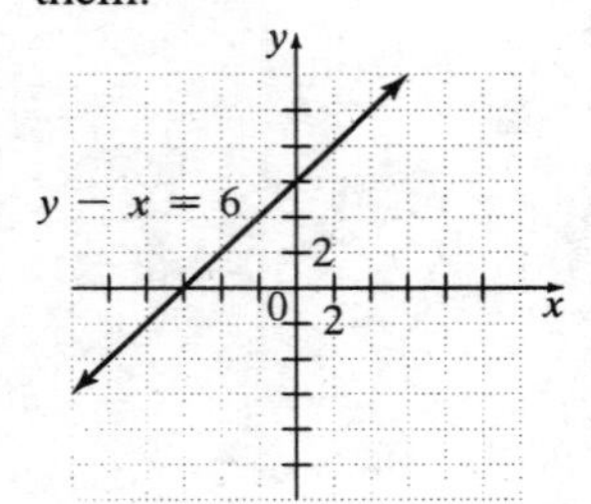

3.

x	$y = -3x + 2$
0	$-3 \cdot 0 + 2 = 2$
1	$-3 \cdot 1 + 2 = -1$
2	$-3 \cdot 2 + 2 = -4$

Plot the three ordered pairs. Then draw the line through them.

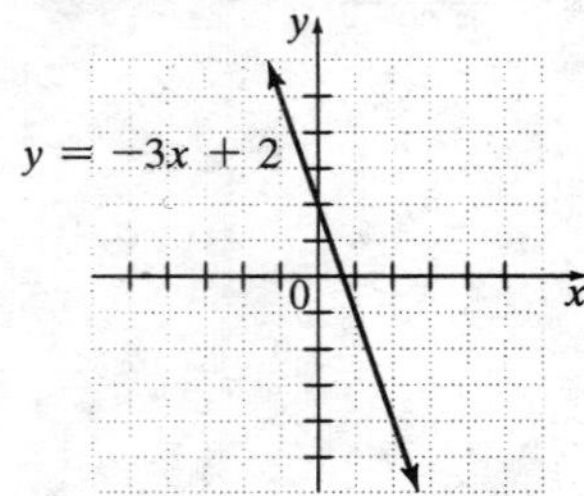

4.

x	$y = -1$
–5	–1
0	–1
5	–1

Plot the three ordered pairs. Then draw the line through them.

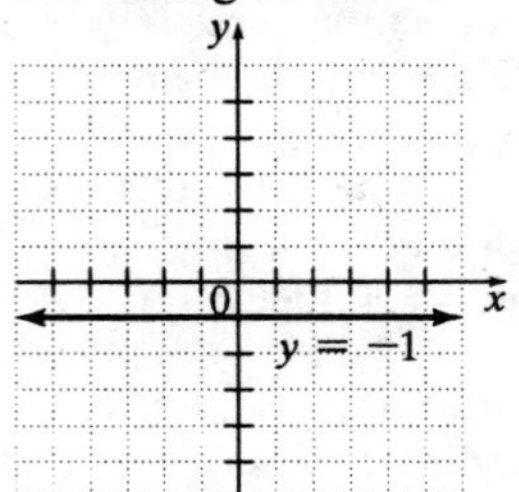

5.

$x = -2$	y
–2	–5
–2	0
–2	5

Plot the three ordered pairs. Then draw the line through them.

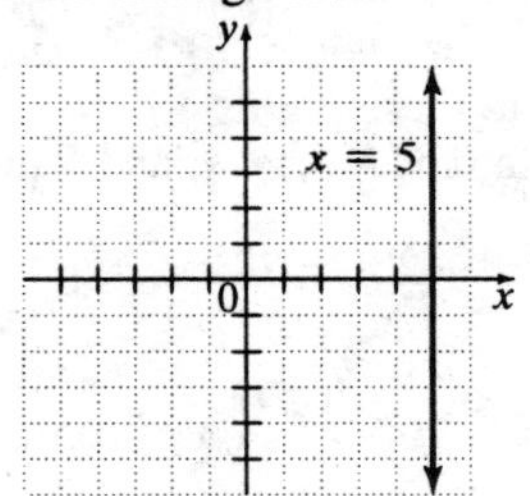

Exercise Set 8.4

1. $x + y = 6$
Find any 3 ordered pair solutions.
Let $x = 0$.
$x + y = 6$
$0 + y = 6$
$y = 6$
(0, 6)
Let $x = 3$.

$$x + y = 6$$
$$3 + y = 6$$
$$3 + y \cdot 3 = 6 - 3$$
$$y = 3$$

(3, 3)

Let $x = 6$.

$$x + y = 6$$
$$6 + y = 6$$
$$6 + y - 6 = 6 - 6$$
$$y = 0$$

(6, 0)

Plot (0, 6), (3, 3), and (6, 0). Then draw the line through them.

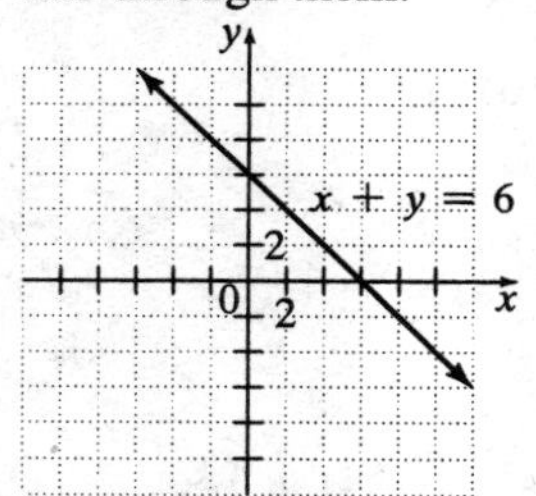

3. $x - y = -2$

Let $x = -2$.

$$x - y = -2$$
$$-2 - y = -2$$
$$-2 - y + 2 = -2 + 2$$
$$-y = 0$$
$$y = 0$$

(–2, 0)

Let $x = 0$.

$$x - y = -2$$
$$0 - y = -2$$
$$-y = -2$$
$$y = 2$$

(0, 2)

Let $x = 2$.

$$x - y = -2$$
$$2 - y = -2$$
$$2 - y - 2 = -2 - 2$$
$$-y = -4$$
$$y = 4$$

(2, 4)

Plot (–2, 0), (0, 2), and (2, 4). Then draw the line through them.

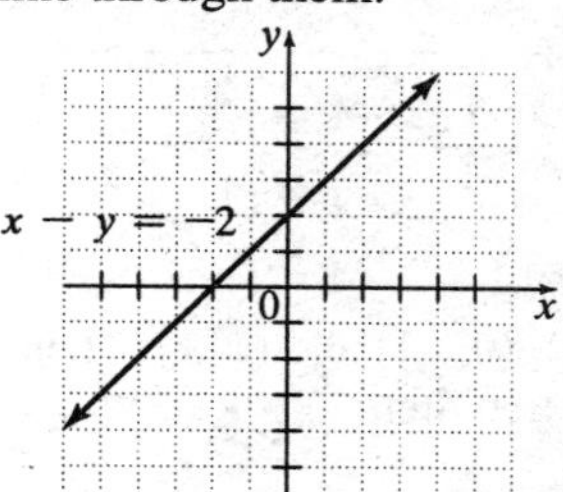

5. $y = 4x$

Find any 3 ordered pair solutions.

Let $x = 0$.

$$y = 4x$$
$$y = 4(0)$$
$$y = 0$$

(0, 0)

Let $x = 1$.

$$y = 4x$$
$$y = 4(1)$$
$$y = 4$$

(1, 4)

Let $x = -1$.

$$y = 4x$$
$$y = 4(-1)$$
$$y = -4$$

(–1, –4)

Plot (0, 0), (1, 4), and (–1, –4). Then draw the line through them.

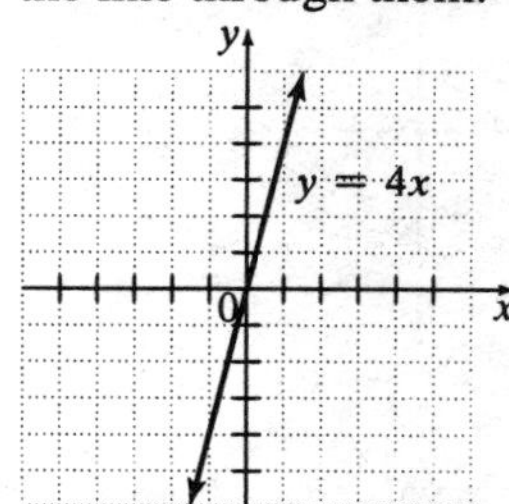

7. $y = 2x - 1$

x	$y = 2x - 1$
−1	$2(-1) - 1 = -2 - 1 = -3$
0	$2(0) - 1 = 0 - 1 = -1$
1	$2(1) - 1 = 2 - 1 = 1$

Plot (−1, −3), (0, −1) and (1, 1). Then draw the line through them.

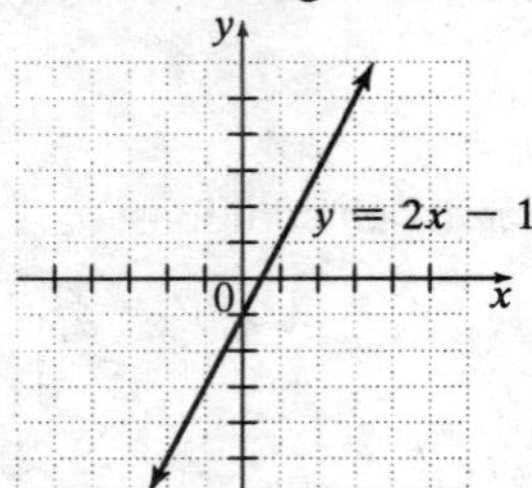

9. $x = 5$

No matter what y-value we choose, x is always 5.

x	y
5	−4
5	0
5	4

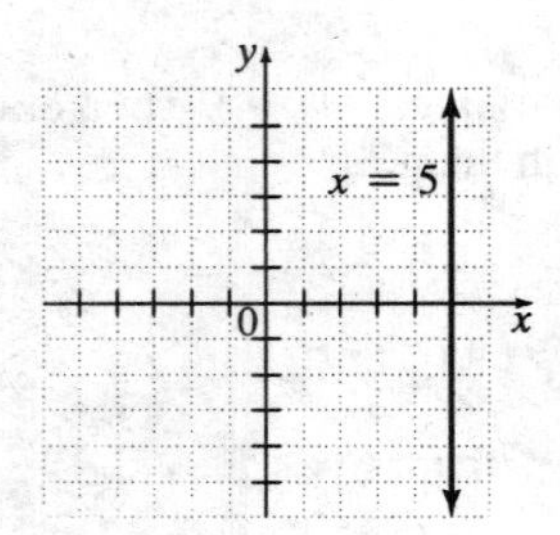

11. $y = -3$

No matter what x-value we choose, y is always −3.

y	y
−5	−3
0	−3
5	−3

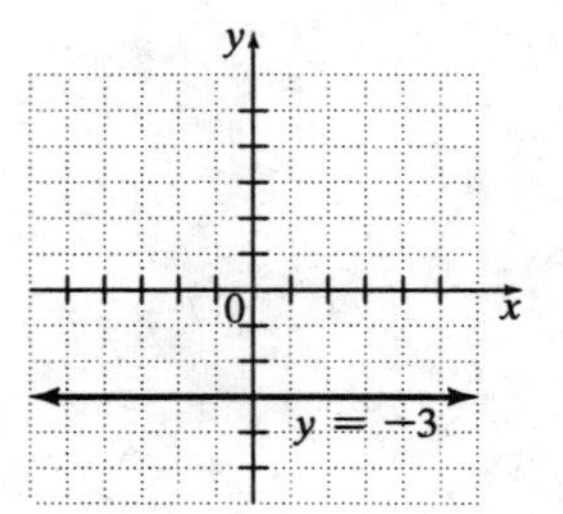

13. $x = 0$

No matter what y-value we choose, x is always 0.

x	y
0	−3
0	0
0	3

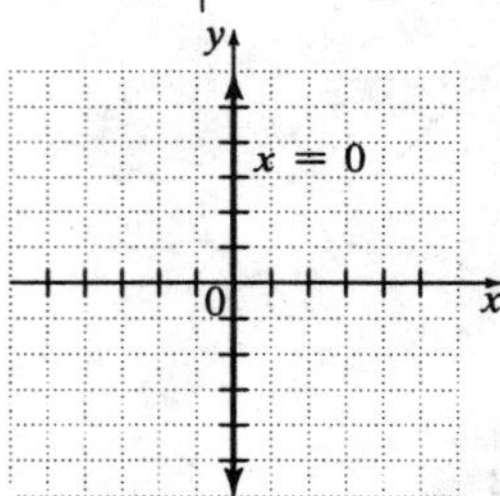

15. $y = -2x$

x	$y = -2x$
−2	$-2(-2) = 4$
0	$-2(0) = 0$
2	$-2(2) = -4$

Plot (−2, 4), (0, 0), and (2, −4). Then draw the line through them.

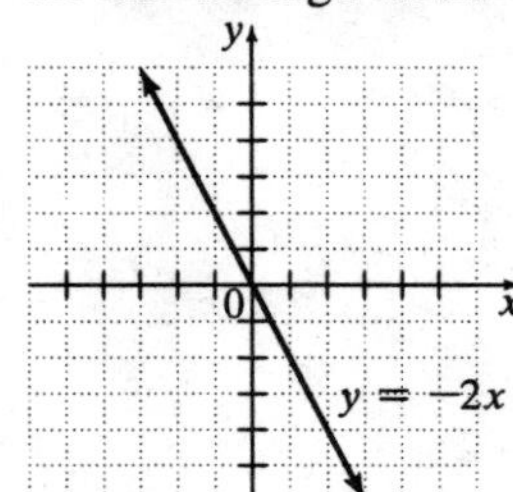

17. $y = -2$
No matter what x-value we choose, y is always -2.

x	y
-4	-2
0	-2
4	-2

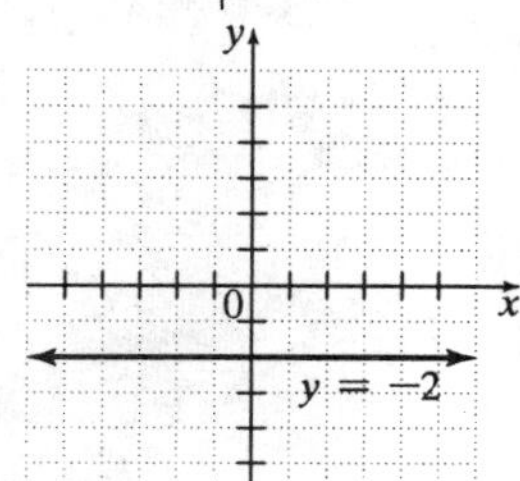

19. $x + 2y = 12$
Let $x = 0$.
$$x + 2y = 12$$
$$0 + 2y = 12$$
$$2y = 12$$
$$y = 6$$
(0, 6)
Let $y = 0$.
$$x + 2y = 12$$
$$x + 2(0) = 12$$
$$x + 0 = 12$$
$$x = 12$$
(12, 0)
Let $x = 6$.
$$x + 2y = 12$$
$$6 + 2y = 12$$
$$6 + 2y - 6 = 12 - 6$$
$$2y = 6$$
$$y = 3$$
(6, 3)
Plot (0, 6), (12, 0), and (6, 3). Then draw the line through them.

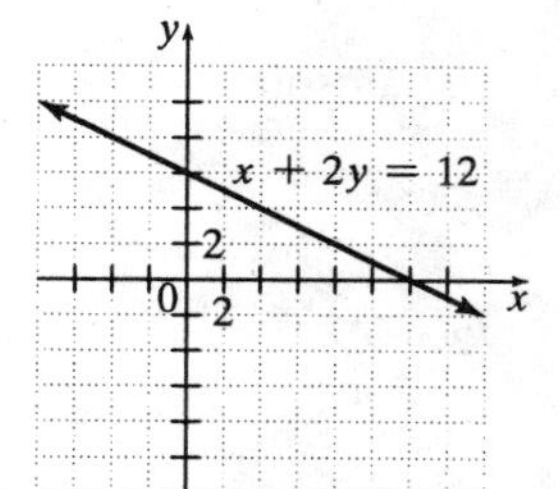

21. $x = 6$
No matter what y-value we choose, x is always 6.

x	y
6	-5
6	0
6	5

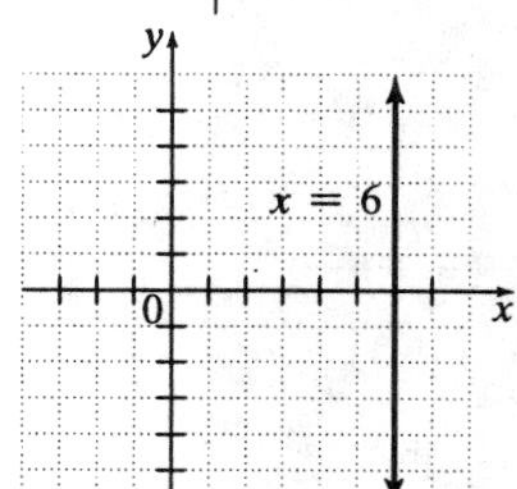

23. $y = x - 3$

x	$y = x - 3$
0	$0 - 3 = -3$
3	$3 - 3 = 0$
6	$6 - 3 = 3$

Plot (0, −3), (3, 0), and (6, 3). Then draw the line through them.

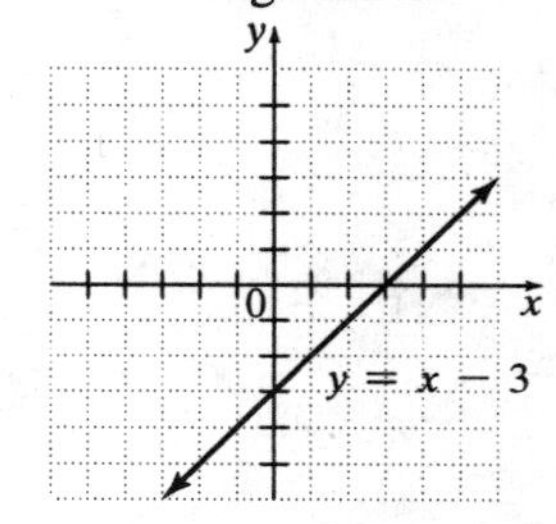

25. $x = y - 4$
Find any 3 ordered pair solutions.
Let $y = 0$.
$x = y - 4$
$x = 0 - 4$
$x = -4$
$(-4, 0)$
Let $y = 4$.
$x = y - 4$
$x = 4 - 4$
$x = 0$
$(0, 4)$
Let $y = 6$.
$x = y - 4$
$x = 6 - 4$
$x = 2$
$(2, 6)$
Plot (–4, 0), (0, 4), and (2, 6). Then draw the line through them.

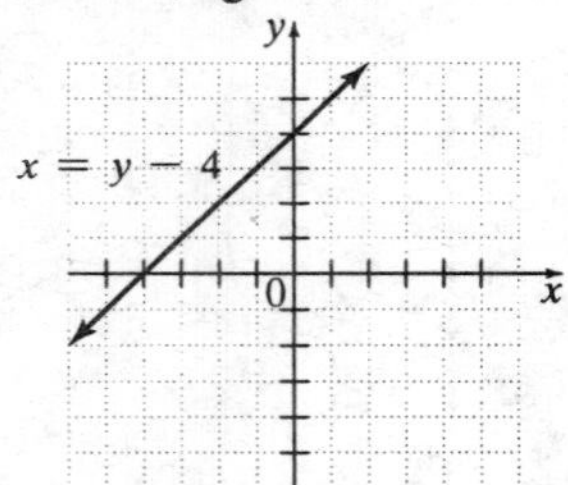

27. $x + 3 = 0$

$$x + 3 = 0$$
$$x + 3 - 3 = 0 - 3$$
$$x = -3$$

No matter what y-value we choose, x is always –3.

x	y
–3	4
–3	0
–3	–4

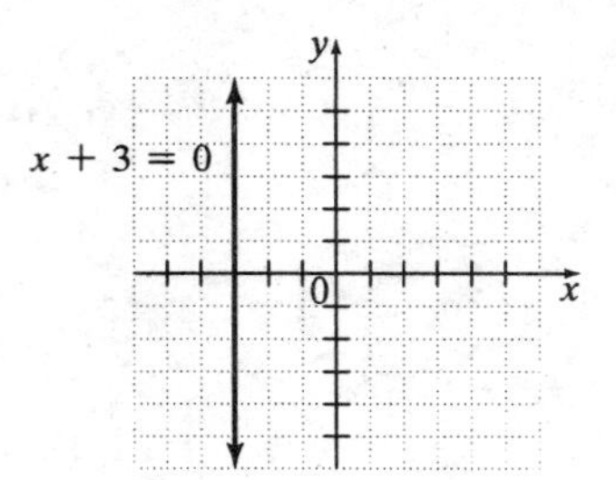

29. $x = 4y$
Find any 3 ordered pair solutions.
Let $y = -1$.
$x = 4y$
$x = 4(-1)$
$x = -4$
$(-4, -1)$
Let $y = 0$.
$x = 4y$
$x = 4(0)$
$x = 0$
$(0, 0)$
Let $y = 1$.
$x = 4y$
$x = 4(1)$
$x = 4$
$(4, 1)$
Plot (–4, –1), (0, 0), and (4, 1). Then draw the line through them.

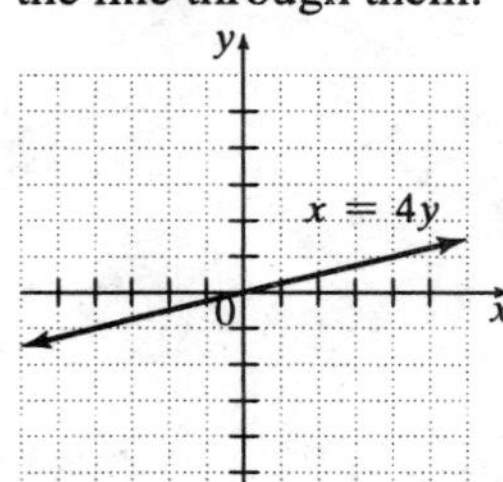

31. $y = \frac{1}{3}x$

x	$y = \frac{1}{3}x$
0	$\frac{1}{3}(0) = 0$
3	$\frac{1}{3}(3) = 1$
6	$\frac{1}{3}(3) = 2$

Plot (0, 0), (3, 1), and (6, 2). Then draw the line thorough them.

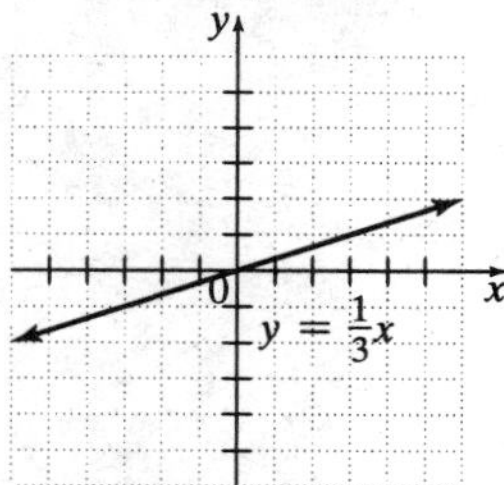

33. $y = 4x + 2$

Find any 3 ordered pair solutions.

$x = -1$
$y = 4x + 2$
$y = 4(-1) + 2$
$y = -4 + 2$
$y = -2$
$(-1, -2)$
$x = 0$
$y = 4x + 2$
$y = 4(0) + 2$
$y = 0 + 2$
$y = 2$
$(0, 2)$
$x = 1$
$y = 4x + 2$
$y = 4(1) + 2$
$y = 4 + 2$
$y = 6$
$(1, 6)$

Plot (–1, –2), (0, 2), and (1, 6). Then draw the line through them.

35. $2x + 3y = 6$

Let $x = -3$.

$$2x + 3y = 6$$
$$2(-3) + 3y = 6$$
$$-6 + 3y = 6$$
$$-6 + 3y + 6 = 6 + 67$$
$$3y = 12$$
$$y = 4$$

$(-3, 4)$

Let $x = 0$.

$$2x + 3y = 6$$
$$2(0) + 3y = 6$$
$$0 + 3y = 6$$
$$3y = 6$$
$$y = 2$$

$(0, 2)$

Let $y = 0$.

$$2x + 3y = 6$$
$$2x + 3(0) = 6$$
$$2x + 0 = 6$$
$$2x = 6$$
$$x = 3$$

$(3, 0)$

Plot (–3, 4), (0, 2), and (3, 0). Then draw the line through them.

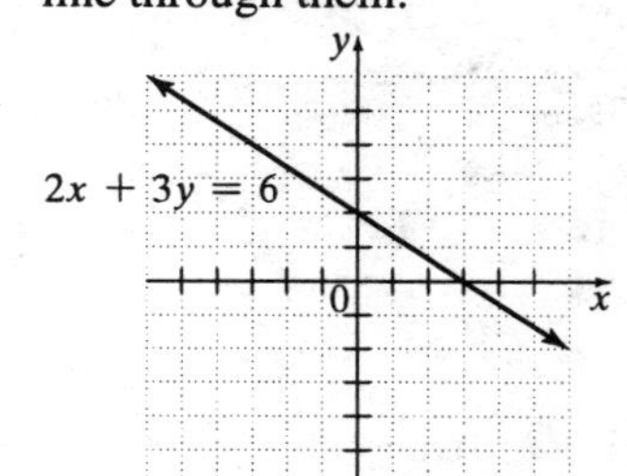

37. $x = -3.5$
No matter what y-value we choose, x is always -3.5.

x	y
-3.5	-3
-3.5	0
-3.5	3

39. $3x - 4y = 24$
Let $x = 0$.

$$3x - 4y = 24$$
$$3(0) - 4y = 24$$
$$0 - 4y = 24$$
$$-4y = 24$$
$$y = -6$$

$(0, -6)$
Let $y = 0$.

$$3x - 4y = 24$$
$$3x - 4(0) = 24$$
$$3x - 0 = 24$$
$$3x = 24$$
$$x = 8$$

$(8, 0)$
Let $x = 4$.

$$3x - 4y = 24$$
$$3(4) - 4y = 24$$
$$12 - 4y = 24$$
$$12 - 4y - 12 = 24 - 12$$
$$-4y = 12$$
$$y = -3$$

$(4, -3)$
Plot $(0, -6)$, $(8, 0)$, and $(4, -3)$. Then draw the line through them.

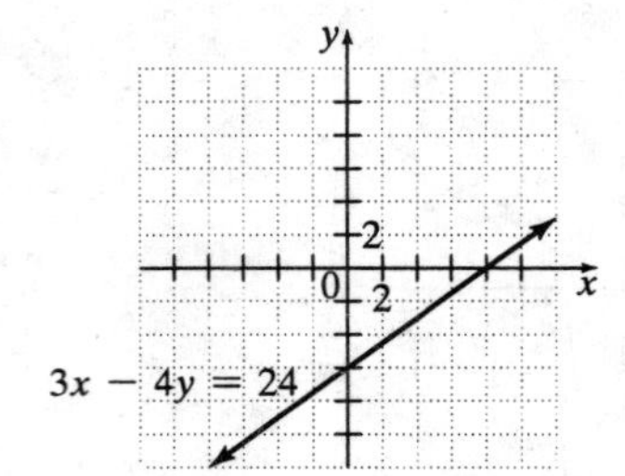

41. $\dfrac{86+94}{2} = \dfrac{180}{2} = 90$

43. $\dfrac{12+28+20}{3} = \dfrac{60}{3} = 20$

45. $\dfrac{30+22+23+33}{4} = \dfrac{108}{4} = 27$

47. $y = |x|$

x	y
-3	3
-2	2
-1	1
0	0
1	1
2	2
3	3

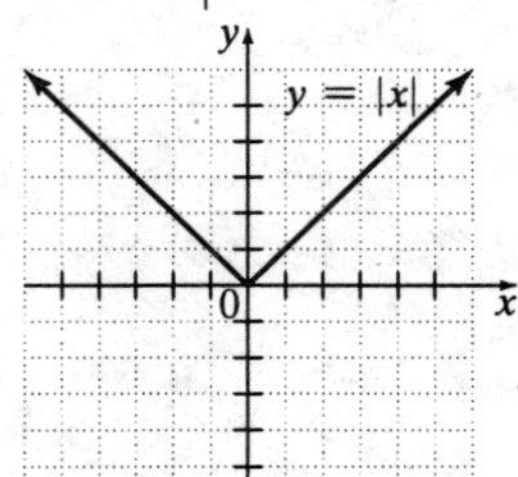

49. Answers may vary.

Section 8.5

Practice Problems

1. $\text{mean} = \frac{77+85+86+91+88}{5} = \frac{427}{5} = 85.4$

2. $\text{grade point average} = \frac{4\cdot 2+2\cdot 4+3\cdot 5+1\cdot 2+4\cdot 2}{2+4+5+2+2} = \frac{41}{15} \approx 2.73$

3. Because the list is in numerical order, the median is the middle number, 24.

4. First list the scores in numerical order. Then find the middle number, which will be the mean of the two middle numbers since there is an even number of scores.
43, 65, 71, 78, 88, 89, 95, 95
$\text{median} = \frac{78+88}{2} = \frac{166}{2} = 83$

5. The number 15 occurs most often. Therefore, 15 is the mode.

6. First write the numbers in numerical order.
15, 15, 15, 16, 18, 26, 26, 30, 31, 35
Since there is an even number of items, the median is the mean of the two middle numbers.
$\text{median} = \frac{18+26}{2} = \frac{44}{2} = 22$
The mode is 15, since 15 occurs most often.

Mental Math

1. $\frac{3+5}{2} = \frac{8}{2} = 4$

2. $\frac{10+20}{2} = \frac{30}{2} = 15$

3. $\frac{1+3+5}{3} = \frac{9}{3} = 3$

4. $\frac{7+7+7}{3} = \frac{21}{3} = 7$

Exercise Set 8.5

1. Mean: $\frac{21+28+16+42+38}{5}=\frac{145}{5}=29$

Median: Write the numbers in order.
16, 21, 28, 38, 42
The middle number is 28.
Mode: There is no mode, since there is no number that occurs more often than the others.

3. Mean: $\frac{7.6+8.2+8.2+9.6+5.7+9.1}{6}=\frac{48.4}{6}$
≈ 8.1

Median: Write the numbers in order.
5.7, 7.6, 8.2, 8.2, 9.1, 9.6
Since there is an even number of items, find the average of the two middle numbers.
$\frac{8.2+8.2}{2}=\frac{16.4}{2}=8.2$
Mode: 8.2 occurs most often.

5. Mean: $\frac{0.2+0.3+0.5+0.6+0.6+0.9+0.2+0.7+1.1}{9}=\frac{5.1}{9}$
≈ 0.6

Median: Write the numbers in order.
0.2, 0.2, 0.3, 0.5, 0.6, 0.6, 0.7, 0.9, 1.1
The middle number is 0.6.
Mode: Since 0.2 and 0.6 occur most often, there are two modes, 0.2 and 0.6.

7. Mean: $\frac{231+543+601+293+588+109+334+268}{8}=\frac{2967}{8}$
≈ 370.9

Median: Write the numbers in order.
109, 231, 268, 293, 334, 543, 588, 601
Since there is an even number of items, find the average of the two middle numbers.
$\frac{293+334}{2}=\frac{627}{2}=313.5$
Mode: There is no mode since there is no number that occurs more often than the others.

9. Mean: $\frac{1450+1368+1362+1250+1136}{5}=\frac{6566}{5}$
$=1313.2$

The mean height of the five tallest buildings is 1313.2 feet.

11. The table presents the numbers in order. Since there is an even number of building heights, find the average of the middle two.
median $=\frac{1136+1127}{2}=\frac{2263}{2}=1131.5$
The median height for the ten tallest buildings is 1131.5 feet.

13.

Grade	Point Value	Credit Hours	$\begin{pmatrix}\text{Point}\\\text{Value}\end{pmatrix}\cdot\begin{pmatrix}\text{Credit}\\\text{Hours}\end{pmatrix}$
B	3	3	9
C	2	3	6
A	4	4	16
C	2	4	8
Totals		14	39

Grade Point Average $= \frac{39}{14} \approx 2.79$

15.

Grade	Point Value	Credit Hours	$\begin{pmatrix}\text{Point}\\\text{Value}\end{pmatrix}\cdot\begin{pmatrix}\text{Credit}\\\text{Hours}\end{pmatrix}$
A	4	3	12
A	4	3	12
B	3	4	12
B	3	1	3
B	3	2	6
Totals		13	45

Grade Point Average $= \frac{45}{13} \approx 3.46$

17. Mean:

$$\frac{7.8+6.9+7.5+4.7+6.9+7.0}{6} = \frac{40.8}{6} = 6.8$$

The mean time is 6.8 seconds.

19. The most common time is 6.9 seconds.

21. Write the numbers in order.
74, 77, 85, 86, 91, 95
Since there is an even number of scores, find the average of the middle two numbers.
median $= \frac{85+86}{2} = \frac{171}{2} = 85.5$

23. $\text{mean} = \frac{78+80+66+68+71+64+82+71+70+65+70+75+77+86+72}{15} = \frac{1095}{15} = 73$

25. Since 70 and 71 occur more often than the other numbers, the modes are 70 and 71.

27. Nine: the rates of 66, 68, 71, 64, 71, 70, 65, 70, and 72 are below the mean of 73.

29. $\frac{6}{18} = \frac{2 \cdot 3}{2 \cdot 3 \cdot 3} = \frac{1}{3}$

31. $\frac{18}{30} = \frac{2 \cdot 3 \cdot 3}{2 \cdot 3 \cdot 5} = \frac{3}{5}$

33. $\frac{55}{75} = \frac{5 \cdot 11}{5 \cdot 15} = \frac{11}{15}$

35. There are 5 numbers, so the median is the middle number, 37. The mode is 35 so it must occur two times since the others occur only once. Now we have 35, 35, 37, 40, ___.
Let x = the unknown number.
Since the mean is 38,

$$38 = \frac{35+35+37+40+x}{5}$$
$$5 \cdot 38 = 147 + x$$
$$190 = 147 + x$$
$$190 - 147 = 147 + x - 147$$
$$43 = x$$

The list is 35, 35, 37, 40, 43.

Section 8.6

Practice Problems

1.

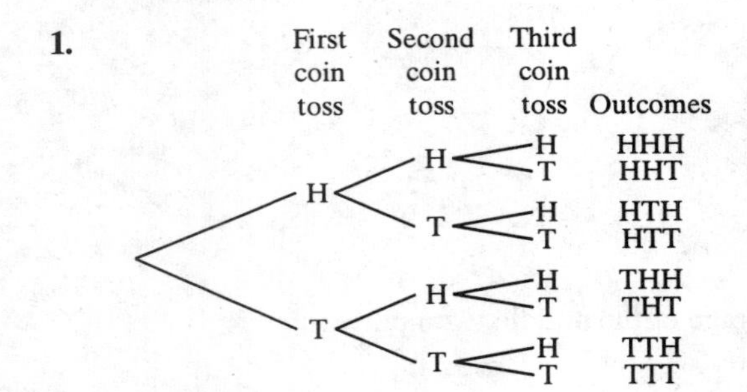

There are 8 outcomes.

2.

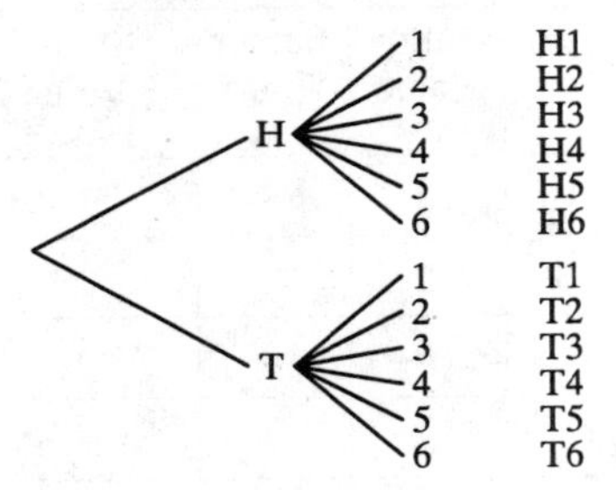

There are 12 outcomes.

3. Possible outcomes: HHH, HHT, HTH, HTT, THH, THT, TTH, TTT
probability
$= \frac{\text{number of ways the event can occur}}{\text{number of possible outcomes}}$
$= \frac{1}{8}$

4. Possible outcomes: 1, 2, 3, 4, 5, 6
probability
$= \frac{\text{number of ways the event can occur}}{\text{number of possible outcomes}}$
$= \frac{2}{6}$
$= \frac{1}{3}$

5. probability
$= \frac{\text{number of ways the event can occur}}{\text{number of possible outcomes}}$
$= \frac{2}{4}$
$= \frac{1}{2}$

Mental Math

1. $\frac{1}{2}$

2. $\frac{1}{2}$

3. $\frac{1}{2}$

4. $\frac{1}{2}$

Exercise Set 8.6

1.

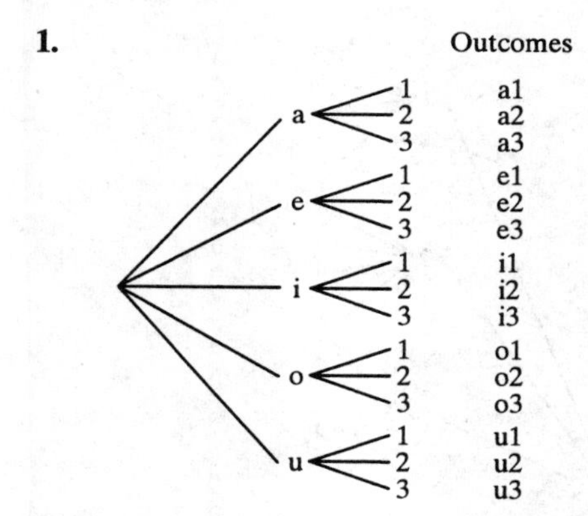

There are 15 outcomes.

3.

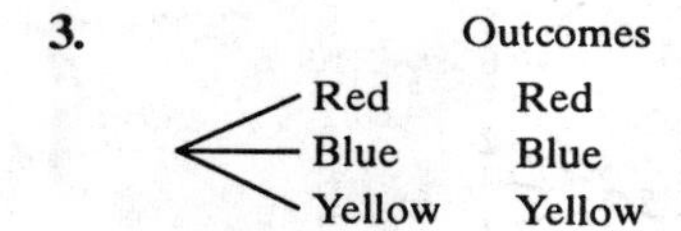

There are 3 outcomes.

5.

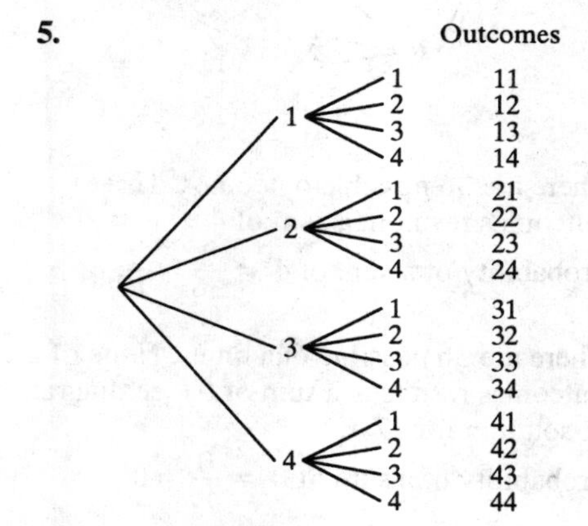

There are 16 outcomes.

7.

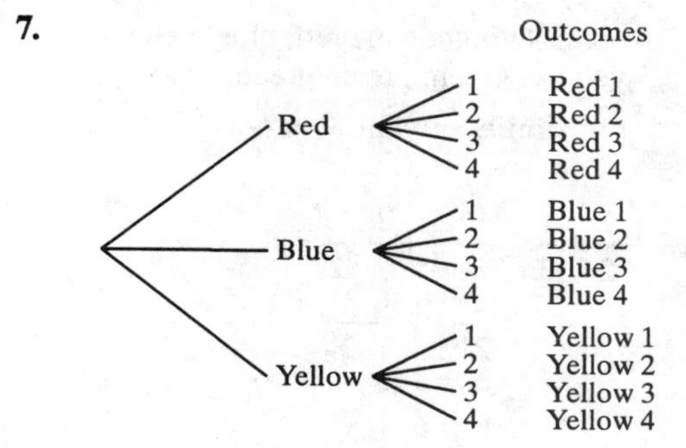

There are 12 outcomes.

9.

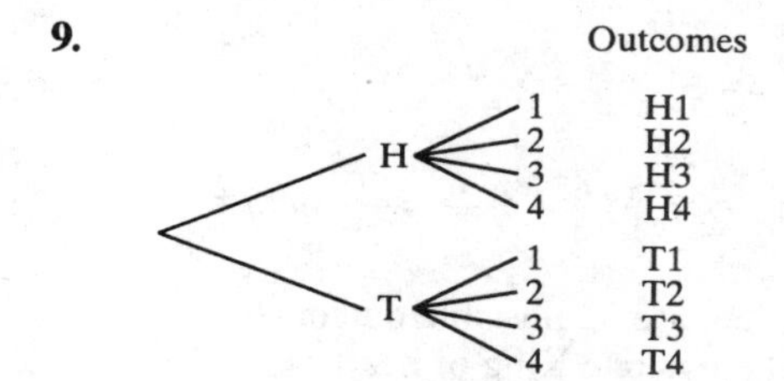

There are 8 outcomes.

11. possible outcomes: 1, 2, 3, 4, 5, 6
probability of a 5 $= \frac{1}{6}$

13. possible outcomes: 1, 2, 3, 4, 5, 6
probability of a 1 or a 4 $= \frac{2}{6} = \frac{1}{3}$

15. possible outcomes: 1, 2, 3, 4, 5, 6
probability of an even number $= \frac{3}{6} = \frac{1}{2}$

17. possible outcomes: 1, 2, 3
probability of a 2 $= \frac{1}{3}$

19. possible outcomes: 1, 2, 3
probability of an odd number $= \frac{2}{3}$

21. possible outcomes: red, blue, yellow, yellow, green, green, green
probability of a red $= \frac{1}{7}$

23. possible outcomes: red, blue, yellow, yellow, green, green, green
probability of yellow $= \frac{2}{7}$

25. $\frac{1}{2}+\frac{1}{3}=\frac{1}{2}\cdot\frac{3}{3}+\frac{1}{3}\cdot\frac{2}{2}$
$=\frac{1\cdot 3}{2\cdot 3}+\frac{1\cdot 2}{3\cdot 2}$
$=\frac{3}{6}+\frac{2}{6}$
$=\frac{5}{6}$

27. $\frac{1}{2}\cdot\frac{1}{3}=\frac{1\cdot 1}{2\cdot 3}=\frac{1}{6}$

29. $5\div\frac{3}{4}=\frac{5}{1}\cdot\frac{4}{3}=\frac{5\cdot 4}{1\cdot 3}=\frac{20}{3}$ or $6\frac{2}{3}$

31. There are 52 possible outcomes.
There is one King of hearts.
probability of King of Hearts $= \frac{1}{52}$

33. There are 52 possible outcomes.
There are four kings.
probability of a King $= \frac{4}{52}=\frac{1}{13}$

35. There are 52 possible outcomes.
There are 13 hearts.
probability of a heart $= \frac{13}{52}=\frac{1}{4}$

37.

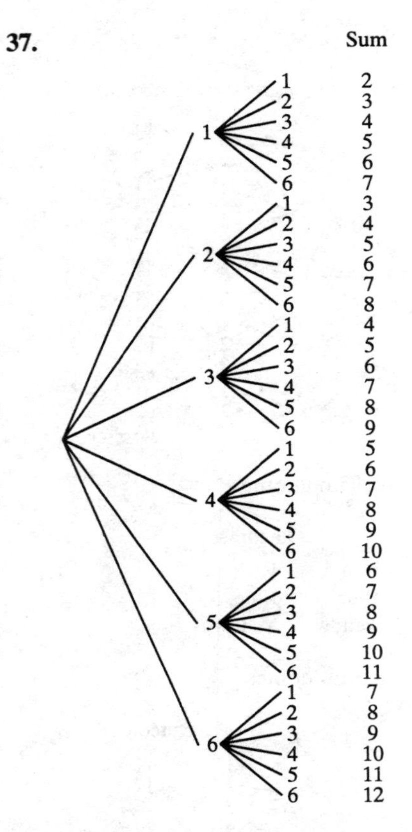

There are 36 possible outcomes. Three outcomes result in a sum of 4.
probability of a sum of 4 $= \frac{3}{36}=\frac{1}{12}$

39. There are 36 possible outcomes. None of the outcomes results is a sum of 0 (see diagram in solution for **37.**)
probability of a sum of 0 $= \frac{0}{36}=0$

41. Answers may vary.

Chapter 8 Review

1. The largest sector of the circle graph represents the largest budget item, which is House Mortgage.

3. The smallest sector of the circle graph represents the smallest budget item, which is Utilities.

3. Add the two amounts of the two sectors: $\$975 + \$250 = \$1225$

4. Add the two amounts of the two sectors: $\$400 + \$300 = \$700$

5. $\text{ratio} = \frac{\text{house mortgage}}{\text{monthly budget}} = \frac{\$975}{\$4000} = \frac{39}{160}$

6. $\text{ratio} = \frac{\text{food}}{\text{monthly budget}} = \frac{\$700}{\$4000} = \frac{7}{40}$

7. $\text{amount} = 40\% \cdot 50 = 20$ states

8. $\text{amount} = 20\% \cdot 50 = 10$ states

9.
$$\begin{aligned}\text{amount} &= (32\% + 20\%) \cdot 50 \\ &= 52\% \cdot 50 \\ &= 26 \text{ states}\end{aligned}$$

10.
$$\begin{aligned}\text{amount} &= (6\% + 40\%) \cdot 50 \\ &= 46\% \cdot 50 \\ &= 23 \text{ states}\end{aligned}$$

11. Oklahoma has about $5\frac{1}{2}$ houses, and each house represents 2000 homes. There were approximately $5.5 \cdot 2000 = 11{,}000$ new homes constructed in Oklahoma in 1999.

12. Kansas has about $5\frac{1}{2}$ houses, and each house represents 2000 homes. There were approximately $5.5 \cdot 2000 = 11{,}000$ new homes constructed in Kansas in 1999.

13. Alabama had the largest number of houses in the pictograph, so Alabama had the most new homes constructed.

14. Delaware has the fewest number of houses in the pictograph, so Delaware had the fewest new homes constructed.

15. A state must have more than $\frac{13{,}000}{2000} = 6.5$ houses in the pictograph. Utah and Alabama have more than 6.5 houses, so Utah and Alabama had more than 13,000 new homes constructed.

16. A state must have fewer than $\frac{8000}{2000} = 4$ houses in the pictograph. Both Maine and Delaware have fewer than 4 houses, so Maine and Delaware had less than 8000 new homes constructed.

17. The bar above 1960 is approximately 8 high, so approximately 8% of persons age 25 or more completed 4 or more years of college in 1960.

18. The tallest bar is for the year 1998, so 1998 had the greatest percentage of persons age 25 or more completing 4 or more years of college.

19. The years 1980, 1990, and 1998 all have bars of height 15 or more, so those years had 15% or more persons age 25 or more completing 4 or more years of college.

20. Answers may vary.

21. The Poultry line graph shows approximately 10 billion pounds of poultry consumed in 1970.

22. The red meats line graph shows approximately 39.5 billion pounds of red meats eaten in 1970.

23. The greatest increase in the number of pounds of poultry eaten occurred from 1980–1990.

24. The smallest increase in the number of pounds of poultry eaten occurred from 1940–1950.

25.
$$\begin{aligned}x &= -7y \\ 0 &= -7y \\ \frac{0}{-7} &= \frac{-7y}{-7} \\ 0 &= y\end{aligned}$$
The ordered pair is (0, 0).

$$\begin{aligned}x &= -7y \\ x &= -7(-1) \\ x &= 7\end{aligned}$$
The ordered pair is (7, –1).

$x = -7y$
$-7 = -7y$
$\frac{-7}{-7} = \frac{-7y}{-7}$
$1 = y$
The ordered pair is (–7, 1).

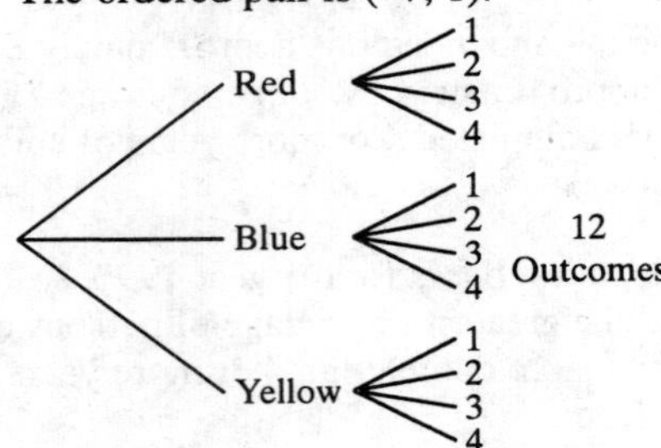

26. $y = 3x - 2$
$y = 3(0) - 2$
$y = -2$
The ordered pair is (0, –2).
$y = 3x - 2$
$y = 3(1) - 2$
$y = 3 - 2$
$y = 1$
The ordered pair is (1, 1).
$y = 3x - 2$
$y = 3(-2) - 2$
$y = -6 - 2$
$y = -8$
The ordered pair is (–2, –8).

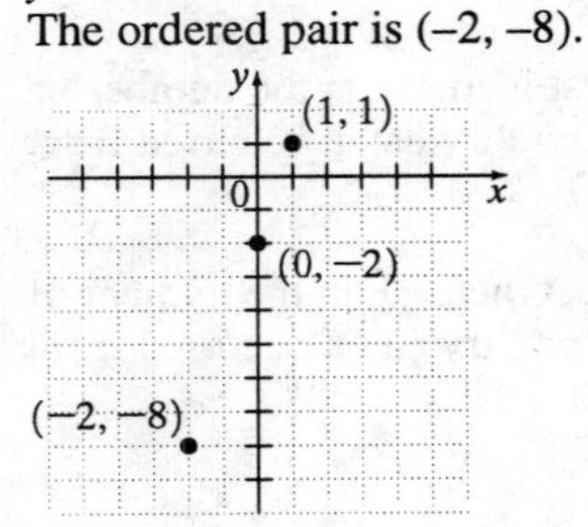

27. $x + y = -9$
$-1 + y = -9$
$-1 + y + 1 = -9 + 1$
$y = -8$
The ordered pair is (–1, –8).
$x + y = -9$
$x + 0 = -9$
$x = -9$
The ordered pair is (–9, 0).
$x + y = -9$
$-5 + y = -9$
$y = -9 + 5$
$y = -4$
The ordered pair is (–5, –4).

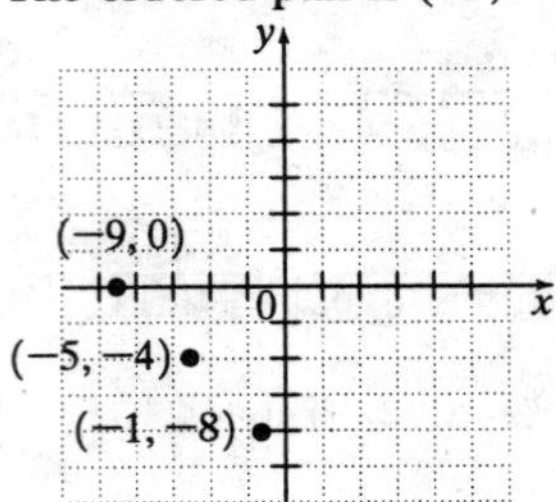

28. $x - y = 3$
$4 - y = 3$
$4 - y - 4 = 3 - 4$
$-y = -1$
$\frac{-y}{-1} = \frac{-1}{-1}$
$y = 1$
The ordered pair is (4, 1).
$x - y = 3$
$0 - y = 3$
$-y = 3$
$\frac{-y}{-1} = \frac{3}{-1}$
$y = -3$
The ordered pair is (0, –3).
$x - y = 3$
$x - 3 = 3$
$x - 3 + 3 = 3 + 3$
$x = 6$
The ordered pair is (6, 3).

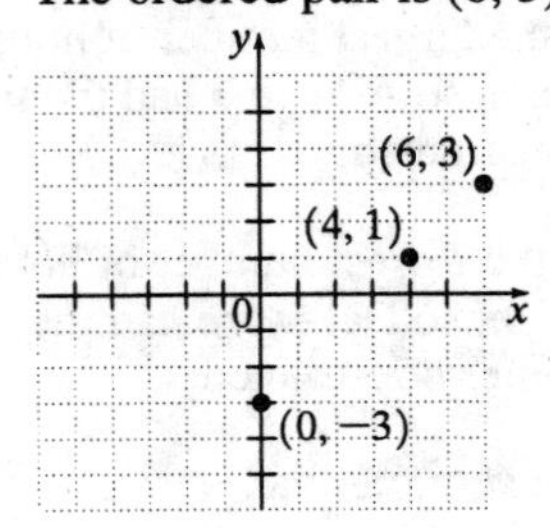

29. $y = 3x$
$y = 3(1)$
$y = 3$
The ordered pair is (1, 3).

$y = 3x$
$y = 3(-2)$
$y = -6$
The ordered pair is (–2, –6).

$y = 3x$
$0 = 3x$
$\frac{0}{3} = \frac{3x}{3}$
$0 = x$
The ordered pair is (0, 0).

30. $x = y + 6$
$1 = y + 6$
$1 - 6 = y + 6 - 6$
$-5 = y$
The ordered pair is (1, –5).
$x = y + 6$
$6 = y + 6$
$6 - 6 = y + 6 - 6$
$0 = y$
The ordered pair is (6, 0).
$x = y + 6$
$-1 = y + 6$
$-1 - 6 = y + 6 - 6$
$-7 = y$
The ordered pair is (–1, –7).

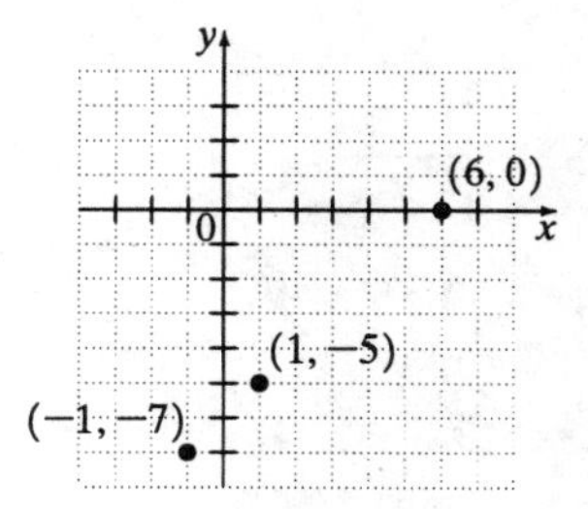

31. $x = -6$

x	y
–6	–5
–6	0
–6	5

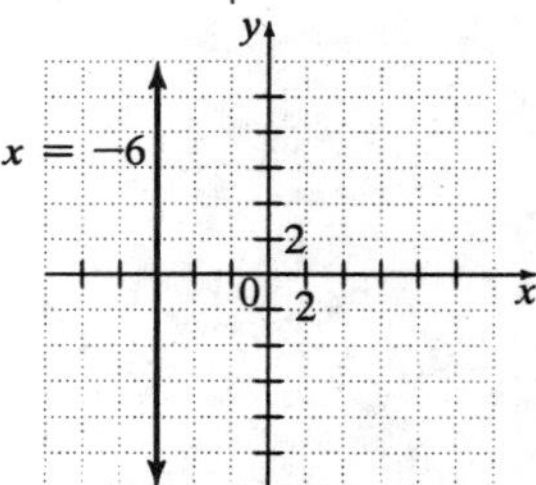

32. $y = 0$

x	y
–5	0
0	0
5	0

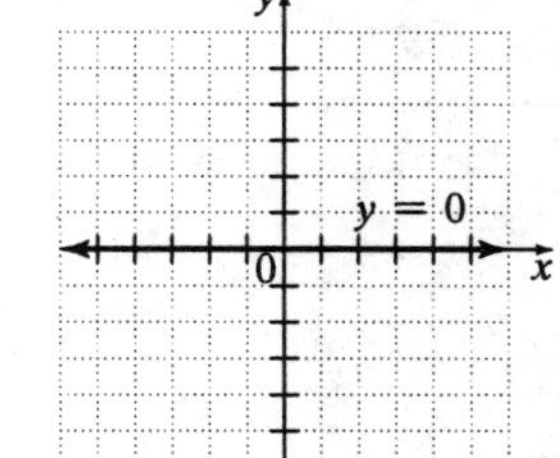

33. $x + y = 11$

x	y
0	11
5	6
11	0

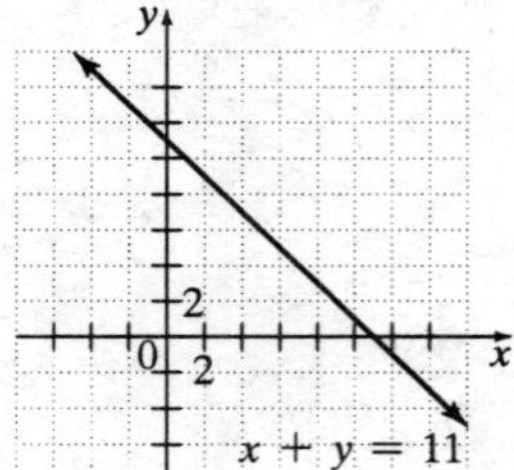

34. $x - y = 11$

x	y
0	–11
5	–6
11	0

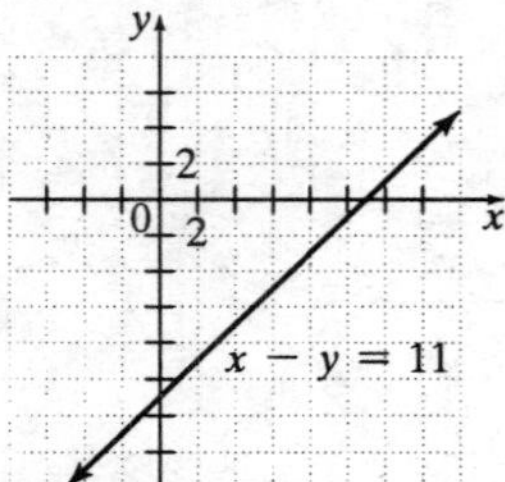

35. $y = 4x - 2$

x	y
–1	–6
0	–2
1	2

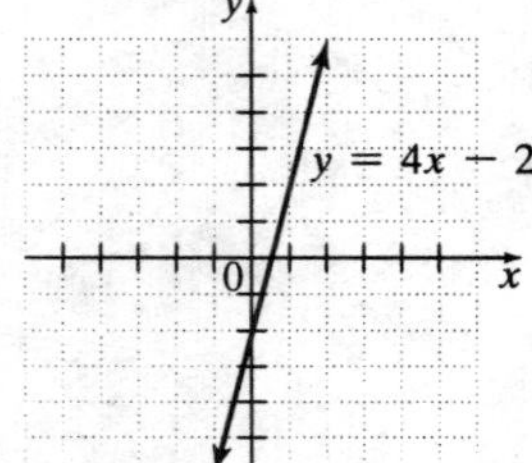

36. $y = 5x$

x	y
–1	–5
0	0
1	5

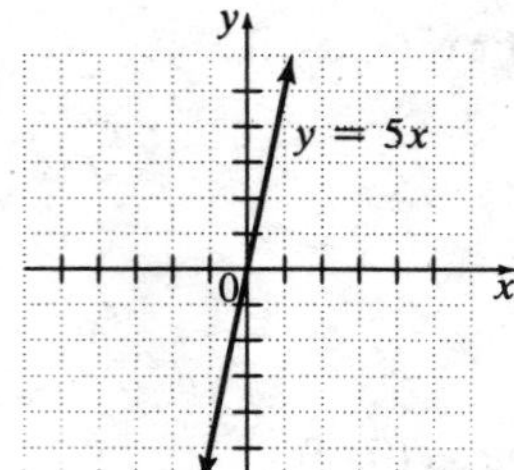

37. $x = -2y$

x	y
2	–1
0	0
–2	1

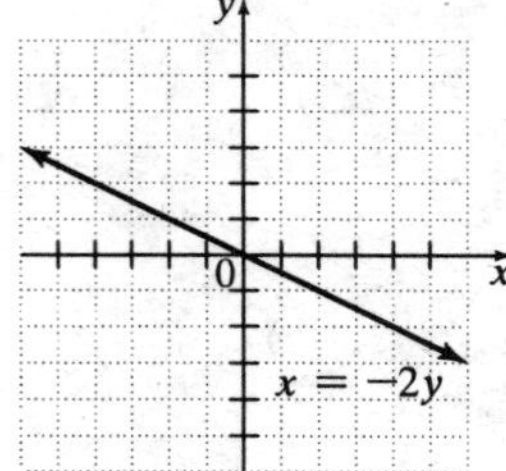

38. $x + y = -1$

x	y
–3	2
–1	0
2	–3

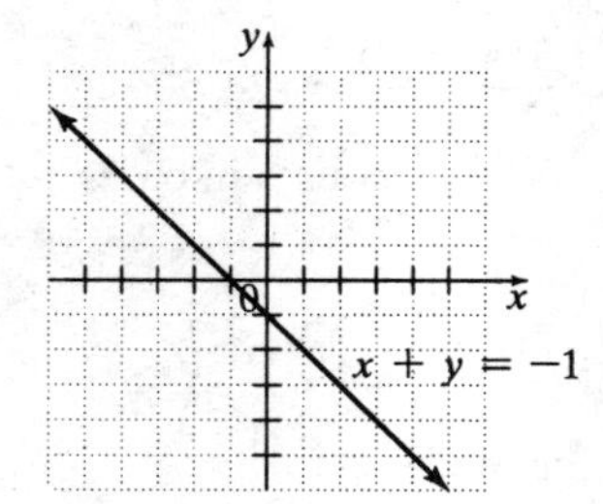

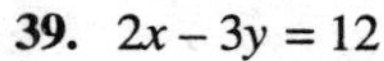

39. $2x - 3y = 12$

x	y
0	−4
3	−2
6	0

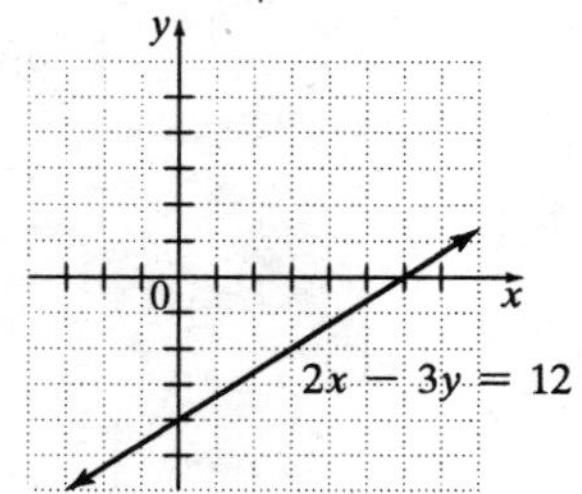

40. $x = \frac{1}{2}y$

x	y
−2	−4
0	0
2	4

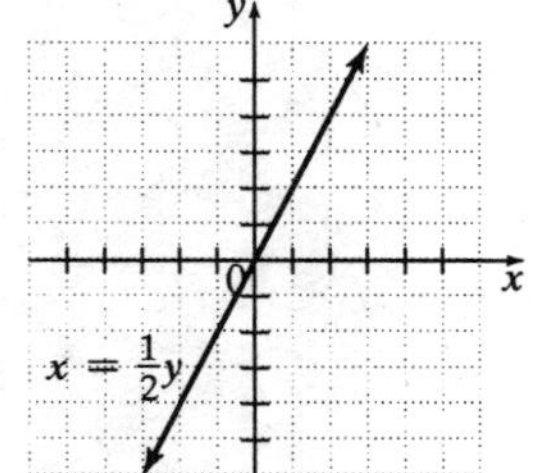

41. $\text{Mean} = \frac{13+23+33+14+6}{5}$

$= \frac{89}{5}$

$= 17.8$

Median: Write the numbers in order to find the middle number. Write the numbers in order.
6, 13, 14, 23, 33
The median is the middle number:
Median = 14.
Mode: Since each number occurs only once, there is no mode.

42. $\text{Mean} = \frac{45+21+60+86+64}{5}$

$= \frac{276}{5}$

$= 55.2$

Median: Write the numbers in order to find the middle number. Write the numbers in order.
21, 45, 60, 64, 86
The median is the middle number:
Median = 60.
Mode: Since each number occurs only once, there is no mode.

43. Mean $= \dfrac{\$14,000 + \$20,000 + \$12,000 + \$20,000 + \$36,000 + \$45,000}{6} = \dfrac{\$147,000}{6} = \$24,500$

Median: Write the numbers in order to find the middle number. Write the dollar amounts in order.
\$12,000; \$14,000; \$20,000; \$20,000; \$36,000; \$45,000
The median is the average of the two middle dollar amounts.

median $= \dfrac{\$20,000 + \$20,000}{2} = \$20,000$

Mode: \$20,000 occurs more often than the other numbers, so \$20,000 is the mode.

44. Average $= \dfrac{560 + 620 + 123 + 400 + 410 + 300 + 400 + 780 + 430 + 450}{10} = \dfrac{4473}{10} = 447.3$

Median: Write the numbers in order to find the middle number. Write the numbers in order.
123, 300, 400, 400, 410, 430, 450, 560, 620, 780
The median is the average of the two middle numbers.

median $= \dfrac{410 + 430}{2} = 420$

Mode: The number 400 occurs more often than the other numbers, so 400 is the mode.

45.

Grade	Point Value	Credit Hours	$\begin{pmatrix}\text{Point}\\\text{Value}\end{pmatrix} \cdot \begin{pmatrix}\text{Credit}\\\text{Hours}\end{pmatrix}$
A	4	3	12
A	4	3	12
C	2	2	4
B	3	3	9
C	2	1	2
Totals		12	39

Grade Point Average $= \dfrac{39}{12} = 3.25$

46.

Grade	Point Value	Credit Hours	$\begin{pmatrix}\text{Point}\\\text{Value}\end{pmatrix} \cdot \begin{pmatrix}\text{Credit}\\\text{Hours}\end{pmatrix}$
B	3	3	9
B	3	4	12
C	2	2	4
D	1	2	2
B	3	3	9
Totals		14	36

Grade Point Average $= \dfrac{36}{14} \approx 2.57$

47.

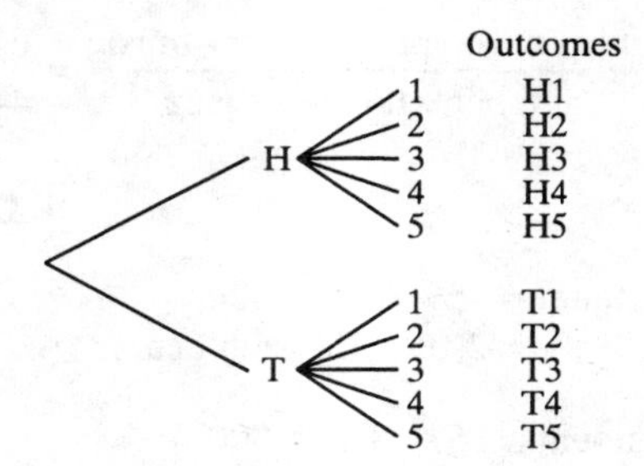

There are 10 outcomes.

48.

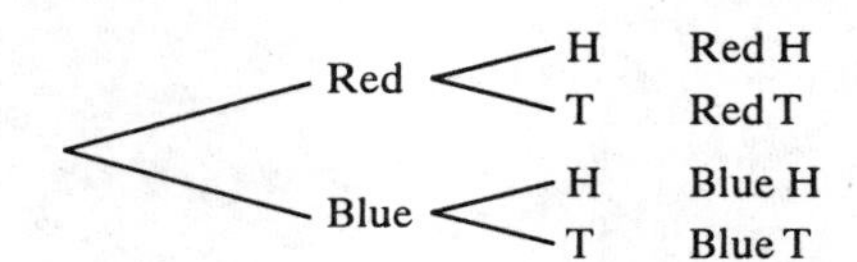

There are 4 outcomes.

49.

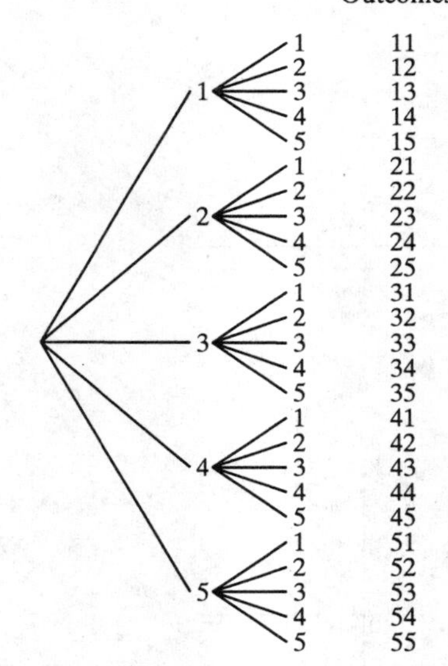

There are 25 outcomes.

50.

Outcomes
Red — Red — Red Red
Red — Blue — Red Blue
Blue — Red — Blue Red
Blue — Blue — Blue Blue

There are 4 outcomes.

51.

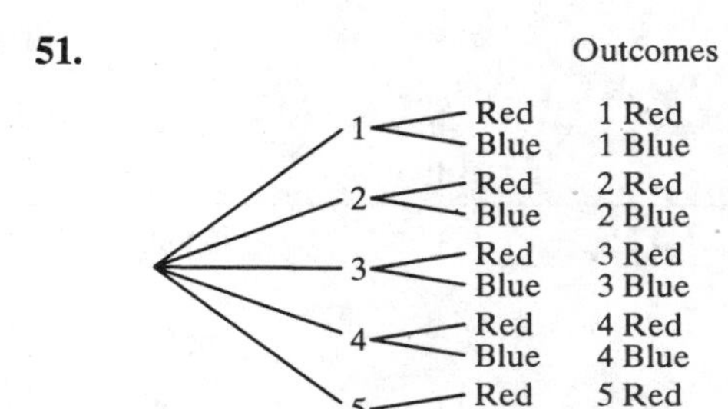

There are 10 outcomes.

52. possible outcomes: 1, 2, 3, 4, 5, 6
probability of a 4 $= \frac{1}{6}$

53. possible outcomes: 1, 2, 3, 4, 5, 6
probability of a 3 $= \frac{1}{6}$

54. possible outcomes: 1, 2, 3, 4, 5
probability of a 4 $= \frac{1}{5}$

55. possible outcomes: 1, 2, 3, 4, 5
probability of a 3 $= \frac{1}{5}$

56. possible outcomes: 1, 2, 3, 4, 5
probability of a 1, 3, or 5 $= \frac{3}{5}$

57. possible outcomes: 1, 2, 3, 4, 5
probability of a 2 or a 4 $= \frac{2}{5}$

Chapter 8 Test

1. The second week has $4\frac{1}{2}$ bills and each bill represents \$50. $4.5 \cdot \$50 = \$225$ was collected during the second week.

2. The most bills are for week 3 with 7 bills. Week 3 took in the most money:
$7 \cdot \$50 = \$350$

3.

Week	Number of Bills	$\begin{pmatrix}\text{Number}\\\text{of Bills}\end{pmatrix}\cdot \$50$
1	3	\$150
2	4.5	\$225
3	7	\$350
4	5.5	\$275
5	2	\$100
	Total	\$1100

4. The bars for the months of June, August, and September have heights above 9. June, August, and September normally have more than 9 centimeters of monthly precipitation in Chicago.

5. The shortest bar is above the month of February. February has the least amount of normal monthly precipitation with 3 centimeters.

6. The months of March and November show a normal monthly precipitation of 7 centimeters.

7. Locate the darkened point representing 1998. Then move over horizontally to the vertical axis. The sales are approximately \$110 million.

8.

Year	Increase
1997	$47 - 18 = 29$
1998	$110 - 47 = 63$
1999	$240 - 110 = 130$

1999 showed the greatest increase.

9. There are two points below the line representing \$50 million. 1996 and 1997 had sales less than \$50 million.

10. $\dfrac{\text{number who prefer rock}}{\text{number surveyed}} = \dfrac{85}{200} = \dfrac{17}{40}$

11. $\dfrac{\text{number who prefer contry music}}{\text{number who prefer jazz}} = \dfrac{62}{44} = \dfrac{31}{22}$

12. amount $= 25\% \cdot 8,139,000$
$= 2,034,750$ small cars

13. amount $= 16\% \cdot 8,139,000$
$= 1,302,240$ luxury cars

14. (4, 0)

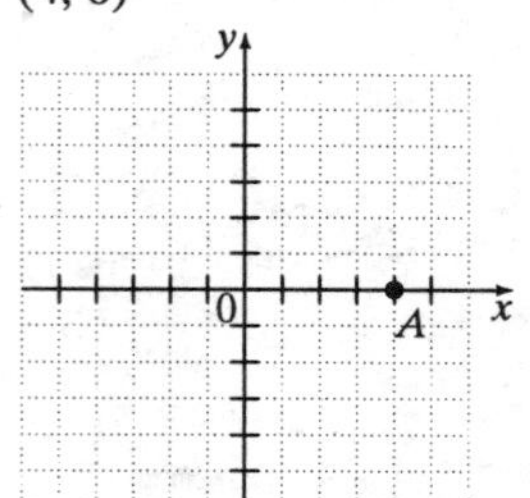

15. (0, –3)

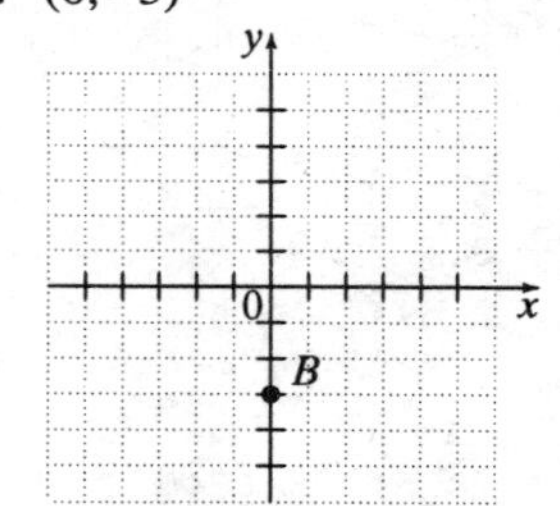

16. (–3, 4)

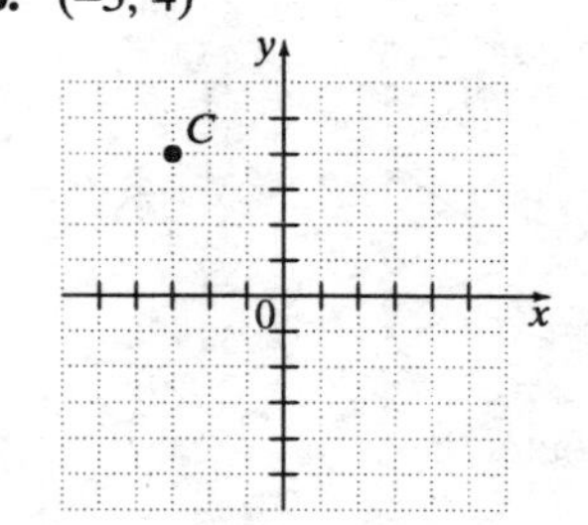

17. (–2, –1)

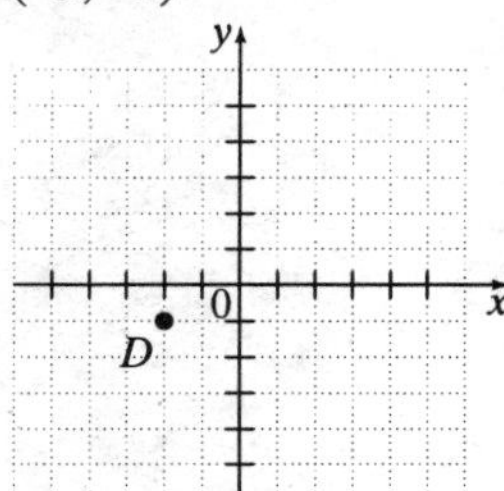

18. $x = -6y$
$0 = -6y$
$0 = y$
The ordered pair is (0, 0).

$x = -6y$
$x = -6(1)$
$x = -6$
The ordered pair is (–6, 1).

$x = -6y$
$12 = -6y$
$-2 = y$
The ordered pair is (12, –2).

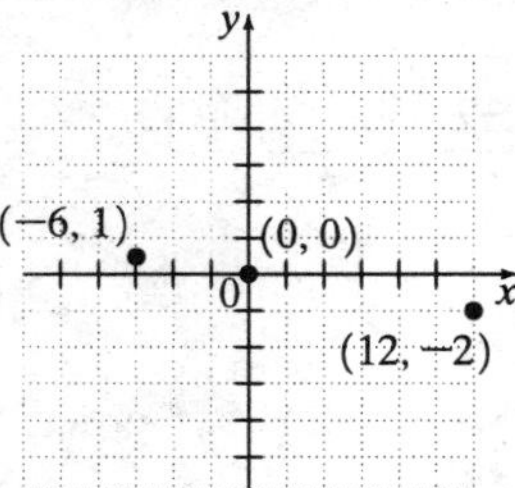

19. $y = 7x - 4$
$y = 7(2) - 4$
$y = 14 - 4$
$y = 10$
The ordered pair is (2, 10).

$y = 7x - 4$
$y = 7(-1) - 4$
$y = -7 - 4$
$y = -11$
The ordered pair is (–1, –11).

$y = 7x - 4$
$y = 7(0) - 4$
$y = 0 - 4$
$y = -4$
The ordered pair is (0, –4).

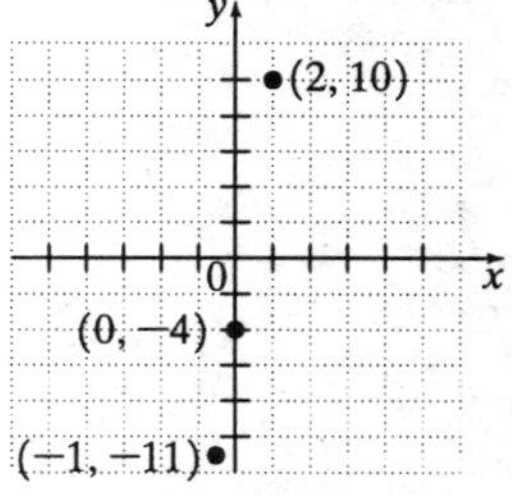

20. $y + x = -4$

x	y
–4	0
–2	–2
0	–4

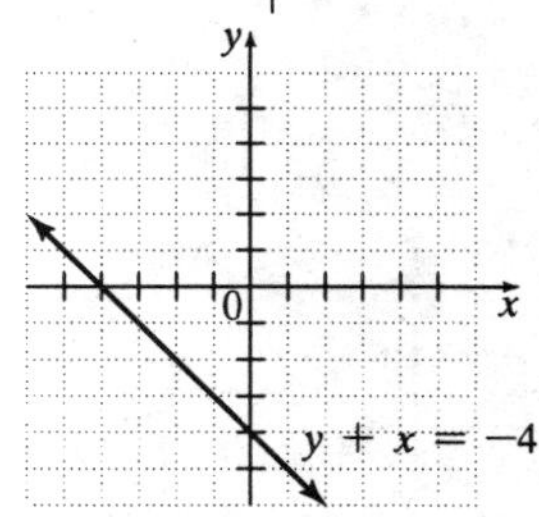

21. $y = -4$

x	y
–5	–4
0	–4
5	–4

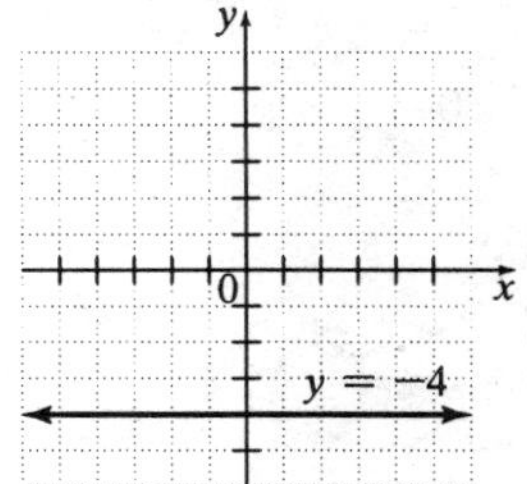

22. $y = 3x - 5$

x	y
0	–5
1	–2
2	1

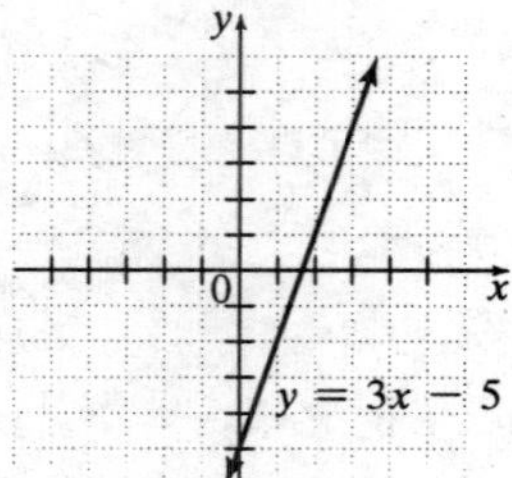

23. $x = 5$

x	y
5	–5
5	0
5	5

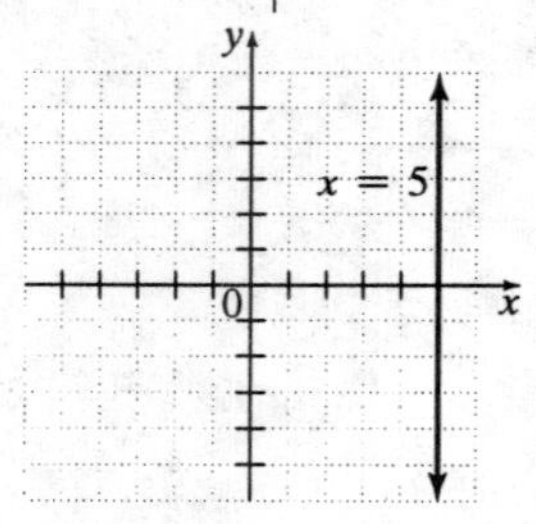

24. $y = -\frac{1}{2}x$

x	y
–4	2
0	0
4	–2

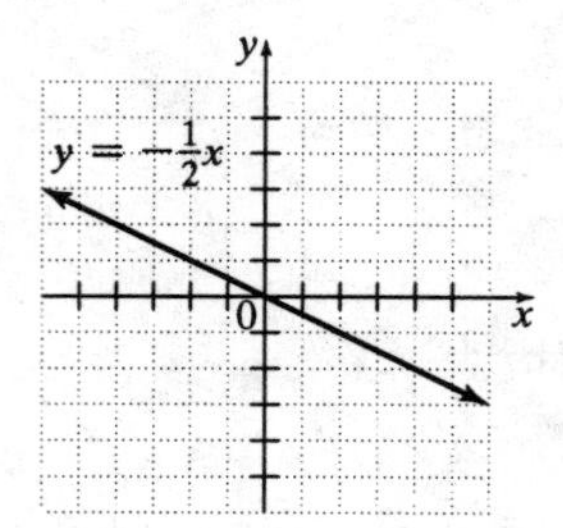

25. $3x - 2y = 12$

x	y
0	–6
2	–3
4	0

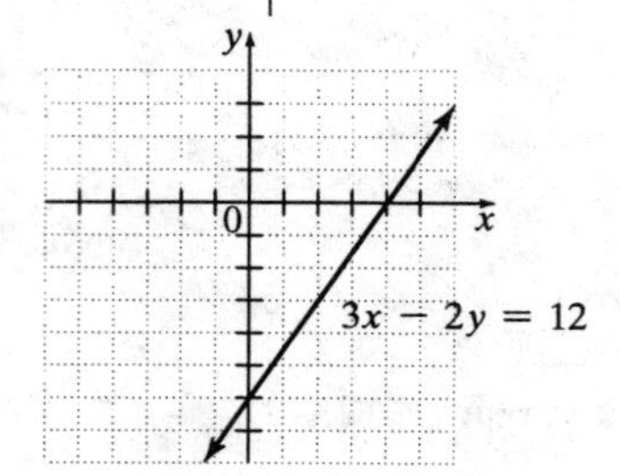

26. Mean $= \frac{26+32+42+43+49}{5}$

$= \frac{192}{5}$

$= 38.4$

Median: Write the numbers in order.
26, 32, 42, 43, 49
The median is the middle number, so median = 42.
Mode: Since each number occurs only once, there is no mode.

27. Mean

$$= \frac{8+10+16+16+14+12+12+13}{8}$$
$$= \frac{101}{8}$$
$$= 12.625$$

Median: Write the numbers in order.
8, 10, 12, 12, 13, 14, 16, 16
The median is the average of the two middle numbers.

$$\text{Median} = \frac{12+13}{2} = 12.5$$

Mode: Since 12 and 16 occur more often than the other numbers, there are two modes: 12 and 16.

28.

Grade	Point Value	Credit Hours	(Point Value)·(Credit Hours)
A	4	3	12
B	3	3	9
C	2	3	6
B	3	4	12
A	4	1	4
	Totals	14	43

$$\text{Grade Point Average} = \frac{43}{14} \approx 3.07$$

29.

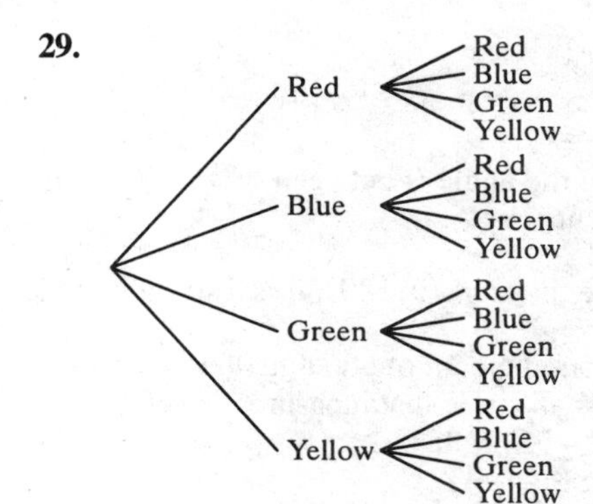

30.

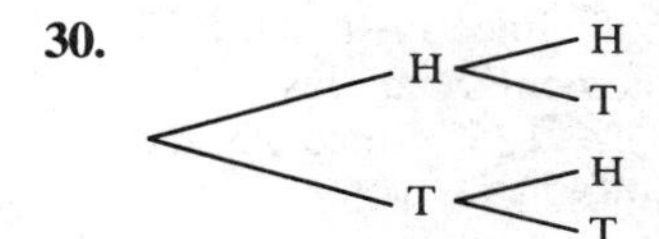

31. possible outcomes: 1, 2, 3, 4, 5, 6, 7, 8, 9, 10
probability of a 6 $= \frac{1}{10}$

33. possible outcomes: 1, 2, 3, 4, 5, 6, 7, 8, 9, 10
probability of a 3 or a 4 $= \frac{2}{10} = \frac{1}{5}$

Chapter 9

Pretest

1. Since the angle is between 0° and 90°, it is an acute angle.

2. Since the angle is 180°, it is a straight angle.

3. The supplement of an angle that measures 92° is an angle that measures $180° - 92° = 88°$.

4. Since $\angle x$ and the 55° angle are vertical angles, $\angle x = 55°$. Since $\angle y$ and the 55° angle are supplementary angles, $\angle y = 180° - 55° = 125°$. Since $\angle y$ and $\angle z$ are vertical angles, $\angle z = 125°$.

5. Since the sum of the measures of the angles of a triangle is 180°, $\angle x = 180° - 40° - 30° = 110°$.

6. $11 \text{ feet} \cdot \frac{1 \text{ yard}}{3 \text{ feet}} = \frac{11 \text{ yards}}{3} = 3\frac{2}{3} \text{ yards}$

7. To convert centimeters to kilometers, move the decimal point 5 places to the left. 6,250,000 cm = 62.5 km

8. $$\begin{aligned} P &= 2l + 2w \\ &= 2 \cdot 24 \text{ inches} + 2 \cdot 6 \text{ inches} \\ &= 48 \text{ inches} + 12 \text{ inches} \\ &= 60 \text{ inches} \end{aligned}$$

9. $$\begin{aligned} C &= 2\pi r \\ &= 2\pi \cdot 9 \text{ yards} \\ &= 18\pi \text{ yards} \\ &\approx 18 \cdot 3.14 \text{ yards} \\ &= 56.52 \text{ yards} \end{aligned}$$

10. $$\begin{aligned} A &= \frac{1}{2}bh \\ &= \frac{1}{2} \cdot 4 \text{ cm} \cdot 3 \text{ cm} \\ &= 6 \text{ square centimeters} \end{aligned}$$

11. $$\begin{array}{r} 7 \text{ lb} \quad 12 \text{ oz} \\ 2\overline{)\ 15 \text{ lb} \quad 8 \text{ oz}} \\ \underline{-14} \qquad\qquad \\ 1 \text{ lb} = \underline{16 \text{ oz}} \\ 24 \text{ oz} \\ \underline{-24 \text{ oz}} \\ 0 \end{array}$$

12. $$\begin{array}{rcr} 9 \text{ gal } 1 \text{ qt} & = & 8 \text{ gal } 5 \text{ qt} \\ \underline{- \quad 2\text{qt}} & & \underline{- \quad 2 \text{ qt}} \\ & & 8 \text{ gal } 3 \text{ qt} \end{array}$$

13. To convert liters to milliliters, move the decimal point 3 places to the right. 25 L = 25,000 ml

14. $$\begin{aligned} C &= \frac{5}{9}(F - 32) \\ &= \frac{5}{9}(118 - 32) \\ &= \frac{5}{9}(86) \\ &\approx 47.8 \end{aligned}$$
$118°F \approx 47.8°C$

Section 9.1

Practice Problems

1. a. The figure has two endpoints. It is line segment RS, or $\overline{RS}$.

   b. The figure is part of a line with one endpoint. It is ray AB, or $\overrightarrow{AB}$.

   c. The figure extends indefinitely in two directions. It is line EF, or $\overleftrightarrow{EF}$.

   d. The figure is two rays with a common endpoint. It is $\angle HVT$, $\angle TVH$, or $\angle V$.

2. Two other ways to name $\angle z$ are $\angle RTS$ or $\angle STR$.

3. a. $\angle M$ is an acute angle. It measures between 0° and 90°.

b. $\angle N$ is a straight angle.

c. $\angle O$ is an obtuse angle. It measures between 90° and 180°.

d. $\angle P$ is a right angle.

4. The complement of an angle that measures 36° is an angle that measures $90° - 36° = 54°$.

5. The supplement of an angle that measures 88° is an angle that measures $180° - 88° = 92°$.

6. $\angle y = 136° - 97° = 39°$

7. Since $\angle a$ and the 112° angle are vertical angles, the measure of $\angle a$ is 112°. $\angle a$ and $\angle b$ are supplementary angles, so $\angle b$ measures $180° - 112° = 68°$. $\angle b$ and $\angle c$ are vertical angles, so $\angle c$ measures 68°.

8. Since $\angle w$ and $\angle x$ are vertical angles and $\angle w = 40°$, $\angle x = 40°$. $\angle y$ and $\angle w$ are corresponding angles, so $\angle y = 40°$. $\angle y$ and $\angle z$ are supplementary angles, so $\angle z = 180° - 40° = 140°$.

9. Since the sum of the measures of the three angles is 180°, we have $\angle x = 180° - 110° - 25° = 45°$.

10. Since the sum of the measures of the three angles is 180°, we have $\angle y = 180° - 90° - 25° = 65°$.

Exercise Set 9.1

1. The figure extends indefinitely in two directions. It is line *yz*, or $\overleftrightarrow{yz}$.

3. The figure has two endpoints. It is line segment *LM*, or $\overline{LM}$.

5. The figure has two endpoints. It is line segment *PQ*, or $\overline{PQ}$.

7. The figure is part of a line with one endpoint. It is ray *UW*, or $\overrightarrow{UW}$.

9. $\angle ABC = 15°$

11. $\angle CBD = 50°$

13. $\angle DBA = 50° + 15° = 65°$

15. $\angle CBE = 50° + 45° = 95°$

17. 90°

19. 0°; 90°

21. $\angle S$ is a straight angle.

23. $\angle R$ is a right angle.

25. Since $\angle Q$ measures between 90° and 180°, it is an obtuse angle.

27. $\angle N$ is a right angle.

29. The complement of an angle that measures 17° is an angle that measures $90° - 17° = 73°$.

31. The supplement of an angle that measures 17° is an angle that measures $180° - 17° = 163°$.

33. The complement of a 45° angle is an angle that measures $90° - 45° = 45°$.

35. The supplement of a 125° angle is an angle that measures $180° - 125° = 55°$.

37. $\angle MNP$ and $\angle RNO$ are complementary angles since $60° + 30° = 90°$.
$\angle PNQ$ and $\angle QNR$ are complementary angles since $52° + 38° = 90°$.

39. $\angle SPT$ and $\angle TPQ$ are supplementary angles since $45° + 135° = 180°$.
$\angle SPR$ and $\angle RPQ$ are supplementary angles since $135° + 45° = 180°$.
$\angle SPT$ and $\angle SPR$ are supplementary angles since $45° + 135° = 180°$.
$\angle TPQ$ and $\angle QPR$ are supplementary angles since $135° + 45° = 180°$.

41. $\angle x = 120° - 88° = 32°$

43. $\angle x = 90° - 15° = 75°$

45. Since $\angle x$ and the 35° angle are vertical angles, $\angle x = 35°$. $\angle x$ and $\angle y$ are supplementary angles, so $\angle y = 180° - 35° = 145°$. $\angle y$ and $\angle z$ are vertical angles, so $\angle z = 145°$.

47. Since $\angle x$ and the 103° angle are supplementary angles, $\angle x = 180° - 103° = 77°$. $\angle y$ and the 103° angle are vertical angles, so $\angle y = 103°$. $\angle x$ and $\angle z$ are vertical angles, so $\angle z = 77°$.

49. $\angle x$ and the 80° angle are adjacent angles, so $\angle x = 180° - 80° = 100°$. $\angle y$ and the 80° angle are alternate interior angles, so $\angle y = 80°$. $\angle x$ and $\angle z$ are corresponding angles, so $\angle z = 100°$.

51. $\angle x$ and the 46° angle are supplementary angles, so $\angle x = 180° - 46° = 134°$. $\angle y$ and the 46° angle are corresponding angles, so $\angle y = 46°$. $\angle x$ and $\angle z$ are corresponding angles, so $\angle z = 134°$.

53. $\angle x = 180° - 70° - 85° = 25°$

55. $\angle x = 180° - 95° - 72° = 13°$

57. $\angle x = 180° - 90° - 50° = 40°$

59. $\frac{7}{8} + \frac{1}{4} = \frac{7}{8} + \frac{2}{8} = \frac{9}{8}$ or $1\frac{1}{8}$

61. $\frac{7}{8} \cdot \frac{1}{4} = \frac{7 \cdot 1}{8 \cdot 4} = \frac{7}{32}$

63. $3\frac{1}{3} - 2\frac{1}{2} = \frac{10}{3} - \frac{5}{2} = \frac{20}{6} - \frac{15}{6} = \frac{5}{6}$

65.
$$\begin{aligned} 3\frac{1}{3} \div 2\frac{1}{2} &= \frac{10}{3} \div \frac{5}{2} \\ &= \frac{10}{3} \cdot \frac{2}{5} \\ &= \frac{10 \cdot 2}{3 \cdot 5} \\ &= \frac{20}{15} \\ &= \frac{4}{3} \text{ or } 1\frac{1}{3} \end{aligned}$$

67. The supplement of a 125.2° angle is $180° - 125.2° = 54.8°$.

69. $\angle a$ and the 60° angle are alternate interior angles, so $\angle a = 60°$.
$\angle a$, $\angle b$, and the 70° angle form a straight angle, so
$$\begin{aligned} \angle b &= 180° - 70° - \angle a \\ &= 180° - 70° - 60° \\ &= 50°. \end{aligned}$$
$\angle c$ and the angle that is $\angle a$ and $\angle b$ together are alternate interior angles, so
$\angle c = 60° + 50° = 110°$.
$\angle c$ and $\angle d$ are adjacent angles, so
$\angle d = 180° - 110° = 70°$.
$\angle e$ and the 60° angle are adjacent angles, so
$\angle e = 180° - 60° = 120°$.

71. The sum of the measures of the angles of a triangle is 180°.

Section 9.2

Practice Problems

1.
$$\begin{aligned} 5 \text{ ft} &= \frac{5 \text{ ft}}{1} \cdot \frac{12 \text{ in.}}{1 \text{ ft}} \\ &= 5 \cdot 12 \text{ in.} \\ &= 60 \text{ in.} \end{aligned}$$

2.
$$\begin{aligned} 7 \text{ yd} &= \frac{7 \text{ yd}}{1} \cdot \frac{3 \text{ ft}}{1 \text{ yd}} \\ &= 7 \cdot 3 \text{ ft} \\ &= 21 \text{ ft} \end{aligned}$$

3.
$$\begin{array}{r} 5 \\ 12\overline{)\ 68} \\ \underline{-60} \\ 8 \end{array}$$
68 in. = 5 ft 8 in.

4. $5 \text{ yd} = \frac{5 \text{ yd}}{1} \cdot \frac{3 \text{ ft}}{1 \text{ yd}} = 15 \text{ ft}$
5 yd 2 ft = 15 ft + 2 ft = 17 ft

5.
$$\begin{array}{r} 4 \text{ ft } \ 8 \text{ in.} \\ \underline{+\ 8 \text{ ft } 11 \text{ in.}} \\ 12 \text{ ft } 19 \text{ in.} \end{array} = 12 \text{ ft} + 1 \text{ ft } 7 \text{ in.} = 13 \text{ ft } 7 \text{ in.}$$

6.
$$\begin{array}{r} 4 \text{ ft } 7 \text{ in.} \\ \times \quad\quad 4 \\ \hline 16 \text{ ft } 28 \text{ in.} \end{array} = 16 \text{ ft} + 2 \text{ ft } 4 \text{ in.} = 18 \text{ ft } 4 \text{ in.}$$

7.
$$\begin{array}{r} 9 \text{ ft } 3 \text{ in.} \\ 2\overline{)\ 18 \text{ ft } 6 \text{ in.}} \\ -18 \text{ ft} \quad\quad \\ \hline 0 \quad 6 \text{ in.} \\ -6 \text{ in.} \\ \hline 0 \end{array}$$

8.
$$\begin{array}{r} 5 \text{ ft } 8 \text{ in.} \\ -\ 1 \text{ ft } 9 \text{ in.} \\ \hline \end{array} = \begin{array}{r} 4 \text{ ft } 20 \text{ in.} \\ -\ 1 \text{ ft } \ 9 \text{ in.} \\ \hline 3 \text{ ft } 11 \text{ in.} \end{array}$$

9. $3.5 \text{ m} = \frac{3.5 \text{ m}}{1} \cdot \frac{0.001 \text{ km}}{1 \text{ m}}$
$= 0.0035 \text{ km}$

10. To convert meters to millimeters, move decimal point 3 places to the right.
2.5 m = 2500 mm

11. 640 m = 0.64 km or 2.1 km = 2100 m
$$\begin{array}{r} 2.1 \text{ km} \\ -\ 0.64 \text{ km} \\ \hline 1.46 \text{ km} \end{array} \quad \text{or} \quad \begin{array}{r} 2100 \text{ m} \\ -\ 640 \text{ m} \\ \hline 1460 \text{ m} \end{array}$$

12.
$$\begin{array}{r} 18.3 \text{ hm} \\ \times\ 5 \quad \\ \hline 91.5 \text{ hm} \end{array}$$

13. 0.8 m = 80 cm or 45 cm = 0.45 m
$$\begin{array}{r} 80 \text{ cm} \\ +\ 45 \text{ cm} \\ \hline 125 \text{ cm} \end{array} \quad \text{or} \quad \begin{array}{r} 0.80 \text{ m} \\ 0.45 \text{ m} \\ \hline 1.25 \text{ m} \end{array}$$
The scarf will be 125 cm or 1.25 m long.

Calculator Explorations

1. $7 \times 3.28 = 22.96$
Therefore 7 meters ≈ 22.96 feet.

2. $11.5 \times 0.914 = 10.511$
Therefore 11.5 yards ≈ 10.511 meters.

3. $8.5 \times 2.54 = 21.59$
Therefore 8.5 inches = 21.59 centimeters.

4. $15 \times 0.62 = 9.3$
Therefore 15 kilometers ≈ 9.3 miles.

5. $5 \times 0.62 = 3.1$
Therefore 5 kilometers ≈ 3.1 miles.

6. $100 \times 1.09 = 109$
Therefore 100 meters ≈ 109 yards.

Mental Math

1. 12 in. = 1 ft

2. 6 ft. = 2 yd

3. 24 in. = 2 ft

4. 36 in. = 3 ft

5. 36 in. = 1 yd

6. 2 yd = 6 ft = 72 in.

7. No; 30 meters is too long.

8. Yes

9. Yes

10. No; 4 kilometers is much too long.

11. No; 50 kilometers is too long.

12. Yes

Exercise Set 9.2

1. $60 \text{ in.} = 60 \text{ in.} \cdot \frac{1 \text{ ft}}{12 \text{ in.}} = \frac{60 \text{ ft}}{12} = 5 \text{ ft}$
60 inches is 5 feet.

3. $12 \text{ yd} = 12 \text{ yd} \cdot \frac{3 \text{ ft}}{1 \text{ yd}} = 12 \cdot 3 \text{ ft} = 36 \text{ ft}$

5. $42,240 \text{ ft} = 42,240 \text{ ft} \cdot \frac{1 \text{ mi}}{5280 \text{ ft}}$
$= \frac{42,240 \text{ mi}}{5280}$
$= 8 \text{ mi}$
42,240 feet is 8 miles.

7. $102 \text{ in.} = 102 \text{ in.} \cdot \frac{1 \text{ ft}}{12 \text{ in.}} = \frac{102 \text{ ft}}{12} = 8\frac{1}{2} \text{ ft}$

9. $10 \text{ ft} = 10 \text{ ft} \cdot \frac{1 \text{ yd}}{3 \text{ ft}} = \frac{10 \text{ yd}}{3} = 3\frac{1}{3} \text{ yd}$

10 feet is $3\frac{1}{3}$ yards.

11. $6.4 \text{ mi} = 6.4 \text{ mi} \cdot \frac{5280 \text{ ft}}{1 \text{ mi}}$
$= 6.4 \cdot 5280 \text{ ft}$
$= 33,792 \text{ ft}$

13. 40 ft ÷ 3 = 13 with remainder 1
40 ft = 13 yd 1 ft

15. 41 in. ÷ 12 = 3 with remainder 5
41 in. = 3 ft 5 in.

17. 10,000 ft ÷ 5280 = 1 with remainder 4720
10,000 ft = 1 mi 4720 ft

19. $5 \text{ ft} = 5 \text{ ft} \cdot \frac{12 \text{ in.}}{1 \text{ ft}} = 60 \text{ in.}$
5 ft 2 in. = 60 in. + 2 in. = 62 in.

21. $5 \text{ yd} = 5 \text{ yd} \cdot \frac{3 \text{ ft}}{1 \text{ yd}} = 15 \text{ft}$
5 yd 2 ft = 15 ft + 2 ft = 17 ft

23. $2 \text{ yd} = 2 \text{ yd} \cdot \frac{36 \text{ in.}}{1 \text{ yd}} = 72 \text{ in.}$
$1 \text{ ft} = 1 \text{ ft} \cdot \frac{12 \text{ in.}}{1 \text{ ft}} = 12 \text{ in.}$
2 yd 1 ft = 72 in. + 12 in. = 84 in.

25.
$$\begin{array}{r} 5 \text{ ft } 8 \text{ in.} \\ +\ 6 \text{ ft } 7 \text{ in.} \\ \hline 11 \text{ ft } 15 \text{ in.} \end{array} = 11 \text{ ft} + 1 \text{ ft } 3 \text{ in.} = 12 \text{ ft } 3 \text{ in.}$$

27.
$$\begin{array}{r} 12 \text{ yd } 2 \text{ ft} \\ +\ 9 \text{ yd } 2 \text{ ft} \\ \hline 21 \text{ yd } 4 \text{ ft} \end{array} = 21 \text{ yd} + 1 \text{ yd } 1 \text{ ft} = 22 \text{ yd } 1 \text{ ft}$$

29.
$$\begin{array}{r} 24 \text{ ft } 8 \text{ in.} \\ -\ 16 \text{ ft } 3 \text{ in.} \\ \hline 8 \text{ ft } 5 \text{ in.} \end{array}$$

31.
$$\begin{array}{r} 16 \text{ ft } 3 \text{ in.} \\ -\ 10 \text{ ft } 9 \text{ in.} \\ \hline \end{array} = \begin{array}{r} 15 \text{ ft } 15 \text{ in.} \\ -\ 10 \text{ ft } 9 \text{ in.} \\ \hline 5 \text{ ft } 6 \text{ in.} \end{array}$$

33.
$$\begin{array}{r} 3 \text{ ft } 4 \text{ in.} \\ 2 \overline{)\ 6 \text{ ft } 8 \text{ in.}} \\ \underline{-6 \text{ ft}} \quad\quad \\ 0 \quad 8 \text{ in.} \\ \underline{-8 \text{ in.}} \\ 0 \end{array}$$

35.
$$\begin{array}{r} 12 \text{ yd } 2 \text{ ft} \\ \times \quad\quad 4 \\ \hline 48 \text{ yd } 8 \text{ ft} \end{array} = 48 \text{ yd} + 2 \text{ yd } 2 \text{ ft}$$
$= 50 \text{ yd } 2 \text{ ft}$

37.
$$\begin{array}{r} 6 \text{ ft } 10 \text{ in.} \\ +\ 3 \text{ ft } 8 \text{ in.} \\ \hline 9 \text{ ft } 18 \text{ in.} \end{array} = 9 \text{ ft} + 1 \text{ ft } 6 \text{ in.}$$
$= 10 \text{ ft } 6 \text{ in.}$

The bamboo is 10 ft 6 in. tall.

39.
$$\begin{array}{r} 42 \text{ ft} \quad\quad \\ -\ 23 \text{ ft } 7 \text{ in.} \\ \hline \end{array} = \begin{array}{r} 41 \text{ ft } 12 \text{ in.} \\ -\ 23 \text{ ft } 7 \text{ in.} \\ \hline 18 \text{ ft } 5 \text{ in.} \end{array}$$

The right arm is 18 ft 5 in. longer than the tablet.

41.
$$\begin{array}{r} 1 \text{ ft } 9 \text{ in.} \\ \times \quad\quad 9 \\ \hline 9 \text{ ft } 81 \text{ in.} \end{array} = 9 \text{ ft} + 6 \text{ ft } 9 \text{ in.}$$
$= 15 \text{ ft } 9 \text{ in.}$

9 stacks would extend 15 ft 9 in. from the wall.

43.
$$\begin{array}{r} 3 \text{ ft } 1 \text{ in.} \\ 3 \overline{)\ 9 \text{ ft } 3 \text{ in.}} \\ \underline{-9 \text{ ft}} \quad\quad \\ 0 \quad 3 \text{ in.} \\ \underline{-3 \text{ in.}} \\ 0 \end{array}$$

Each cut piece will be 3 ft 1 in. long.

45. $\text{Perimeter} = 24\text{ ft }9\text{ in.} + 18\text{ ft }6\text{ in.} + 24\text{ ft }9\text{ in.} + 18\text{ ft }6\text{ in.}$
$= 84\text{ ft }30\text{ in.}$
$= 84\text{ ft} + 2\text{ ft }6\text{ in.}$
$= 86\text{ ft }6\text{ in.}$
She must purchase 86 ft 6 in. of fencing material.

47. $79\text{ ft} = 79\text{ ft} \cdot \frac{1\text{ yd}}{3\text{ ft}} = \frac{79}{3}\text{ yd}$
$\frac{79}{3} \cdot 4 = \frac{316}{3} = 105\frac{1}{3}$
4 trucks are $105\frac{1}{3}$ yd long.

49. To convert meters to centimeters, move the decimal 2 places to the right.
40 m = 4000 cm

51. To convert millimeters to centimeters, move the decimal 1 place to the right.
40 mm = 4.0 cm = 4 cm

53. To convert meters to kilometers, move the decimal 3 places to the left.
300 m = 0.300 km = 0.3 km

55. To convert millimeters to meters, move the decimal 3 places to the left.
1400 mm = 1.4 m

57. To convert centimeters to meters, move the decimal 2 places to the left.
1500 cm = 15 m

59. To convert centimeters to millimeters, move the decimal 1 place to the right.
8.3 cm = 83 mm

61. To convert millimeters to decimeters, move the decimal 2 places to the left.
20.1 mm = 0.201 dm

63. To convert meters to millimeters, move the decimal 3 places to the right.
0.04 m = 0040. mm = 40 mm

65.
$$\begin{array}{r} 8.60\text{ m} \\ +\ 0.34\text{ m} \\ \hline 8.94\text{ m} \end{array}$$

67.
$$\begin{array}{rcl} 2.9\text{ m} & = & 2.90\text{ m} \\ +\ 40.0\text{ mm} & = & 0.04\text{ m} \\ \hline & & 2.94\text{ m or } 2940\text{ mm} \end{array}$$

69.
$$\begin{array}{rcr} 24.8\text{ mm} & = & 2.48\text{ cm} \\ -\ 1.19\text{ cm} & = & -\ 1.19\text{ cm} \\ \hline & & 1.29\text{ cm or } 12.9\text{ mm} \end{array}$$

71.
$$\begin{array}{rcl} 15\text{ km} & = & 15.000\text{ km} \\ -\ 2360\text{ m} & = & -\ 2.360\text{ km} \\ \hline & & 12.640\text{ km or } 12{,}640\text{ m} \end{array}$$

73.
$$\begin{array}{r} 18.3\text{ m} \\ \times\ 3 \\ \hline 5.49\text{ m} \end{array}$$

75.
$$\begin{array}{r} 1.55 \\ 4\overline{)\ 6.2} \\ -4 \\ \hline 2.2 \\ -2\ 0 \\ \hline 20 \\ -20 \\ \hline 0 \end{array}$$
6.2 km ÷ 4 = 1.55 km

77. 3.4 m + 5.8 m – 8 cm
= 3.4 m + 5.8 m – 0.08 m
= 9.12 m
The tied ropes are 9.12 m long.

79. 80 mm – 5.33 cm
= 80 mm – 53.3 mm
= 26.7 mm
The ice must be 26.7 mm thicker.

81. 25(1 m + 65 cm) = 25(100 cm + 65 cm)
= 25(165 cm)
= 4125 cm or 41.25 m
4125 cm or 41.25 m of wood must be ordered.

83.
$$\begin{array}{r} 3.35 \\ 20\overline{)\ 67.00} \\ -60 \\ \hline 70 \\ -60 \\ \hline 100 \\ -100 \\ \hline 0 \end{array}$$
67 m ÷ 20 = 3.35 m
Each piece will be 3.35 m long.

85. 5.988 km + 21 m = 5988 m + 21 m
= 6009 m or 6.009 km
The elevation is 6009 m or 6.009 km.

87. 3.429 m ÷ 22.86 cm = 3.429 m ÷ 0.2286 m
$0.2286\overline{)3.429}$
Move the decimal points 4 places.
$$\begin{array}{r} 15 \\ 2286.\overline{)\ 34290.} \\ -2286 \\ \hline 11430 \\ -11430 \\ \hline 0 \end{array}$$
15 tiles are needed.

89. perimeter $= 10 + 6 + 10 + 6$
$= 32$
The perimeter is 32 yards.

91. perimeter $= 7 + 7 + 7 + 7 + 7$
$= 35$
The perimeter is 35 meters.

93. Answers may vary.

95. A square has 4 equal sides.
$$\begin{array}{r} 6.575 \\ 4\overline{)\ 26.300} \\ -24 \\ \hline 2\ 3 \\ -2\ 0 \\ \hline 30 \\ -28 \\ \hline 20 \\ -20 \\ \hline 0 \end{array}$$
Each side will be 6.575 m long.

Section 9.3

Practice Problems

1. Perimeter
= 60 feet + 60 feet + 80 feet + 80 feet
= 280 feet

2. $P = 2l + 2w$
$= 2 \cdot 22$ centimeters $+ 2 \cdot 10$ centimeters
= 44 centimeters + 20 centimeters
= 64 centimeters

3. $P = 4s = 4 \cdot 50$ yards = 200 yards

4. $P = a + b + c = 5 \text{ cm} + 9 \text{ cm} + 7 \text{ cm} = 21 \text{ cm}$

5. $\text{Perimeter} = 3 \text{ km} + 7 \text{ km} + 3 \text{ km} + 4 \text{ km}$
$= 17 \text{ km}$

6. There are two sides of the room that are unknown. One side is $20 \text{ m} - 15 \text{ m} = 5 \text{ m}$, and the other side is $26 \text{ m} - 7 \text{ m} = 19 \text{ m}$. The perimeter of the room is
$5 \text{ m} + 19 \text{ m} + 15 \text{ m} + 26 \text{ m} + 20 \text{ m} + 7 \text{ m} = 92 \text{ m}$.

7. First find the perimeter of the room.
$P = 2l + 2w$
$= 2 \cdot 60 \text{ feet} + 2 \cdot 120 \text{ feet}$
$= 120 \text{ feet} + 240 \text{ feet}$
$= 360 \text{ feet}$
The cost for the fencing is
$\text{cost} = 1.90 \cdot 360 = 684$
The cost of the fencing is \$684.

8. $C = \pi d = \pi \cdot 20 \text{ yards} = 20\pi \text{ yards}$
If we use 3.14 as an approximation of π, then the circumference is approximately
$20 \text{ yards} \cdot 3.14 = 62.8 \text{ yards}$.

Exercise Set 9.3

1. $P = 2l + 2w$
$= 2 \cdot 17 \text{ feet} + 2 \cdot 15 \text{ feet}$
$= 34 \text{ feet} + 30 \text{ feet}$
$= 64 \text{ feet}$

3. $P = 4s = 4(9 \text{ centimeters}) = 36 \text{ centimeters}$

5. $P = a + b + c$
$= 5 \text{ inches} + 7 \text{ inches} + 9 \text{ inches}$
$= 21 \text{ inches}$

7. $P = 2l + 2w$
$= 2(35 \text{ centimeters}) + 2(25 \text{ centimeters})$
$= 70 \text{ centimeters} + 50 \text{ centimeters}$
$= 120 \text{ centimeters}$

9. $P = 10 \text{ feet} + 8 \text{ feet} + 8 \text{ feet} + 15 \text{ feet} + 7 \text{ feet}$
$= 48 \text{ feet}$

11. $P = 12 \text{ inches} + 3 \text{ inches} + 15 \text{ inches} + 24 \text{ inches} + 12 \text{ inches}$
$= 66 \text{ inches}$

13. $P = 5 \text{ feet} + 3 \text{ feet} + 2 \text{ feet} + 7 \text{ feet} + 4 \text{ feet}$
$= 21 \text{ feet}$

15. $P = 4s = 4(15 \text{ feet}) = 60 \text{ feet}$

17. $P = 2l + 2w$
$= 2(120 \text{ yards}) + 2(53 \text{ yards})$
$= 240 \text{ yards} + 106 \text{ yards}$
$= 346 \text{ yards}$

19. $P = 2l + 2w$
$= 2(8 \text{ feet}) + 2(3 \text{ feet})$
$= 16 \text{ feet} + 6 \text{ feet}$
$= 22 \text{ feet}$

21. Since 22 feet of stripping is needed,
cost = \$3(22) = \$66.

23. $P = 6(6 \text{ inches}) = 36 \text{ inches}$

25. $P = 4s = 4(7 \text{ inches}) = 28 \text{ inches}$

27. First find the perimeter of the room.
$P = 2l + 2w$
$= 2(6 \text{ feet}) + 2(8 \text{ feet})$
$= 12 \text{ feet} + 16 \text{ feet}$
$= 28 \text{ feet}$
The cost of the wallpaper is
$c = 0.86(28) = \$24.08$.

29. The missing lengths are:
28 meters – 20 meters = 8 meters
20 meters – 17 meters = 3 meters
$P = 8 \text{ meters} + 3 \text{ meters} + 20 \text{ meters} + 20 \text{ meters} + 28 \text{ meters} + 17 \text{ meters}$
$= 96 \text{ meters}$

31. The missing length is: 3 feet + 6 feet + 4 feet = 13 feet
$P = 13 \text{ feet} + 15 \text{ feet} + 3 \text{ feet} + 5 \text{ feet} + 6 \text{ feet} + 5 \text{ feet} + 4 \text{ feet} + 15 \text{ feet}$
$= 66 \text{ feet}$

33. The missing lengths are:
12 miles + 10 miles = 22 miles
8 miles + 34 miles = 42 miles
$P = 12 \text{ miles} + 34 \text{ miles} + 10 \text{ miles} + 8 \text{ miles} + 22 \text{ miles} + 42 \text{ miles}$
$= 128 \text{ miles}$

35. $C = \pi d$
$= \pi 17 \text{ centimeters}$
$\approx 3.14 \cdot 17 \text{ centimeters}$
$= 53.38 \text{ centimeters}$

37. $C = 2\pi r$
$= 2\pi \cdot 8$ miles
$= 16\pi$ miles
$\approx 16 \cdot 3.14$ miles
$= 50.24$ miles

39. $C = \pi d$
$= \pi \cdot 26$ meters
$= 26\pi$ meters
$\approx 26 \cdot 3.14$ meters
$= 81.64$ meters

41. $C = 2\pi r$
$= 2\pi \cdot 5$ feet
$= 10\pi$ feet
$\approx 10 \cdot \frac{22}{7}$ feet
$= \frac{220}{7}$ feet
$= 31\frac{3}{7}$ feet

43. $C = \pi d$
$= \pi \cdot 4000$ feet
$= 4000\pi$ feet
$\approx 4000 \cdot 3.14$ feet
$= 12,560$ feet

45. $5 + 6 \cdot 3 = 5 + 18 = 23$

47. $(20 - 16) \div 4 = 4 \div 4 = 1$

49. $(18 + 8) - (12 + 4) = 26 - 16 = 10$

51. $(72 \div 2) \cdot 6 = 36 \cdot 6 = 216$

53. Since a fence goes along the edge of the yard, we are concerned with the perimeter of the yard.

55. Since carpet covers the surface of the floor, we are concerned with the area of the room.

57. Since we paint the surface of the wall, we are concerned with the area of the wall.

59. Since the wallpaper border goes along the edge of the room, we are concerned with the perimeter of the wall.

61. **a.** Circumference of the smaller circle:
$C = 2\pi r$
$= 2\pi \cdot 10$ meters
$= 20\pi$ meters
$\approx 20 \cdot 3.14$ meters
$= 62.8$ meters
Circumference of the larger circle:
$C = 2\pi r$
$= 2\pi \cdot 20$ meters
$= 40\pi$ meters
$\approx 40 \cdot 3.14$ meters
$= 125.6$ meters

b. Yes, the circumference is doubled.

63. The perimeter of the skating rink is equal to the circumference of a circle with radius 5 m plus two sides of a rectangle, each of length 22 m.
Perimeter $= 2\pi \cdot 5 \text{ m} + 2 \cdot 22 \text{ m}$
$= 10\pi \text{ m} + 44 \text{ m}$
$\approx 10 \cdot 3.14 \text{ m} + 44 \text{ m}$
$= 75.4 \text{ m}$

Section 9.4

Practice Problems

1. $A = \frac{1}{2} byh$
$= \frac{1}{2} \cdot 8 \text{ in.} \cdot 6\frac{1}{4} \text{ in.}$
$= \frac{1}{2} \cdot \frac{8}{1} \cdot \frac{25}{4}$ square inches
$= 25$ square inches

2. $A = \frac{1}{2}(b + B)h$
$= \frac{1}{2} \cdot (4 \text{ yd} + 10 \text{ yd}) \cdot 4.1 \text{ yd}$
$= \frac{1}{2} \cdot 14 \text{ yd} \cdot 4.1 \text{ yd}$
$= 28.7$ square yards

3. Split the figure into two rectangles, and find the sum of the areas of the rectangles.

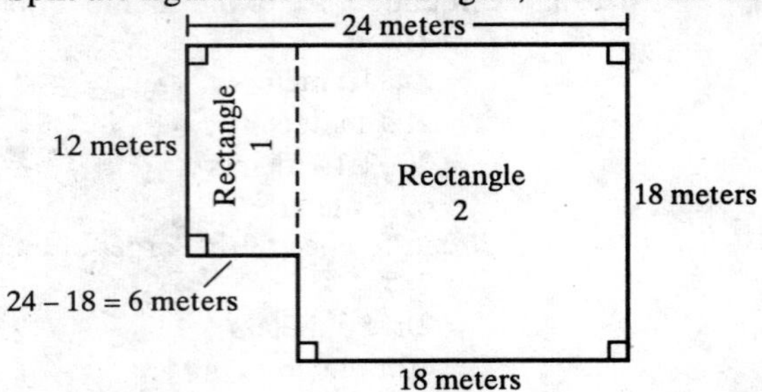

The area of rectangle 1
= (6 meters)(12 meters)
= 72 square meters

The area of rectangle 2
= (18 meters)(18 meters)
= 324 square meters

The area of the figure
= 324 square meters + 72 square meters
= 396 square meters

4. $A = \pi r^2$
$= \pi \cdot (7 \text{ centimeters})^2$
$= 49\pi$ square centimeters
$\approx 49 \cdot 3.14$ square centimeters
$= 153.86$ square centimeters

5.

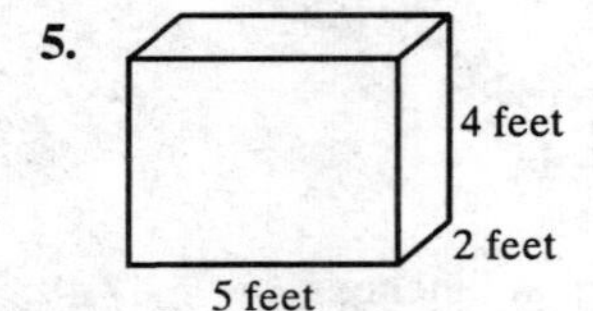

$V = lwh$
$= 5 \text{ feet} \cdot 2 \text{ feet} \cdot 4 \text{ feet}$
$= 40$ cubic feet

6. $\frac{1}{2}$ centimeter

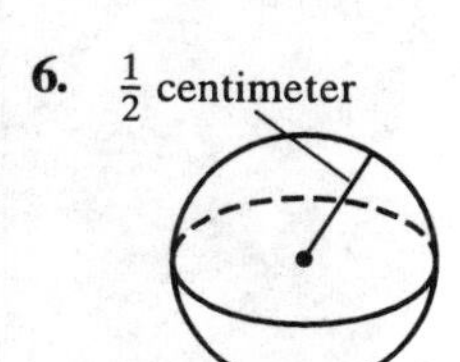

$$V = \frac{4}{3}\pi r^3$$
$$= \frac{4}{3}\pi \cdot \left(\frac{1}{2} \text{ centimeter}\right)^3$$
$$= \frac{1}{8} \cdot \frac{4}{3}\pi \text{ cubic centimeters}$$
$$= \frac{1}{6}\pi \text{ cubic centimeters}$$
$$\approx \frac{1}{6} \cdot \frac{22}{7} \text{ cubic centimeters}$$
$$= \frac{11}{21} \text{ cubic centimeters}$$

7. $V = \pi r^2 h$
$$= \pi \cdot (5 \text{ inches})^2 \cdot 7 \text{ inches}$$
$$= 25 \cdot 7 \cdot \pi \text{ cubic inches}$$
$$= 175\pi \text{ cubic inches}$$
$$\approx 175 \cdot 3.14 \text{ cubic inches}$$
$$= 549.5 \text{ cubic inches}$$

8. $V = \frac{1}{3}s^2 h$
$$= \frac{1}{3} \cdot (3 \text{ meters})^2 \cdot 5.1 \text{ meters}$$
$$= \frac{1}{3} \cdot 9 \cdot 5.1 \text{ cubic meters}$$
$$= 15.3 \text{ cubic meters}$$

Exercise Set 9.4

1. $A = lw$
$$= 3.5 \text{ meters} \cdot 2 \text{ meters}$$
$$= 7 \text{ square meters}$$

3. $A = \frac{1}{2}bh$
$$= \frac{1}{2} \cdot 6\frac{1}{2} \text{ yards} \cdot 3 \text{ yards}$$
$$= \frac{1}{2} \cdot \frac{13}{2} \cdot \frac{3}{1} \text{ square yards}$$
$$= \frac{39}{4} \text{ or } 9\frac{3}{4} \text{ square yards}$$

5. $A = \frac{1}{2}bh$
$$= \frac{1}{2} \cdot 6 \text{ yards} \cdot 5 \text{ yards}$$
$$= \frac{1}{2} \cdot 6 \cdot 5 \text{ square yards}$$
$$= 15 \text{ square yards}$$

7. $r = \frac{d}{2}$
$$= \frac{3 \text{ inches}}{2}$$
$$= 1.5 \text{ inches}$$
$$A = \pi r^2$$
$$= \pi(1.5 \text{ inches})^2$$
$$= 2.25\pi \text{ square inches}$$
$$\approx 2.25(3.14) \text{ square inches}$$
$$= 7.065 \text{ square inches}$$

9. $A = bh$
$$= 7 \text{ feet} \cdot 5.25 \text{ feet}$$
$$= 36.75 \text{ square feet}$$

11. $A = \frac{1}{2}(b + B)h$
$$= \frac{1}{2} \cdot (5 \text{ meters} + 9 \text{ meters}) \cdot 4 \text{ meters}$$
$$= \frac{1}{2} \cdot 14 \cdot 4 \text{ square meters}$$
$$= 28 \text{ square meters}$$

13. $A = \frac{1}{2}(b + B)h$
$= \frac{1}{2} \cdot (4 \text{ yards} + 7 \text{ yards}) \cdot 4 \text{ yards}$
$= \frac{1}{2} \cdot 11 \text{ yards} \cdot 4 \text{ yards}$
$= \frac{1}{2} \cdot 11 \cdot 4 \text{ square yards}$
$= 22 \text{ square yards}$

15. $A = bh$
$= 7 \text{ feet} \cdot 5\frac{1}{4} \text{ feet}$
$= 7 \cdot \frac{21}{4} \text{ square feet}$
$= \frac{147}{4} \text{ or } 36\frac{3}{4} \text{ square feet}$

17. $A = bh$
$= 5 \text{ inches} \cdot 4\frac{1}{2} \text{ inches}$
$= 5 \cdot \frac{9}{2} \text{ square inches}$
$= \frac{45}{2} \text{ or } 22\frac{1}{2} \text{ square inches}$

19. Area of rectangle $= bh$
$= 7 \text{ centimeters} \cdot 3 \text{ centimeters}$
$= 21 \text{ square centimeters}$

Area of triangle $= \frac{1}{2}bh$
$= \frac{1}{2} \cdot \left(7 - 1\frac{1}{2} - 1\frac{1}{2}\right) \text{ centimeters} \cdot 2 \text{ centimeters}$
$= \frac{1}{2} \cdot \left(\frac{14}{2} - \frac{3}{2} - \frac{3}{2}\right) \cdot 2 \text{ square centimeters}$
$= \frac{1}{2} \cdot \frac{8}{2} \cdot 2 \text{ square centimeters}$
$= 4 \text{ square centimeters}$

Total area $= 21 \text{ square centimeters} + 4 \text{ square centimeters}$
$= 25 \text{ square centimeters}$

21.

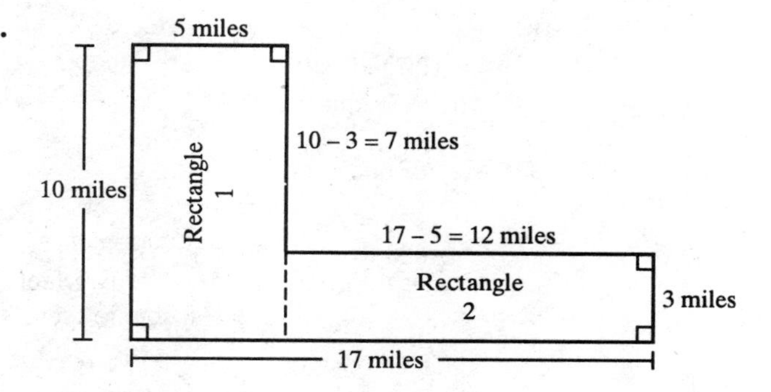

Area of rectangle 1 $= bh$
$= 5 \text{ miles} \cdot 10 \text{ miles}$
$= 50$ square miles

Area of rectangle 2 $= bh$
$= 12 \text{ miles} \cdot 3 \text{ miles}$
$= 36$ square miles

Total area = 50 square miles + 36 square miles
= 86 square miles

23. Area of square $= s^2$
$= (3 \text{ centimeters})^2$
$= 9$ square centimeters

Area of parallelogram $= b \cdot h$
$= (3 \text{ centimeters})(5 \text{ centimeters})$
$= 15$ square centimeters

Total area = 9 square centimeters + 15 square centimeters = 24 square centimeters

25. $A = \pi r^2$
$= \pi(6 \text{ inches})^2$
$= 36\pi$ square inches
$\approx 36 \cdot \frac{22}{7}$ square inches
≈ 113.1 square inches

27. $V = lwh$
$= 6 \text{ inches} \cdot 4 \text{ inches} \cdot 3 \text{ inches}$
$= 72$ cubic inches

29. $V = s^3$
$= (8 \text{ centimeters})^3$
$= 512$ cubic centimeters

31. $V = \frac{1}{3}\pi r^2 h$
$= \frac{1}{3}\cdot\pi\cdot(2 \text{ yards})^2\cdot(3 \text{ yards})$
$= 4\pi$ cubic yards
$\approx 4\cdot\frac{22}{7}$ cubic yards
$= \frac{88}{7}$ or $12\frac{4}{7}$ cubic yards

33. $V = \frac{4}{3}\pi r^3$
$= \frac{4}{3}\cdot\pi\cdot\left(\frac{10 \text{ inches}}{2}\right)^3$
$= \frac{4}{3}\cdot 125\cdot\pi$ cubic inches
$= \frac{500}{3}\pi$ cubic inches
$\approx \frac{500}{3}\cdot\frac{22}{7}$ cubic inches
$= \frac{11,000}{21}$ cubic inches
$= 523\frac{17}{21}$ cubic inches

35. $V = \frac{1}{3}s^2 h$
$= \frac{1}{3}\cdot(5 \text{ centimeters})^2\cdot 9$ centimeters
$= 75$ cubic centimeters

37. $V = s^3$
$= \left(1\frac{1}{3} \text{ inches}\right)^3$
$= \left(\frac{4}{3}\right)^3$ cubic inches
$= \frac{64}{27}$ cubic inches
$= 2\frac{10}{27}$ cubic inches

39. $V = lwh$
$= 2 \text{ feet}\cdot 1.4 \text{ feet}\cdot 3$ feet
$= 8.4$ cubic feet

41. $A = lw$
$= 100 \text{ feet}\cdot 50$ feet
$= 5000$ square feet

43. Area of one panel $= lw$
$= 6 \text{ feet}\cdot 7$ feet
$= 42$ square feet
Area of four panels $= 4\cdot 42$ square feet
$= 168$ square feet

45. $V = \frac{1}{3}s^2 h$
$= \frac{1}{3}\cdot(12 \text{ centimeters})^2\cdot 20$ centimeters
$= \frac{2880}{3}$ cubic centimeters
$= 960$ cubic centimeters

47. $A = \frac{1}{2}(b+B)h$
$= \frac{1}{2}\cdot(90 \text{ feet} + 140 \text{ feet})\cdot 80$ feet
$= \frac{1}{2}\cdot 230\cdot 80$ square feet
$= 9200$ square feet

49. Area of shaded part
$= \frac{1}{2}(b+B)h$
$= \frac{1}{2}\cdot(25 \text{ feet} + 36 \text{ feet})\cdot 12\frac{1}{2}$ feet
$= \frac{1}{2}\cdot 61 \text{ feet}\cdot\frac{25}{2}$ feet
$= \frac{1}{2}\cdot\frac{61}{1}\cdot\frac{25}{2}$ square feet
$= \frac{1525}{4}$ square feet
$= 381\frac{1}{4}$ square feet
The area of the shaded part to the nearest foot, is 381 square feet.

51. $5^2 = 25$

53. $3^2 = 9$

55. $1^2 + 2^2 = 1 + 4 = 5$

57. $4^2 + 2^2 = 16 + 4 = 20$

59. Area of a 12-inch pizza

$= \pi r^2$

$= \pi\left(\frac{12}{2} \text{ inches}\right)^2$

$= \pi(6 \text{ inches})^2$

$= 36\pi$ square inches

Area of two 8-inch pizzas

$= 2(\pi r^2)$

$= 2 \cdot \pi\left(\frac{8}{2} \text{ inches}\right)^2$

$= 2\pi(4 \text{ inches})^2$

$= 2\pi \cdot 16$ square inches

$= 32\pi$ square inches

The 12-inch pizza is

$\frac{\$10}{36\pi \text{ square inches}} \approx \$0.08$ per square inch.

The two 8-inch pizzas are $\frac{\$9}{32\pi \text{ square inches}} \approx \$0.09$ per square inch.

The 12-inch pizza is the better buy.

61. 8 inches $= \frac{8}{12}$ feet $= \frac{2}{3}$ feet

$A = lw$

$= 2 \text{ feet} \cdot \frac{2}{3} \text{ feet}$

$= \frac{4}{3}$ or $1\frac{1}{3}$ square feet

2 feet = 24 inches

$A = lw$

$= 24 \text{ inches} \cdot 8 \text{ inches}$

$= 192$ square inches

63. $V = \frac{1}{3}s^2h$

$= \frac{1}{3} \cdot (704 \text{ feet})^2 \cdot 471 \text{ feet}$

$= \frac{233{,}435{,}136}{3}$ cubic feet

$= 77{,}811{,}712$ cubic feet

65. $V = \frac{1}{3}s^2h$
$= \frac{1}{3}\cdot(227 \text{ meters})^2 \cdot 137 \text{ meters}$
$= \frac{7{,}059{,}473}{3} \text{ cubic meters}$
$\approx 2{,}353{,}158 \text{ cubic meters}$
The original volume was approximately 2,583,283 cubic meters, so the volume has decreased 2,583,283 cubic meters – 2,353,158 cubic meters = 230,125 cubic meters.

67. a. $V = lwh$
$= (2 \text{ feet } 1 \text{ inch})\cdot(1 \text{ foot } 8 \text{ inches})\cdot(1 \text{ foot } 7 \text{ inches})$
$= \left(2\frac{1}{12} \text{ feet}\right)\cdot\left(1\frac{8}{12} \text{ feet}\right)\cdot\left(1\frac{7}{12} \text{ feet}\right)$
$= \frac{25}{12}\cdot\frac{20}{12}\cdot\frac{19}{12} \text{ cubic feet}$
$= \frac{9500}{1728} \text{ cubic feet}$
$\approx 5 \text{ cubic feet}$

b. $V = lwh$
$= (1 \text{ foot } 1 \text{ inch})\cdot(2 \text{ feet})\cdot(8 \text{ inches})$
$= \left(1\frac{1}{12} \text{ feet}\right)\cdot(2 \text{ feet})\cdot\left(\frac{8}{12} \text{ feet}\right)$
$= \frac{13}{12}\cdot\frac{2}{1}\cdot\frac{8}{12} \text{ cubic feet}$
$= \frac{208}{144} \text{ cubic feet}$
$\approx 1 \text{ cubic foot}$

Kennel (a) is larger.

69. Answers may vary.

Integrated Review

1. The supplement of a 27° angle is 180° – 27° = 153°. The complement of a 27° angle is 90° – 27° = 63°.

2. Since $\angle x$ and the 105° angle are supplementary angles, $\angle x = 180° - 105° = 75°$. $\angle y$ and the 105° angle are vertical angles, so $\angle y = 105°$. $\angle x$ and $\angle z$ are vertical angles, so $\angle z = 75°$.

3. Since $\angle x$ and the 52° angle are supplementary angles, $\angle x = 180° - 52° = 128°$. $\angle y$ and the 52° angle are corresponding angles, so $\angle y = 52°$. $\angle y$ and $\angle z$ are supplementary angles, so $\angle z = 180° - 52° = 128°$.

4. $\angle x = 180° - 90° - 38° = 52°$

5. $36 \text{ in.} = \frac{36 \text{ in.}}{1}\cdot\frac{1 \text{ ft}}{12 \text{ in.}} = 3 \text{ ft}$

6. $10{,}560 \text{ ft} = \frac{10{,}560 \text{ ft}}{1} \cdot \frac{1 \text{ mi}}{5280 \text{ ft}} = 2 \text{ mi}$

7. $20 \text{ ft} = \frac{20 \text{ ft}}{1} \cdot \frac{1 \text{ yd}}{3 \text{ ft}} = \frac{20}{3} \text{ yd} = 6\frac{2}{3} \text{ yd}$

8. $6 \text{ yd} = \frac{6 \text{ yd}}{1} \cdot \frac{3 \text{ ft}}{1 \text{ yd}} = 18 \text{ ft}$

9. $2.1 \text{ mi} = \frac{2.1 \text{ mi}}{1} \cdot \frac{5280 \text{ ft}}{1 \text{ mi}} = 11{,}088 \text{ ft}$

10. $3.2 \text{ ft} = \frac{3.2 \text{ ft}}{1} \cdot \frac{12 \text{ in.}}{1 \text{ ft}} = 38.4 \text{ in.}$

11. 30 m = 3000 cm

12. 24 mm = 2.4 cm

13. 2000 mm = 2 m

14. 18 m = 1800 cm

15. 7.2 cm = 72 mm

16. 600 m = 0.6 km

17. $P = 4s = 4 \cdot (5 \text{ meters}) = 20 \text{ meters}$
$A = s^2 = (5 \text{ meters})^2 = 25 \text{ square meters}$

18.
$$\begin{aligned} P &= a + b + c \\ &= 3 \text{ feet} + 4 \text{ feet} + 5 \text{ feet} \\ &= 12 \text{ feet} \\ A &= \frac{1}{2} bh \\ &= \frac{1}{2} \cdot 4 \text{ feet} \cdot 3 \text{ feet} \\ &= 6 \text{ square feet} \end{aligned}$$

19.
$$\begin{aligned} C &= 2\pi r \\ &= 2 \cdot \pi \cdot (3 \text{ centimeters}) \\ &= 6\pi \text{ centimeters} \\ &\approx 6 \cdot 3.14 \text{ centimeters} \\ &= 18.84 \text{ centimeters} \\ A &= \pi r^2 \\ &= \pi \cdot (3 \text{ centimeters})^2 \\ &= 9\pi \text{ square centimeters} \\ &\approx 9 \cdot 3.14 \text{ square centimeters} \\ &= 28.26 \text{ square centimeters} \end{aligned}$$

20. Perimeter
= 11 miles + 5 miles + 11 miles + 5 miles
= 32 miles

$$\begin{aligned} A &= bh \\ &= 11 \text{ miles} \cdot 4 \text{ miles} \\ &= 44 \text{ square miles} \end{aligned}$$

21.
$$\begin{aligned} P &= 2l + 2w \\ &= 2 \cdot 14 \text{ feet} + 2 \cdot 17 \text{ feet} \\ &= 28 \text{ feet} + 34 \text{ feet} \\ &= 62 \text{ feet} \\ A &= lw \\ &= 14 \text{ feet} \cdot 17 \text{ feet} \\ &= 238 \text{ square feet} \end{aligned}$$

22.
$$\begin{aligned} V &= s^3 \\ &= (4 \text{ inches})^3 \\ &= 64 \text{ cubic inches} \end{aligned}$$

23.
$$\begin{aligned} V &= lwh \\ &= 2 \text{ feet} \cdot 3 \text{ feet} \cdot 5.1 \text{ feet} \\ &= 30.6 \text{ cubic feet} \end{aligned}$$

24.
$$\begin{aligned} V &= \frac{1}{3} s^2 h \\ &= \frac{1}{3} \cdot (10 \text{ centimeters})^2 \cdot 12 \text{ centimeters} \\ &= 400 \text{ cubic centimeters} \end{aligned}$$

25.
$$\begin{aligned} V &= \frac{4}{3} \pi r^3 \\ &= \frac{4}{3} \cdot \pi \cdot \left(\frac{16 \text{ miles}}{2}\right)^3 \\ &= \frac{4}{3} \cdot 512 \cdot \pi \text{ cubic miles} \\ &= \frac{2048}{3} \pi \text{ cubic miles} \\ &\approx \frac{2048}{3} \cdot \frac{22}{7} \text{ cubic miles} \\ &= \frac{45{,}056}{21} \text{ or } 2145\frac{11}{21} \text{ cubic miles} \end{aligned}$$

Section 9.5

Practice Problems

1. $$\begin{aligned} 4500\text{ lb} &= \frac{4500\text{ lb}}{1}\cdot\frac{1\text{ ton}}{2000\text{ lb}} \\ &= \frac{4500\text{ tons}}{2000} \\ &= \frac{9}{4}\text{ tons} \\ &= 2\frac{1}{4}\text{ tons} \end{aligned}$$

2. $$\begin{aligned} 56\text{ oz} &= \frac{56\text{ oz}}{1}\cdot\frac{1\text{ lb}}{16\text{ oz}} \\ &= \frac{56\text{ lb}}{16} \\ &= \frac{7}{2}\text{ lb} \\ &= 3\frac{1}{2}\text{ lb} \end{aligned}$$

3. $$\begin{array}{rr} 8\text{ tons } 100\text{ lb} \\ -\ 5\text{ tons } 1200\text{ lb} \\ \hline \end{array} = \begin{array}{rr} 7\text{ tons } 2100\text{ lb} \\ -\ 5\text{ tons } 1200\text{ lb} \\ \hline 2\text{ tons } 900\text{ lb} \end{array}$$

4. $$\begin{array}{r} 4\text{ lb } 11\text{ oz} \\ \times \quad 8 \\ \hline 32\text{ lb } 88\text{ oz} \end{array} \begin{aligned} &= 32\text{ lb} + 5\text{ lb } 8\text{ oz} \\ &= 37\text{ lb } 8\text{ oz} \end{aligned}$$

$$\begin{array}{r} 5 \\ 16\overline{)\ 88} \\ -80 \\ \hline 8 \end{array}$$

5. $$\begin{array}{r} 1\text{ lb } 6\text{ oz} \\ 4\overline{)\ 5\text{ lb } 8\text{ oz}} \\ -4 \quad\quad \\ \hline 1 = 16\text{ oz} \\ 24\text{ oz} \\ -24 \\ \hline 0 \end{array}$$

6. $$\begin{array}{r} 5\text{ lb } 14\text{ oz} \\ +\quad 6\text{ oz} \\ \hline 5\text{ lb } 20\text{ oz} \end{array} \begin{aligned} &= 5\text{ lb} + 1\text{ lb } 4\text{ oz} \\ &= 6\text{ lb } 4\text{ oz} \end{aligned}$$

7. $$\begin{aligned} 3.41\text{ g} &= \frac{3.41\text{ g}}{1}\cdot\frac{1000\text{ mg}}{1\text{ g}} \\ &= 3140\text{ mg} \end{aligned}$$

8. To convert centigrams to grams, move the decimal point 2 places to the left.

$$\begin{aligned} 56.2\text{ cg} &= \frac{56.2\text{ cg}}{1}\cdot\frac{1\text{ g}}{100\text{ cg}} \\ &= 0.562\text{ g} \end{aligned}$$

9. $3.1\text{ dg} = 0.31\text{ g}$ or $2.5\text{ g} = 25\text{ dg}$

$$\begin{array}{r} 2.50\text{ g} \\ -\ 0.31\text{ g} \\ \hline 2.19\text{ g} \end{array} \quad \text{or} \quad \begin{array}{r} 25.0\text{ dg} \\ -\ 3.1\text{ dg} \\ \hline 21.9\text{ dg} \end{array}$$

10. $$\begin{array}{r} 12.6\text{ kg} \\ \times\ 4 \\ \hline 50.4\text{ kg} \end{array}$$

11. $$\begin{array}{r} 22.9 \approx 23 \\ 24\overline{)\ 550.0} \\ -48 \quad \\ \hline 70 \quad \\ -48 \quad \\ \hline 22\,0 \\ -21\,6 \\ \hline 4 \end{array}$$

Each bag weighs approximately 23 kg.

Calculator Explorations

1. $15 \times 28.35 = 425.25$
Therefore 15 ounces ≈ 425. 25 grams

2. $11.2 \times 0.035 = 0.392$
Therefore 11.2 gaphs ≈ 0.392 ounces

3. $7 \times 2.20 = 15.4$
Therefore 7 kilograms ≈ 15.4 pounds

4. $23 \times 0.454 = 10.442$
Therefore 23 pounds ≈ 10.442 kilograms

5. $5 \times 0.035 = 0.175$
Therefore 5 graphs ≈ 0.175 ounces

6. $82 \times 2.20 = 180.4$
Therefore 82 kilograms ≈ 180.4 pounds

Mental Math

1. 16 ounces = 1 lb
2. 32 ounces = 2 lb
3. 1 ton = 2000 lb
4. 2 tons = 4000 lb
5. 1 pound = 16 oz
6. 3 pounds = 48 oz
7. 2000 pounds = 1 ton
8. 4000 pounds = 2 tons
9. No
10. No
11. Yes
12. Yes
13. No
14. No

Exercise Set 9.5

1. $2 \text{ lb} = 2 \text{ lb} \cdot \dfrac{16 \text{ oz}}{1 \text{ lb}} = 2 \cdot 16 \text{ oz} = 32 \text{ oz}$

3. $$\begin{aligned} 5 \text{ tons} &= 5 \text{ tons} \cdot \frac{2000 \text{ lb}}{1 \text{ ton}} \\ &= 5 \cdot 2000 \text{ lb} \\ &= 10,000 \text{ lb} \end{aligned}$$

5. $$\begin{aligned} 12,000 \text{ lb} &= 12,000 \text{ lb} \cdot \frac{1 \text{ ton}}{2000 \text{ lb}} \\ &= \frac{12,000 \text{ tons}}{2000} \\ &= 6 \text{ tons} \end{aligned}$$

7. $$\begin{aligned} 60 \text{ oz} &= 60 \text{ oz} \cdot \frac{1 \text{ lb}}{16 \text{ oz}} \\ &= \frac{60 \text{ lb}}{16} \\ &= \frac{15}{4} \text{ lb} \\ &= 3\frac{3}{4} \text{ lb} \end{aligned}$$

9. $$\begin{aligned} 3500 \text{ lb} &= 3500 \text{ lb} \cdot \frac{1 \text{ ton}}{2000 \text{ lb}} \\ &= \frac{3500 \text{ tons}}{2000} \\ &= \frac{7}{4} \text{ tons} \\ &= 1\frac{3}{4} \text{ tons} \end{aligned}$$

11. $$\begin{aligned} 16.25 \text{ lb} &= 16.25 \text{ lb} \cdot \frac{16 \text{ oz}}{1 \text{ lb}} \\ &= 16.25 \cdot 16 \text{ oz} \\ &= 260 \text{ oz} \end{aligned}$$

13. $$\begin{aligned} 4.9 \text{ tons} &= 4.9 \text{ tons} \cdot \frac{2000 \text{ lb}}{1 \text{ ton}} \\ &= 4.9 \cdot 2000 \text{ lb} \\ &= 9800 \text{ lb} \end{aligned}$$

15. $$\begin{aligned} 4\frac{3}{4} \text{ lb} &= 4\frac{3}{4} \text{ lb} \cdot \frac{16 \text{ oz}}{1 \text{ lb}} \\ &= \frac{19}{4} \cdot 16 \text{ oz} \\ &= 19 \cdot 4 \text{ oz} \\ &= 76 \text{ oz} \end{aligned}$$

17. $$\begin{aligned} 2950 \text{ lb} &= 2950 \text{ lb} \cdot \frac{1 \text{ ton}}{2000 \text{ lb}} \\ &= \frac{2950 \text{ tons}}{2000} \\ &= 1.475 \text{ tons} \\ &\approx 1.5 \text{ tons} \end{aligned}$$

19. $$\begin{array}{r} 34 \text{ lb } 12 \text{ oz} \\ +\ 18 \text{ lb } 14 \text{ oz} \\ \hline 52 \text{ lb } 26 \text{ oz} \end{array} = 52 \text{ lb} + 1 \text{ lb } 10 \text{ oz} = 53 \text{ lb } 10 \text{ oz}$$

21.
$$\begin{array}{r} 6 \text{ tons } 1540 \text{ lb} \\ +\ 2 \text{ tons } \ \ 850 \text{ lb} \\ \hline 8 \text{ tons } 2390 \text{ lb} \end{array} = 8 \text{ tons} + 1 \text{ ton } 390 \text{ lb} = 9 \text{ tons } 390 \text{ lb}$$

23.
$$\begin{array}{r} 5 \text{ tons } 1050 \text{ lb} \\ -\ 2 \text{ tons } \ \ 875 \text{ lb} \\ \hline 3 \text{ tons } \ \ 175 \text{ lb} \end{array}$$

25.
$$\begin{array}{rcr} 12 \text{ lb } 4 \text{ oz} & = & 11 \text{ lb } 20 \text{ oz} \\ -\ 3 \text{ lb } 9 \text{ oz} & = & -\ 3 \text{ lb } \ \ 9 \text{ oz} \\ \hline & & 8 \text{ lb } 11 \text{ oz} \end{array}$$

27.
$$\begin{array}{r} 5 \text{ lb } 3 \text{ oz} \\ \times \qquad 6 \\ \hline 30 \text{ lb } 18 \text{ oz} \end{array} = 30 \text{ lb} + 1 \text{ lb } 2 \text{ oz} = 31 \text{ lb } 2 \text{ oz}$$

29.
$$\begin{array}{r} 1 \text{ ton } \quad 700 \text{ lb} \\ 5\overline{)\ 6 \text{ tons } 1500 \text{ lb}} \\ -5 \text{ tons} \qquad\qquad \\ \hline 1 \text{ ton} = 2000 \text{ lb} \\ 3500 \text{ lb} \\ -\ 3500 \text{ lb} \\ \hline 0 \end{array}$$

31.
$$\begin{array}{r} 1 \text{ lb } 10 \text{ oz} \\ +\ 3 \text{ lb } 14 \text{ oz} \\ \hline 4 \text{ lb } 24 \text{ oz} \end{array} = 4 \text{ lb} + 1 \text{ lb } 8 \text{ oz} = 5 \text{ lb } 8 \text{ oz}$$

She has 5 lb 8 oz of rice.

33.
$$\begin{array}{rcr} 64 \text{ lb } \ \ 8 \text{ oz} & = & 63 \text{ lb } 24 \text{ oz} \\ -\ 28 \text{ lb } 10 \text{ oz} & = & -\ 28 \text{ lb } 10 \text{ oz} \\ \hline & & 35 \text{ lb } 14 \text{ oz} \end{array}$$

His zucchini was 35 lb 14 oz below the record.

35. 4 boxes · 10 packages = 40 total packages

$$\begin{array}{r} 3 \text{ lb } \quad 4 \text{ oz} \\ \times \qquad\quad 40 \\ \hline 120 \text{ lb } 160 \text{ oz} \end{array} = 120 \text{ lb} + 10 \text{ lb} = 130 \text{ lb}$$

4 boxes weigh 130 lb.

37.
$$\begin{array}{rcr} 55 \text{ lb } 4 \text{ oz} & = & 54 \text{ lb } 20 \text{ oz} \\ -\ 2 \text{ lb } 8 \text{ oz} & = & -\ 2 \text{ lb } \ \ 8 \text{ oz} \\ \hline & & 52 \text{ lb } 12 \text{ oz} \end{array}$$

$$\begin{array}{r} 52 \text{ lb } 12 \text{ oz} \\ \times \qquad\quad 4 \\ \hline 208 \text{ lb } 48 \text{ oz} \end{array} = 208 \text{ lb} + 3 \text{ lb} = 211 \text{ lb}$$

There are 211 lb of pineapple in 4 boxes.

39. $6\frac{3}{4} \text{ oz} \cdot 12 = \frac{27}{4} \text{oz} \cdot 12 = 81 \text{ oz}$

$81 \text{ oz} \cdot \frac{1 \text{ lb}}{16 \text{ oz}} = 5 \text{ lb } 1 \text{ oz}$

A dozen bags will weigh 5 lb 1 oz.

41. To convert grams to kilograms, move the decimal 3 places to the left.
500 g = 0.5 kg

43. To convert grams to milligrams, move the decimal 3 places to the right.
4 g = 4000 mg

45. To convert kilograms to grams, move the decimal 3 places to the right.
25 kg = 25,000 g

47. To convert milligrams to grams, move the decimal 3 places to the left.
48 mg = 0.048 g

49. To convert grams to kilograms, move the decimal 3 places to the left.
6.3 g = 0.0063 kg

51. To convert grams to milligrams, move the decimal 3 places to the right.
15.14 g = 15,140 mg

53. To convert kilograms to grams, move the decimal 3 places to the right.
4.01 kg = 4010 g

55.
$$\begin{array}{r} 3.8 \text{ mg} \\ +\ 9.7 \text{ mg} \\ \hline 13.5 \text{ mg} \end{array}$$

57. 205 mg = 0.205 g or 5.61 g = 5610 mg

$$\begin{array}{rcr} 0.205\text{ g} & = & 205\text{ mg} \\ +\ 5.610\text{ g} & = & +\ 5610\text{ mg} \\ \hline 5.815\text{ g} & \text{or} & 5815\text{ mg} \end{array}$$

59. 9 g = 9000 mg or 7150 mg = 7.15 g

$$\begin{array}{rcr} 9000\text{ mg} & = & 9.00\text{ g} \\ -7150\text{ mg} & = & -7.15\text{ g} \\ \hline 1850\text{ mg} & \text{or} & 1.85\text{ g} \end{array}$$

61. 1.61 kg = 1610 g or 250 g = 0.25 kg

$$\begin{array}{rr} 1610\text{ g} & 1.61\text{ kg} \\ -250\text{ g} & -0.25\text{ kg} \\ \hline 1360\text{ g} & 1.36\text{ kg} \end{array}$$

63.

$$\begin{array}{r} 5.2\text{ kg} \\ \times\ 2.6 \\ \hline 312 \\ 1040 \\ \hline 13.52\text{ kg} \end{array}$$

65.

$$\begin{array}{r} 2.125\text{ kg} \\ 8\overline{)\ 17.000\text{ kg}} \\ -16 \\ 1\ 0 \\ -\ 8 \\ \hline 20 \\ -16 \\ \hline 40 \\ -40 \\ \hline 0 \end{array}$$

67.

$$\begin{array}{r} 336\text{ g} \\ \times\ 24 \\ \hline 1344 \\ 6720 \\ \hline 8064\text{ g} \end{array} = 8.064\text{ kg}$$

24 cans weigh 8.064 kg.

69. 0.09 g – 60 mg = 90 mg – 60 mg
= 30 mg
The extra-strength tablet contains 30 mg more medication.

71. 0.6 g – 350 mg = 600 mg – 350 mg
= 250 mg
He can have 250 mg more sodium.

73. 3 · 16 · 3 mg = 144 mg
3 cartons contain 144 mg of perservatives.

75. 6.432 kg – 12 · 26 g = 6.432 kg – 312 g
= 6.432 kg – 0.312 kg
= 6.12 kg
There is 6.12 kg of oatmeal in the carton.

77. 0.3 kg + 0.15 kg + 400 g
= 0.3 kg + 0.15 kg + 0.4 kg
= 0.85 kg or 850 g
The package weighs 0.85 kg or 850 g.

79.

$$\begin{array}{r} 198\text{ g} \\ 12 \\ \hline 396 \\ 1980 \\ \hline 2376\text{ g} \end{array} = 2.376\text{ kg}$$

A dozen bags weigh about 2.38 kg.

81. $\frac{1}{4} = \frac{1 \cdot 25}{4 \cdot 25} = \frac{25}{100} = 0.25$

83. $\frac{4}{25} = \frac{4 \cdot 4}{25 \cdot 4} = \frac{16}{100} = 0.16$

85. $\frac{7}{8} = \frac{7 \cdot 125}{8 \cdot 125} = \frac{875}{1000} = 0.875$

87. Answers may vary.

Section 9.6

Practice Problems

1. $43\text{ pints} = \frac{43\text{ pints}}{1} \cdot \frac{1\text{ qt}}{2\text{ pints}}$
$= \frac{43}{2}\text{ qt}$
$= 21\frac{1}{2}\text{ qt}$

2. $26\text{qt} = \frac{26\text{ qt}}{1} \cdot \frac{4\text{ c}}{1\text{ qt}}$
$= 104\text{ c}$

3.

$$\begin{array}{r} 1\text{ gal }1\text{ qt} \\ -\quad 2\text{ qt} \\ \hline \end{array} \qquad \begin{array}{r} 0\text{ gal }5\text{ qt} \\ -\quad 2\text{ qt} \\ \hline 3\text{ qt} \end{array}$$

4. $\begin{array}{r} 2 \text{ gal } 3 \text{ qt} \\ \times \quad\quad 2 \\ \hline 4 \text{ gal } 6 \text{ qt} \end{array} = 4 \text{ gal} + 1 \text{ gal } 2 \text{ qt} = 5 \text{ gal } 2 \text{ qt}$

5. $2\overline{)6 \text{ gal } 1 \text{ qt}}$ = 3 gal 1 pt
-6, 0; $1 \text{ qt} = 2 \text{ pt}$, -2 pt, 0

6. $\begin{array}{r} 15 \text{ gal } 3 \text{ qt} \\ +\ 4 \text{ gal } 3 \text{ qt} \\ \hline 19 \text{ gal } 6 \text{ qt} \end{array} = 19 \text{ gal} + 1 \text{ gal } 2 \text{ qt} = 20 \text{ gal } 2 \text{ qt}$

7. $2100 \text{ ml} = \frac{2100 \text{ ml}}{1} \cdot \frac{1 \text{ L}}{1000 \text{ ml}} = \frac{2100}{1000} \text{ L} = 2.1 \text{ L}$

8. To convert decaliters to liters, move the decimal one place to the right.
2.13 dal = 21.3 L

9. 1250 ml = 1.25 L or 2.9 L = 2900 ml

$\begin{array}{r} 1250 \text{ ml} \\ +\ 2900 \text{ ml} \\ \hline 4150 \text{ ml} \end{array} \qquad \begin{array}{r} 1.25 \text{ L} \\ +\ 2.90 \text{ L} \\ \hline 4.15 \text{ L} \end{array}$

10. $13\overline{)146.9 \text{ L}}$ = 11.3 L
-13, 16, -13, 39, -39, 0

11. $\begin{array}{r} 28.6 \\ \times\ 8.5 \\ \hline 14\ 30 \\ 228\ 80 \\ \hline 243.10 \end{array}$

243.1 L can be pumped in 8.5 minutes.

Calculator Explorations

1. $5 \times 0.946 = 4.73$
Therefore 5 quarts ≈ 4.73 liters.
2. $26 \times 3.785 = 98.41$
Therefore 26 gallons ≈ 98.41 liters.
3. $17.5 \times 0.264 = 4.62$
Therefore 17.5 liters ≈ 4.62 gallons.
4. $7.8 \times 1.06 = 8.268$
Therefore 7.8 liters ≈ 8.268 quarts.
5. $1 \times 3.785 = 3.785$
Therefore 1 gallon ≈ 3.785 liters.
6. $3 \times 1.06 = 3.18$
Therefore 3 liters ≈ 3.18 quarts.

Mental Math

1. 2 c = 1 pt
2. 4 c = 2 pt
3. 4 qt = 1 gal
4. 8 qt = 2 gal
5. 2 pt = 1 qt
6. 6 pt = 3 qt
7. 8 fl oz = 1 c
8. 24 fl oz = 3 c
9. 1 pt = 2 c
10. 3 pt = 6 c
11. 1 gal = 4 qt
12. 2 gal = 8 qt
13. No
14. Yes
15. No

16. Yes

Exercise Set 9.6

1. $32 \text{ fl oz} = 32 \text{ fl oz} \cdot \frac{1 \text{ c}}{8 \text{ fl oz}} = \frac{32 \text{ c}}{8} = 4 \text{ c}$

3. $8 \text{ qt} = 8 \text{ qt} \cdot \frac{2 \text{ pt}}{1 \text{ qt}} = 8 \cdot 2 \text{ pt} = 16 \text{ pt}$

5. $10 \text{ qt} = 10 \text{ qt} \cdot \frac{1 \text{ gal}}{4 \text{ qt}} = \frac{10 \text{ gal}}{4} = 2\frac{1}{2} \text{ gal}$

7. $80 \text{ fl oz} = 80 \text{ fl oz} \cdot \frac{1 \text{ pt}}{16 \text{ fl oz}} = \frac{80 \text{ pt}}{16} = 5 \text{ pt}$

9. $2 \text{ qt} = 2 \text{ qt} \cdot \frac{4 \text{ c}}{1 \text{ qt}} = 2 \cdot 4 \text{ c} = 8 \text{ c}$

11.
$$\begin{aligned} 120 \text{ fl oz} &= 120 \text{ fl oz} \cdot \frac{1 \text{ c}}{8 \text{ fl oz}} \cdot \frac{1 \text{ pt}}{2 \text{ c}} \cdot \frac{1 \text{ qt}}{2 \text{ pt}} \\ &= \frac{120 \text{ qt}}{8 \cdot 2 \cdot 2} \\ &= \frac{120 \text{ qt}}{32} \\ &= \frac{15 \text{ qt}}{4} \\ &= 3\frac{3}{4} \text{ qt} \end{aligned}$$

13.
$$\begin{aligned} 6 \text{ gal} &= 6 \text{ gal} \cdot \frac{4 \text{ qt}}{1 \text{ gal}} \cdot \frac{2 \text{ pt}}{1 \text{ qt}} \cdot \frac{2 \text{ c}}{1 \text{ pt}} \cdot \frac{8 \text{ fl oz}}{1 \text{ c}} \\ &= 6 \cdot 4 \cdot 2 \cdot 2 \cdot 8 \text{ fl oz} \\ &= 768 \text{ fl oz} \end{aligned}$$

15. $4\frac{1}{2} \text{ pt} = \frac{9}{2} \text{ pt} \cdot \frac{2 \text{ c}}{1 \text{ pt}} = 9 \text{ c}$

17.
$$\begin{aligned} 2\frac{3}{4} \text{gal} &= \frac{11}{4} \text{ gal} \cdot \frac{4 \text{ qt}}{1 \text{ gal}} \cdot \frac{2 \text{ pt}}{1 \text{ qt}} \\ &= \frac{11 \cdot 4 \cdot 2 \text{ pt}}{4} \\ &= 22 \text{ pt} \end{aligned}$$

19.
$$\begin{array}{l} \quad\ 4 \text{ gal } 3 \text{ qt} \\ \underline{+\ 5 \text{ gal } 2 \text{ qt}} \\ \quad\ 9 \text{ gal } 5 \text{ qt} = 9 \text{ gal} + 1 \text{ gal } 1 \text{ qt} = 10 \text{ gal } 1 \text{ qt} \end{array}$$

21.
$$\begin{array}{l} \quad\ 1 \text{ c} \quad 5 \text{ fl oz} \\ \underline{+\ 2 \text{ c} \quad 7 \text{ fl oz}} \\ \quad\ 3 \text{ c } 12 \text{ fl oz} = 3 \text{ c} + 1 \text{ c } 4 \text{ fl oz} = 4 \text{ c } 4 \text{ fl oz} \end{array}$$

23.
$$\begin{array}{lcl} \quad 3 \text{ gal} & = & \quad 2 \text{ gal } 4 \text{ qt} \\ \underline{-\ 1 \text{ gal } 3 \text{ qt}} & = & \underline{-\ 1 \text{ gal } 3 \text{ qt}} \\ & & \quad 1 \text{ gal } 1 \text{ qt} \end{array}$$

25.
$$\begin{array}{lcr} 3 \text{ gal } 1 \text{ qt} & = & 2 \text{ gal } 4 \text{ qt } 2 \text{ pt} \\ \underline{-\qquad 1 \text{ qt } 1 \text{ pt}} & = & \underline{-\qquad 1 \text{ qt } 1 \text{ pt}} \\ & & 2 \text{ gal } 3 \text{ qt } 1 \text{ pt} \end{array}$$

27.
$$\begin{array}{l} 1 \text{ pt } 1 \text{ c} \\ \underline{\times \quad 3} \\ 3 \text{ pt } 3 \text{ c} = 3 \text{ pt} + 1 \text{ pt } 1 \text{ c} \\ \qquad\qquad = 4 \text{ pt } 1 \text{ c} = 2 \text{ qt } 1 \text{ c} \end{array}$$

29.
$$\begin{array}{l} 8 \text{ gal } 2 \text{ qt} \\ \underline{\times \qquad\quad 2} \\ 16 \text{ gal } 4 \text{ qt} = 16 \text{ gal} + 1 \text{ gal} = 17 \text{ gal} \end{array}$$

31.
$$\begin{array}{r} 4 \text{ gal} \quad 3 \text{ qt} \\ 2\overline{)\ 9 \text{ gal} \quad 2 \text{ qt}} \\ \underline{-8 \text{ gal}} \qquad\quad \\ 1 \text{ gal} = \underline{4 \text{qt}} \\ 6 \text{ qt} \\ \underline{-6 \text{ qt}} \\ 0 \end{array}$$

33.
$$\begin{aligned} 1\frac{1}{2} \text{ qt} &= \frac{3}{2} \text{qt} \cdot \frac{2 \text{ pt}}{1 \text{ qt}} \cdot \frac{2 \text{ c}}{1 \text{ qt}} \cdot \frac{8 \text{ fl oz}}{1 \text{ c}} \\ &= \frac{3 \cdot 2 \cdot 2 \cdot 8 \text{ fl oz}}{2} \\ &= 48 \text{ fl oz} \end{aligned}$$

One can holds 48 fl oz.

35.
$$\begin{aligned} 64 \text{ fl oz} &= 64 \text{ fl oz} \cdot \frac{1 \text{ c}}{8 \text{ fl oz}} \cdot \frac{1 \text{ pt}}{2 \text{ c}} \cdot \frac{1 \text{ qt}}{2 \text{ pt}} \\ &= \frac{64 \text{ qt}}{8 \cdot 2 \cdot 2} \\ &= 2 \text{ qt} \end{aligned}$$

Individuals should drink 2 qt of water daily.

37. $\begin{array}{r} 5 \text{ pt } 1 \text{ c} \\ +\ 2 \text{ pt } 1 \text{ c} \\ \hline 7 \text{ pt } 2 \text{ c} \end{array} = 7 \text{ pt} + 1 \text{ pt} = 8 \text{ pt} = 4 \text{ qt} = 1 \text{ gal}$

Yes, the fruit punch can be poured into the container.

39. $\begin{array}{l} 1 \text{ qt } 1 \text{ pt} \\ +\quad 1 \text{ pt } 1 \text{ c} \\ \hline 1 \text{ qt } 2 \text{ pt } 1 \text{ c} \end{array} = 1 \text{ qt} + 1 \text{ qt} + 1 \text{ c}$

$= 2 \text{ qt } 1 \text{ c}$

2 qt 1 c of punch has been prepared.

41. $12 \text{ fl oz} \cdot 24 = 288 \text{ fl oz}$

$$\begin{aligned} 288 \text{ fl oz} &= 288 \text{ fl oz} \cdot \frac{1 \text{ c}}{8 \text{ fl oz}} \cdot \frac{1 \text{ pt}}{2 \text{ c}} \cdot \frac{1 \text{ qt}}{2 \text{ pt}} \\ &= \frac{288 \text{ qt}}{8 \cdot 2 \cdot 2} \\ &= 9 \text{ qt} \end{aligned}$$

There are 9 quarts in a case of Pepsi.

43. To convert liters to milliliters move the decimal 3 places to the right.
5 L = 5000 ml

45. To convert milliliters to liters, move the decimal 3 places to the left.
4500 ml = 4.5 L

47. To convert liters to kiloliters move the decimal 3 places to the left.
410 L = 0.41 kl

49. To convert milliliters to liters move the decimal 3 places to the left.
64 ml = 0.064 L

51. To convert kiloliters to liters, move the decimal 3 places to the right.
0.16 kl = 160 L

53. To convert liters to milliliters move the decimal 3 places to the right.
3.6 L = 3600 ml

55. To convert liters to kiloliters, move the decimal 3 places to the left.
0.16 L = 0.00016 kl

57. $\begin{array}{r} 2.9 \text{ L} \\ +\ 19.6 \text{ L} \\ \hline 22.5 \text{ L} \end{array}$

59. 2700 ml = 2.7 L or 1.8 L = 1800 ml

$\begin{array}{r} 2.7 \text{ L} \\ +\ 1.8 \text{ L} \\ \hline 4.5 \text{ L} \end{array} \begin{array}{c} = \\ = \\ \text{or} \end{array} \begin{array}{r} 2700 \text{ ml} \\ +\ 1800 \text{ ml} \\ \hline 4500 \text{ ml} \end{array}$

61. 8.6 L = 8600 ml or 190 ml = 0.19 L

$\begin{array}{r} 8600 \text{ ml} \\ -190 \text{ ml} \\ \hline 8410 \text{ ml} \end{array} \begin{array}{c} = \\ = \\ \text{or} \end{array} \begin{array}{r} 8.60 \text{ L} \\ -0.19 \text{ L} \\ \hline 8.41 \text{ L} \end{array}$

63. 11,400 ml = 11.4 L or 0.8 L = 800 ml

$\begin{array}{r} 11.4 \text{ L} \\ -0.8 \text{ L} \\ \hline 10.6 \text{ L} \end{array} \begin{array}{c} = \\ = \\ \text{or} \end{array} \begin{array}{r} 11,400 \text{ ml} \\ -\quad 800 \text{ ml} \\ \hline 10,600 \text{ ml} \end{array}$

65. $\begin{array}{r} 480 \text{ ml} \\ \times\ 8 \\ \hline 3840 \text{ ml} \end{array}$

67. $0.5\overline{)81.2}$

Move decimal points 1 place.

$$\begin{array}{r} 162.4 \\ 5.\overline{)\ 812.0} \\ \underline{-5} \\ 31 \\ \underline{-30} \\ 12 \\ \underline{-10} \\ 20 \\ \underline{-20} \\ 0 \end{array}$$

$81.2 \text{ L} \div 0.5 = 162.4 \text{ L}$

69. $\begin{array}{r} 2 \text{ L} \\ -410 \text{ ml} \\ \hline \end{array} \begin{array}{c} = \\ = \\ \end{array} \begin{array}{r} 2.00 \text{ L} \\ -0.41 \text{ L} \\ \hline 1.59 \text{ L} \end{array}$

1.59 liters remain in the bottle.

71. 354 ml = 0.354 L

$\begin{array}{l} \text{gasoline} \\ +\ \text{dry gas} \\ \hline \end{array} \qquad \begin{array}{r} 18.600 \text{ L} \\ +\ 0.354 \text{ L} \\ \hline 18.954 \text{ L} \end{array}$

There are 18.954 L of gasoline in the tank.

73. $44.3\overline{)\$14.00} \rightarrow 443\overline{)140.0000}$

$$\begin{array}{r} 0.3160 \\ 443\overline{)140.0000} \\ -132\ 9 \\ \hline 7\ 10 \\ -4\ 43 \\ \hline 2\ 670 \\ -2\ 658 \\ \hline 120 \end{array}$$

The cost is about \$0.316 per liter.

75.

$$\begin{array}{rcr} 1.42\text{ L} & = & 1420\text{ mL} \\ -946\text{ mL} & = & -946\text{ mL} \\ \hline & & 474\text{ mL} \end{array}$$

The larger bottle contains 474 ml more.

77. $0.7 = \frac{7}{10}$

79. $0.03 = \frac{3}{100}$

81. $0.006 = \frac{6}{1000} = \frac{3}{500}$

83. Answers may vary.

Section 9.7

Practice Problems

1.
$$\begin{aligned} F &= \frac{9}{5}C + 32 \\ &= \frac{9}{5} \cdot 50 + 32 \\ &= 90 + 32 \\ &= 122 \end{aligned}$$
50°C is 122°F

2.
$$\begin{aligned} F &= 1.8C + 32 \\ &= 1.8 \cdot 18 + 32 \\ &= 32.4 + 32 \\ &= 64.4 \end{aligned}$$
18°C is 64.4°F

3.
$$\begin{aligned} C &= \frac{5}{9}(F - 32) \\ &= \frac{5}{9} \cdot (86 - 32) \\ &= \frac{5}{9} \cdot 54 \\ &= 30 \end{aligned}$$
86°F is 30°C

4.
$$\begin{aligned} C &= \frac{5}{9}(F - 32) \\ &= \frac{5}{9} \cdot (113 - 32) \\ &= \frac{5}{9} \cdot 81 \\ &= 45 \end{aligned}$$
113°F is 45°C

5.
$$\begin{aligned} C &= \frac{5}{9}(F - 32) \\ &= \frac{5}{9} \cdot (102.8 - 32) \\ &= \frac{5}{9} \cdot 70.8 \\ &= 39.\overline{3} \end{aligned}$$
102.8°C is approximately 39.3°F

Mental Math

1. Yes

2. No

3. No

4. Yes

5. No

6. No

7. Yes

8. No

Exercise Set 9.7

1. $C = \frac{5}{9}(F - 32)$

$C = \frac{5}{9} \cdot (41 - 32)$

$= \frac{5}{9} \cdot (9)$

$= 5$

$41° F = 5° C$

3. $C = \frac{5}{9}(F - 32)$

$C = \frac{5}{9} \cdot (104 - 32)$

$= \frac{5}{9} \cdot (72)$

$= 5(8)$

$= 40$

$104°F = 40°C$

5. $F = \frac{9}{5}C + 32$

$F = \frac{9}{5} \cdot (115) + 32$

$= 108 + 32$

$= 140$

$60°C = 140°F$

7. $F = \frac{9}{5}C + 32$

$F = \frac{9}{5} \cdot (115) + 32$

$= 9(23) + 32$

$= 207 + 32$

$= 239$

$115°C = 239°F$

9. $C = \frac{5}{9}(F - 32)$

$C = \frac{5}{9} \cdot (62 - 32)$

$= \frac{5}{9} \cdot (30)$

$= \frac{150}{9}$

≈ 16.7

$62°F \approx 16.7°C$

11. $C = \frac{5}{9}(F - 32)$

$C = \frac{5}{9} \cdot (142.1 - 32)$

$= \frac{5}{9} \cdot (110.1)$

$= \frac{550.5}{9}$

≈ 61.2

$142.1°F \approx 61.2°C$

13. $F = 1.8C + 32$

$F = 1.8(92) + 32$

$= 165.6 + 32$

$= 197.6$

$92°C = 197.6°F$

15. $F = 1.8C + 32$

$F = 1.8(16.3) + 32$

$= 29.34 + 32$

$= 61.34$

$16.3°C \approx 61.3°F$

17. $C = \frac{5}{9}(F - 32)$

$C = \frac{5}{9} \cdot (134 - 32)$

$= \frac{5}{9} \cdot (102)$

$= \frac{510}{9}$

≈ 56.7

$134°F \approx 56.7°C$

19. $F = 1.8C + 32$

$F = 1.8(27) + 32$

$= 48.6 + 32$

$= 80.6$

$27°C = 80.6°F$

21. $C = \frac{5}{9}(F - 32)$

$C = \frac{5}{9} \cdot (70 - 32)$

$= \frac{5}{9} \cdot (38)$

$= \frac{190}{9}$

≈ 21.1

$70°F \approx 21.1°C$

23. $C = \frac{5}{9}(F - 32)$

$C = \frac{5}{9} \cdot (100.2 - 32)$

$= \frac{5}{9} \cdot (68.2)$

$= \frac{341}{9}$

≈ 37.9

$100.2°F \approx 37.9°C$

25. $F = 1.8C + 32$

$F = 1.8 \cdot 118 + 32$

$= 212.4 + 32$

$= 244.4$

$118°C = 244.4°F$

27. $C = \frac{5}{9}(F - 32)$

$C = \frac{5}{9} \cdot (500 - 32)$

$= \frac{5}{9} \cdot (468)$

$= 5(52)$

$= 260$

$500°F = 260°C$

29. $C = \frac{5}{9}(F - 32)$

$C = \frac{5}{9} \cdot (864 - 32)$

$= \frac{5}{9} \cdot (832)$

$= \frac{4160}{9}$

≈ 462.2

$864°F \approx 462.2°C$

31. $P = 2l + 2w$

$= 2(25 \text{ m}) + 2(6 \text{ m})$

$= 50 \text{ m} + 12 \text{ m}$

$= 62 \text{ m}$

33. $P = 3 \text{ ft} + 3 \text{ ft} + 3 \text{ ft} + 3 \text{ ft} + 3 \text{ ft}$

$= 15 \text{ ft}$

35. $P = 4s$

$= 4 \cdot 2.6 \text{ m}$

$= 10.4 \text{ m}$

37. $C = \frac{5}{9}(F - 32)$

$C = \frac{5}{9} \cdot (9010 - 32)$

$= \frac{5}{9} \cdot (8978)$

$= \frac{44,890}{9}$

≈ 4988

9010°F is about 4988°C.

39. a. $C = \frac{5}{9}(F - 32)$

$= \frac{5}{9}(92 - 32)$

$= \frac{5}{9}(60)$

≈ 33.3

Therefore 92°F is approximately 33.3°C.

b. $F = 1.8C + 32$

$= 1.8(-10) + 32$

$= -18 + 32$

$= 14$

Therefore –10°C is the same as 14°F.

Chapter 9 Review

1. $\angle A$ is a right angle.

2. $\angle B$ is a straight angle.

3. $\angle C$ is an acute angle.

4. $\angle D$ is an obtuse angle.

5. The complement of a 25° angle is $90° - 25° = 65°$.

6. The supplement of a 105° angle is $180° - 105° = 75°$.

7. The supplement of a 72° angle is $180° - 72° = 108°$.

8. The complement of a 1° angle is $90° - 1° = 89°$.

9. $\angle x = 90° - 32° = 58°$

10. $\angle x = 180° - 82° = 98°$

11. $\angle x = 105° - 15° = 90°$

12. $\angle x = 45° - 20° = 25°$

13. 47° and 133° are supplementary angles.

14. 47° and 43° are complementary angles.
58° and 32° are complementary angles.

15. $\angle x$ and the 80° angle are vertical angles, so $\angle x = 80°$. $\angle x$ and $\angle y$ are supplementary angles, so $\angle y = 180° - 80° = 100°$. $\angle y$ and $\angle z$ are vertical angles, so $\angle z = 100°$.

16. $\angle x$ and the 25° angle are supplementary angles, so $\angle x = 180° - 25° = 155°$. $\angle x$ and $\angle y$ are vertical angles, so $\angle y = 155°$. $\angle z$ and the 25° angle are vertical angles, so $\angle z = 25°$.

17. $\angle x$ and the 53° angle are vertical angles, so $\angle x = 53°$. $\angle y$ and the 53° angle are corresponding angles, so $\angle y = 53°$. $\angle y$ and $\angle z$ are supplementary angles, so $\angle z = 180° - 53° = 127°$.

18. $\angle x$ and the 42° angle are vertical angles, so $\angle x = 42°$. $\angle y$ and the 42° angle are corresponding angles, so $\angle y = 42°$. $\angle y$ and $\angle z$ are supplementary angles, so $\angle z = 180° - 42° = 138°$.

19. $\angle x = 180° - 32° - 45° = 103°$

20. $\angle x = 180° - 62° - 58° = 60°$

21. $\angle x = 180° - 90° - 30° = 60°$

22. $\angle x = 180° - 90° - 25° = 65°$

23. $108 \text{ in.} = 108 \text{ in.} \cdot \frac{1 \text{ ft}}{12 \text{ in.}} = 9 \text{ ft}$

24. $$\begin{aligned} 72 \text{ ft} &= 72 \text{ ft} \cdot \frac{1 \text{ yd}}{3 \text{ ft}} \\ &= 24 \text{ yd} \end{aligned}$$

25. $2.5 \text{ mi} = 2.5 \text{ mi} \cdot \frac{5280 \text{ ft}}{1 \text{ mi}} = 13,200 \text{ ft}$

26. $$\begin{aligned} 6.25 \text{ ft} &= 6.25 \text{ ft} \cdot \frac{12 \text{ in.}}{1 \text{ ft}} \\ &= 75 \text{ in.} \end{aligned}$$

27. $$\begin{array}{r} 17 \\ 3\overline{)\,52} \\ -3 \\ \hline 22 \\ -21 \\ \hline 1 \end{array}$$
52 ft = 17 yd 1 ft

28. $$\begin{array}{r} 3 \\ 12\overline{)\,46} \\ -36 \\ \hline 10 \end{array}$$
46 in. = 3 ft 10 in.

29. 42 m = 4200 cm

30. 82 cm = 820 mm

31. 12.18 mm = 0.01218 m

32. 2.31 m = 0.00231 km

33. $$\begin{array}{r} 4 \text{ yd } 2 \text{ ft} \\ +\ 16 \text{ yd } 2 \text{ ft} \\ \hline 20 \text{ yd } 4 \text{ ft} \end{array} = 20 \text{ yd} + 1 \text{ yd } 1 \text{ ft} = 21 \text{ yd } 1 \text{ ft}$$

34. $$\begin{array}{r} 12 \text{ ft } 1 \text{ in.} \\ -\ 4 \text{ ft } 8 \text{ in.} \\ \hline \end{array} = \begin{array}{r} 11 \text{ ft } 13 \text{ in.} \\ -\ 4 \text{ ft } \ 8 \text{ in.} \\ \hline 7 \text{ ft } \ 5 \text{ in.} \end{array}$$

35.
$$\begin{array}{r} 8 \text{ ft } 3 \text{ in.} \\ \times \quad\quad 5 \\ \hline 40 \text{ ft } 15 \text{ in.} \end{array} = 40 \text{ ft} + 1 \text{ ft } 3 \text{ in.}$$
$$= 41 \text{ ft } 3 \text{ in.}$$

36.
$$\begin{array}{rl} & 3 \text{ ft} \quad 8 \text{ in.} \\ 2) & \overline{7 \text{ ft} \quad 4 \text{ in.}} \\ & \underline{-6 \text{ ft}} \\ & 1 \text{ ft} = \underline{12 \text{ in.}} \\ & \quad\quad 16 \text{ in.} \\ & \quad \underline{-16 \text{ in.}} \\ & \quad\quad 0 \end{array}$$

37. 8 cm = 80 mm or 15 mm = 1.5 cm
$$\begin{array}{r} 80 \text{ mm} \\ \underline{+15 \text{ mm}} \\ 95 \text{ mm} \end{array} \qquad \begin{array}{r} 8.0 \text{ cm} \\ \underline{+\ 1.5 \text{ cm}} \\ 9.5 \text{ cm} \end{array}$$

38. 4 m = 400 cm or 126 cm = 1.26 m
$$\begin{array}{r} 400 \text{ cm} \\ \underline{+126 \text{ cm}} \\ 526 \text{ cm} \end{array} \qquad \begin{array}{r} 4.00 \text{ m} \\ \underline{+\ 1.26 \text{ m}} \\ 5.26 \text{ m} \end{array}$$

39. 9.3 km = 9300 m or 183 m = 0.183 km
$$\begin{array}{r} 9300 \text{ m} \\ \underline{-183 \text{ m}} \\ 9117 \text{ m} \end{array} \qquad \begin{array}{r} 9.300 \text{ km} \\ \underline{-\ 0.183 \text{ km}} \\ 9.117 \text{ km} \end{array}$$

40. 4100 mm = 4.1 m or 3 m = 3000 mm
$$\begin{array}{r} 4.1 \text{ m} \\ \underline{-3.0 \text{ m}} \\ 1.1 \text{ m} \end{array} \qquad \begin{array}{r} 4100 \text{ mm} \\ \underline{-3000 \text{ mm}} \\ 1100 \text{ mm} \end{array}$$

41.
$$\begin{array}{r} 333 \text{ yd } 1 \text{ ft} \\ \underline{-\ 163 \text{ yd } 2 \text{ ft}} \end{array} \qquad \begin{array}{r} 332 \text{ yd } 4 \text{ ft} \\ \underline{-\ 163 \text{ yd } 2 \text{ ft}} \\ 169 \text{ yd } 2 \text{ ft} \end{array}$$

169 yards 2 feet of cloth remain.

42.
$$\begin{array}{r} 6 \text{ ft } 4 \text{ in.} \\ \times \quad\quad 20 \\ \hline 120 \text{ ft } 80 \text{ in.} \end{array} = 120 \text{ ft} + 6 \text{ ft } 8 \text{ in.}$$
$$= 126 \text{ ft } 8 \text{ in.}$$

126 feet 8 inches of framing material is needed.

43. The round-trip distance is
2(217 km) = 434 km.
$$\frac{434}{4} = 108.5$$
Each must drive 108.5 kilometers.

44. 30 cm = 0.3 m

$$\begin{array}{r} 0.8 \text{ m} \\ \times\ 0.3 \text{ m} \\ \hline 0.24 \end{array} \text{ square meters}$$

45. Perimeter = 27 meters + 17 meters + 27 meters + 17 meters
= 88 meters

46. $P = a + b + c$
= 11 centimeters + 7 centimeters + 12 centimeters
= 30 centimeters

47. The missing lengths are:
10 meters − 7 meters = 3 meters
8 meters − 5 meters = 3 meters
Perimeter = 8 meters + 7 meters + 3 meters + 3 meters + 5 meters + 10 meters
= 36 meters

48. The missing length is: 5 feet + 4 feet + 11 feet = 20 feet
Perimeter = 22 feet + 20 feet + 22 feet + 11 feet + 3 feet + 4 feet + 3 feet + 5 feet
= 90 feet

49. $P = 2l + 2w$
$= 2 \cdot 6$ feet $+ 2 \cdot 10$ feet
= 12 feet + 20 feet
= 32 feet

50. $P = 4s = 4 \cdot 110$ feet = 440 feet

51. $C = \pi d$
$= \pi \cdot 1.7$ inches
$= 1.7\pi$ inches
$\approx 1.7 \cdot 3.14$ inches
= 5.338 inches

52. $C = 2\pi r$
$= 2 \cdot \pi \cdot 5$ yards
$= 10\pi$ yards
$\approx 10 \cdot 3.14$ yards
= 31.4 yards

53. $A = \frac{1}{2}(b + B)h$
$= \frac{1}{2} \cdot (12 \text{ feet} + 36 \text{ feet}) \cdot 10 \text{ feet}$
= 240 square feet

54. $A = \frac{1}{2}bh$
$= \frac{1}{2} \cdot 20 \text{ meters} \cdot 14 \text{ meters}$
$= 140 \text{ square meters}$

55. $A = lw$
$= 40 \text{ centimeters} \cdot 15 \text{ centimeters}$
$= 600 \text{ square centimters}$

56. $A = bh$
$= 21 \text{ yards} \cdot 9 \text{ yards}$
$= 189 \text{ square yards}$

57. $A = \pi r^2$
$= \pi \cdot (7 \text{ feet})^2$
$= 49\pi \text{ square feet}$
$\approx 49 \cdot 3.14 \text{ square feet}$
$= 153.86 \text{ square feet}$

58. $A = \pi r^2$
$= \pi \cdot (2 \text{ inches})^2$
$= 4\pi \text{ square inches}$
$\approx 4 \cdot 3.14 \text{ square inches}$
$= 12.56 \text{ square inches}$

59. $A = \frac{1}{2}bh$
$= \frac{1}{2} \cdot 34 \text{ inches} \cdot 7 \text{ inches}$
$= 119 \text{ square inches}$

60. $A = \frac{1}{2}(b + B)h$
$= \frac{1}{2} \cdot (32 \text{ centimeters} + 64 \text{ centimeters}) \cdot 26 \text{ centimeters}$
$= \frac{1}{2} \cdot 96 \cdot 26 \text{ square centimeters}$
$= 1248 \text{ square centimeters}$

61.

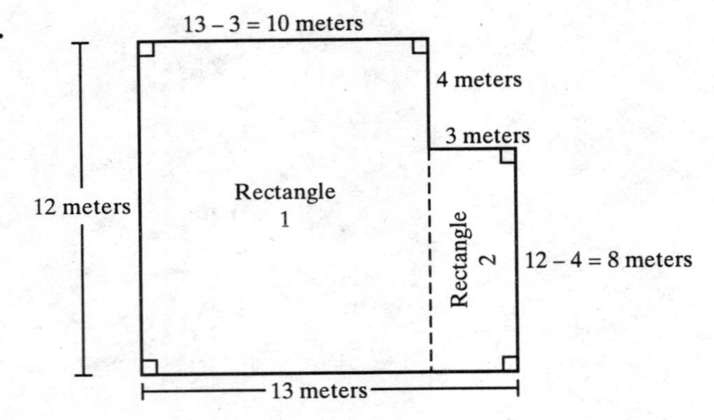

$$\text{Area of rectangle } 1 = 12 \text{ meters} \cdot 10 \text{ meters} = 120 \text{ square meters}$$

$$\text{Area of rectangle } 2 = 8 \text{ meters} \cdot 3 \text{ meters} = 24 \text{ square meters}$$

$$\text{Total area} = 120 \text{ square meters} + 24 \text{ square meters} = 144 \text{ square meters}$$

62. $A = lw$
$= 36 \text{ feet} \cdot 12 \text{ feet}$
$= 432 \text{ square feet}$

63. $A = lw$
$= 13 \text{ feet} \cdot 10 \text{ feet}$
$= 130 \text{ square feet}$

64. $V = s^3$
$= \left(2\frac{1}{2} \text{ inches}\right)^3$
$= \left(\frac{5}{2} \text{ inches}\right)^3$
$= \frac{125}{8} \text{ or } 15\frac{5}{8} \text{ cubic inches}$

65. $V = lwh$
$= 7 \text{ feet} \cdot 2 \text{ feet} \cdot 6 \text{ feet}$
$= 84 \text{ cubic feet}$

66. $V = \pi r^2 h$
$= \pi \cdot (20 \text{ centimeters})^2 \cdot 50 \text{ centimeters}$
$= 20{,}000\pi$ cubic centimeters
$\approx 20{,}000 \cdot \frac{22}{7}$ cubic centimeters
$= 62{,}857\frac{1}{7}$ cubic centimeters

67. $V = \frac{1}{3}\pi r^2 h$
$= \frac{1}{3} \cdot \pi \cdot \left(5\frac{1}{4} \text{ inches}\right)^2 \cdot 12 \text{ inches}$
$= \frac{1}{3} \cdot \left(\frac{21}{4}\right)^2 \cdot 12 \cdot \pi$ cubic inches
$= \frac{441}{4}\pi$ cubic inches
$\approx \frac{441}{4} \cdot \frac{22}{7}$ cubic inches
$= \frac{693}{2}$ or $346\frac{1}{2}$ cubic inches

68. $V = \frac{1}{3}s^2 h$
$= \frac{1}{3} \cdot (2 \text{ feet})^2 \cdot 2 \text{ feet}$
$= \frac{1}{3} \cdot 4 \cdot 2$ cubic feet
$= \frac{8}{3}$ or $2\frac{2}{3}$ cubic feet

69. $V = \pi r^2 h$
$= \pi \cdot (3.5 \text{ inches})^2 \cdot 8 \text{ inches}$
$= 98\pi$ cubic inches
$\approx 98 \cdot 3.14$ cubic inches
$= 307.72$ cubic inches

70. Volume $= 3lwh$
$= 3 \cdot 2\frac{1}{2} \text{ feet} \cdot 1\frac{1}{2} \text{ feet} \cdot \frac{2}{3} \text{ feet}$
$= 3 \cdot \frac{5}{2} \cdot \frac{3}{2} \cdot \frac{2}{3}$ cubic feet
$= \frac{15}{2}$ or $7\frac{1}{2}$ cubic feet

71. $V = \pi r^2 h$
$= \pi \cdot \left(\frac{1 \text{ foot}}{2}\right)^2 \cdot 2 \text{ feet}$
$= \frac{1}{4} \cdot 2 \cdot \pi$ cubic feet
$= \frac{1}{2}\pi$ or 0.5π cubic feet

72. $V = lwh$
$= 15 \text{ feet} \cdot 12 \text{ feet} \cdot 7 \text{ feet}$
$= 1260$ cubic feet

73. Volume of first box $= (3 \text{ feet})^3$
$= 27$ cubic feet
Volume of second box $= (1.2 \text{ feet})^3$
$= 1.728$ cubic feet
Total volume
$= 27 \text{ cubic feet} + 1.728 \text{ cubic feet}$
$= 28.728$ cubic feet

74. $66 \text{ oz} = 66 \text{ oz} \cdot \frac{1 \text{ lb}}{16 \text{ oz}} = 4.125 \text{ lb}$

75. $2.3 \text{ tons} = 2.3 \text{ tons} \cdot \frac{2000 \text{ lb}}{1 \text{ ton}}$
$= 4600 \text{ lb}$

76.
$$\begin{array}{r} 3 \\ 16\overline{)\,52} \\ -\underline{48} \\ 4 \end{array}$$
52 oz = 3 lb 4 oz

77.
$$\begin{array}{r} 4 \\ 2000\overline{)\,8200} \\ \underline{-8000} \\ 200 \end{array}$$
8200 lb = 4 tons 200 lb

78. 1400 mg = 1.4 g

79. 40 kg = 40,000 g

80. 2.1 hg = 21 dag

81. 0.03 mg = 0.0003 dg

82.
$$\begin{array}{r} 6 \text{ lb } 5 \text{ oz} \\ -\,2 \text{ lb } 12 \text{ oz} \\ \hline \end{array} \qquad \begin{array}{r} 5 \text{ lb } 21 \text{ oz} \\ -\,2 \text{ lb } 12 \text{ oz} \\ \hline 3 \text{ lb } 9 \text{ oz} \end{array}$$

83.
$$\begin{array}{r} 5 \text{ tons } 1600 \text{ lb} \\ +\,4 \text{ tons } 1200 \text{ lb} \\ \hline 9 \text{ tons } 2800 \text{ lb} \end{array} = 9 \text{ tons} + 1 \text{ ton } 800 \text{ lb} = 10 \text{ tons } 800 \text{ lb}$$

84.
$$\begin{array}{r} 2 \text{ tons } \;\; 750 \text{ lb} \\ 3\overline{)\;6 \text{ tons } 2250 \text{ lb}} \\ -6 \text{ tons} \qquad\qquad \\ \hline 0 \qquad 2250 \text{ lb} \\ -2250 \text{ lb} \\ \hline 0 \end{array}$$

85.
$$\begin{array}{r} 8 \text{ lb } 6 \text{ oz} \\ \times \qquad\quad 4 \\ \hline 32 \text{ lb } 24 \text{ oz} \end{array} = 32 \text{ lb} + 1 \text{ lb } 8 \text{ oz} = 33 \text{ lb } 8 \text{ oz}$$

86. 1300 mg = 1.3 g or 3.6 g = 3600 mg
$$\begin{array}{r} 1.3 \text{ g} \\ +\,3.6 \text{ g} \\ \hline 4.9 \text{ g} \end{array} \qquad \begin{array}{r} 1300 \text{ mg} \\ +\,3600 \text{ mg} \\ \hline 4900 \text{ mg} \end{array}$$

87. 4.8 kg = 4800 g or 4200 g = 4.2 kg
$$\begin{array}{r} 4800 \text{ g} \\ +\,4200 \text{ g} \\ \hline 9000 \text{ g} \end{array} \qquad \begin{array}{r} 4.8 \text{ kg} \\ +\,4.2 \text{ kg} \\ \hline 9.0 \text{ kg} \end{array} = 9 \text{ kg}$$

88. 9.3 g = 9300 mg or 1200 mg = 1.2 g
$$\begin{array}{r} 9300 \text{ mg} \\ -\,1200 \text{ mg} \\ \hline 8100 \text{ mg} \end{array} \qquad \begin{array}{r} 9.3 \text{ g} \\ -\,1.2 \text{ g} \\ \hline 8.1 \text{ g} \end{array}$$

89.
$$\begin{array}{r} 6.3 \text{ kg} \\ \times\; 8 \quad \\ \hline 50.4 \text{ kg} \end{array}$$

90.
$$\begin{array}{r} 1 \text{ lb } 12 \text{ oz} \\ +\,2 \text{ lb } 8 \text{ oz} \\ \hline 3 \text{ lb } 20 \text{ oz} \end{array} = 3 \text{ lb} + 1 \text{ lb } 4 \text{ oz} = 4 \text{ lb } 4 \text{ oz}$$

He ordered 4 pounds 4 ounces of candy.

91.
$$\begin{array}{r} 9 \text{ tons} \quad 1075 \text{ lb} \\ 4\overline{)\;38 \text{ tons} \quad\; 300 \text{ lb}} \\ -36 \text{ tons} \qquad\qquad\quad \\ \hline 2 \text{ tons} = 4000 \text{ lb} \\ 4300 \text{ lb} \\ -\,4300 \text{ lb} \\ \hline 0 \end{array}$$

Each township receives 9 tons 1075 pounds of cinders.

92. 450 g = 0.45 kg
$$\begin{array}{r} 8.30 \text{ kg} \\ -\,0.45 \text{ kg} \\ \hline 7.85 \text{ kg} \end{array}$$

She received 7.85 kilograms of flour.

93.
$$\begin{array}{r} 1.1625 \text{ kg} \\ 8\overline{)\;9.3000 \text{ kg}} \\ -8 \qquad\quad \\ \hline 13 \qquad\quad \\ -\,8 \qquad\quad \\ \hline 50 \qquad \\ -48 \qquad \\ \hline 20 \qquad \\ -16 \qquad \\ \hline 40 \qquad \\ -40 \qquad \\ \hline 0 \qquad \end{array}$$

Each receives 1.1625 kilograms of syrup.

94. $16 \text{ pt} = 16 \text{ pt} \cdot \frac{1 \text{ qt}}{2 \text{ pt}} = 8 \text{ qt}$

95. $40 \text{ fl oz} = 40 \text{ fl oz} \cdot \frac{1 \text{ c}}{8 \text{ fl oz}} = 5 \text{ c}$

96. $6.75 \text{ gal} = 6.75 \text{ gal} \cdot \frac{4 \text{ qt}}{1 \text{ gal}} = 27 \text{ qt}$

97. $8.5 \text{ pt} = 8.5 \text{ pt} \cdot \frac{2 \text{ c}}{1 \text{ pt}} = 17 \text{ c}$

98.
$$\begin{array}{r} 4 \\ 2\overline{)\;9} \\ -8 \\ \hline 1 \end{array}$$

9 pt = 4 qt 1 pt

99. $\begin{array}{r} 3 \\ 4\overline{)\,15} \\ -12 \\ \hline 3 \end{array}$

15 qt = 3 gal 3 qt

100. 3.8 L = 3800 ml

101. 4.2 ml = 0.042 dl

102. 14 hl = 1.4 kl

103. 30.6 L = 3060 cl

104. $\begin{array}{r} 1 \text{ qt } 1 \text{ pt} \\ +\ 3 \text{ qt } 1 \text{ pt} \\ \hline 4 \text{ qt } 2 \text{ pt} \end{array} = 4 \text{ qt} + 1 \text{ qt}$
$= 5 \text{ qt}$
$= 1 \text{ gal } 1 \text{ qt}$

105. $\begin{array}{r} 3 \text{ gal } 2 \text{ qt } 1 \text{ pt} \\ \times \qquad\qquad 2 \\ \hline 6 \text{ gal } 4 \text{ qt } 2 \text{ pt} \end{array} = 6 \text{ gal} + 1 \text{ gal} + 1 \text{qt}$
$= 7 \text{ gal } 1 \text{ qt}$

106. 0.946 L = 946 ml or 210 ml = 0.210 L

$\begin{array}{r} 946 \text{ ml} \\ -210 \text{ ml} \\ \hline 736 \text{ ml} \end{array} \qquad \begin{array}{r} 0.946 \text{ L} \\ -\ 0.210 \text{ L} \\ \hline 0.736 \text{ L} \end{array}$

107. 6.1 L = 6100 ml or 9400 ml = 9.4 L

$\begin{array}{r} 6100 \text{ ml} \\ +9400 \text{ ml} \\ \hline 15{,}500 \text{ ml} \end{array} \qquad \begin{array}{r} 6.1 \text{ L} \\ +9.4 \text{ L} \\ \hline 15.5 \text{ L} \end{array}$

108. $\begin{array}{r} 4 \text{ gal } 2 \text{ qt} \\ -\ 1 \text{ gal } 3 \text{ qt} \\ \hline \end{array} = \begin{array}{r} 3 \text{ gal } 6 \text{ qt} \\ -\ 1 \text{ gal } 3 \text{ qt} \\ \hline 2 \text{ gal } 3 \text{ qt} \end{array}$

2 gal 3 qt of tea remains.

109. 1 c 4 fl oz = 8 fl oz + 4 fl oz
= 12 fl oz
12 fl oz ÷ 2 = 6 fl oz
6 fluid ounces of beef broth should be used.

110. 85 ml · 8 · 16 = 10,880 ml = 10.88 L
8 boxes contain 10.88 liters of polish.

111. 6 L + 1300 ml + 2.6 L
= 6 L + 1.3 L + 2.6 L
= 9.9 L
Yes, since 9.9 L < 10 L, it will fit.

112. $F = \frac{9}{5} \cdot 245 + 32 = 441 + 32 = 473$

245°C = 473°F

113. $F = \frac{9}{5} \cdot 160 + 32$
$= 288 + 32$
$= 320$
160°C = 320°F

114. $F = 1.8(42) + 32$
$= 75.6 + 32$
$= 107.6$
42°C = 107.6°F

115. $F = 1.8 \cdot 86 + 32$
$= 154.8 + 32$
$= 186.8°\text{F}$
86°C = 186.8°F

116. $C = \frac{5}{9} \cdot (93.2 - 32)$
$= \frac{5}{9} \cdot 61.2$
$= \frac{306.0}{9}$
$= 34$
93.2°F = 34°C

117. $C = \frac{5}{9} \cdot (51.8 - 32)$
$= \frac{5}{9} \cdot 19.8$
$= \frac{99}{9}$
$= 11$
51.8°F = 11°C

118. $C = \frac{5}{9} \cdot (41.3 - 32)$
$= \frac{5}{9} \cdot 9.3$
$= \frac{46.5}{9}$
$= 5.1\overline{6}$
≈ 5.2
41.3°F ≈ 5.2°C

119. $C = \frac{5}{9} \cdot (80 - 32)$
$= \frac{5}{9} \cdot 48$
$= \frac{240}{9}$
$= 26.\overline{6}$
≈ 26.7
80°F ≈ 26.7°C

120. $C = \frac{5}{9} \cdot (35 - 32)$
$= \frac{5}{9} \cdot 3$
$= \frac{15}{9}$
$= 1.\overline{6}$
≈ 1.7
35°F ≈ 1.7°C

121. $F = \frac{9}{5} \cdot 165 + 32$
$= 297 + 32$
$= 329$
165°C = 329°F

Chapter 9 Test

1. The complement of a 78° angle is 90° – 78° = 12°.

2. The supplement of a 124° angle is 180° – 124° = 56°.

3. $\angle x = 90° - 40° = 50°$

4. Since $\angle x$ and the 62° angle are supplementary angles, $\angle x = 180° - 62° = 118°$. $\angle y$ and the 62° angle are vertical angles, so $\angle y = 62°$. $\angle x$ and $\angle z$ are vertical angles, so $\angle z = 118°$.

5. $\angle x$ and the 73° angle are vertical angles, so $\angle x = 73°$. $\angle y$ and the 73° angle are corresponding angles, so $\angle y = 73°$. $\angle z$ and $\angle y$ are vertical angles, so $\angle z = 73°$.

6. $\angle x = 180° - 92° - 62° = 26°$

7. $C = 2\pi r$
$= 2 \cdot \pi \cdot 9$ inches
$= 18\pi$ inches
$\approx 18 \cdot 3.14$ inches
$= 56.52$ inches

$A = \pi r^2$
$= \pi \cdot (9 \text{ inches})^2$
$= 81\pi$ square inches
$\approx 81 \cdot 3.14$ square inches
$= 254.34$ square inches

8. $P = 2l + 2w$
$= 2 \cdot 7$ yards $+ 2 \cdot 5.3$ yards
$= 14$ yards $+ 10.6$ yards
$= 24.6$ yards

$A = lw$
$= 7$ yards $\cdot 5.3$ yards
$= 37.1$ square yards

9.

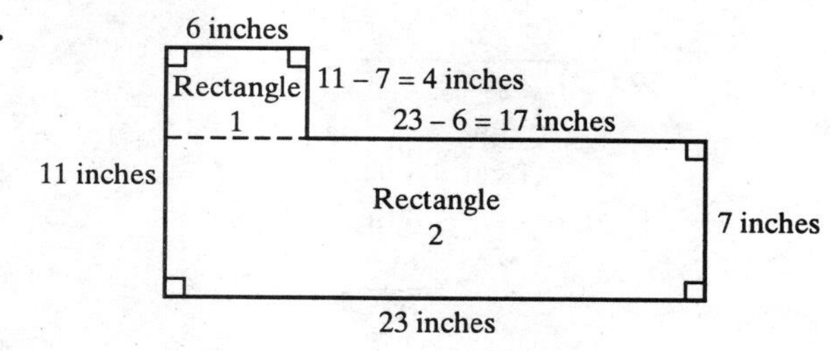

$$\begin{aligned}\text{Perimeter} &= 6 \text{ inches} + 4 \text{ inches} + 17 \text{ inches} + 7 \text{ inches} + 23 \text{ inches} + 11 \text{ inches}\\ &= 68 \text{ inches}\end{aligned}$$

$$\begin{aligned}\text{Area of rectangle } 1 &= 6 \text{ inches} \cdot 4 \text{ inches}\\ &= 24 \text{ square inches}\end{aligned}$$

$$\begin{aligned}\text{Area of rectangle } 2 &= 7 \text{ inches} \cdot 23 \text{ inches}\\ &= 161 \text{ square inches}\end{aligned}$$

$$\begin{aligned}\text{Total area} &= 161 \text{ square inches} + 24 \text{ square inches}\\ &= 185 \text{ square inches}\end{aligned}$$

10. $$\begin{aligned}V &= \pi r^2 h\\ &= \pi \cdot (2 \text{ inches})^2 \cdot 5 \text{ inches}\\ &= 20\pi \text{ cubic inches}\\ &\approx 20 \cdot \frac{22}{7} \text{ cubic inches}\\ &= \frac{440}{7} \text{ or } 62\frac{6}{7} \text{ cubic inches}\end{aligned}$$

11. $$\begin{aligned}V &= lwh\\ &= 3 \text{ feet} \cdot 5 \text{ feet} \cdot 2 \text{ feet}\\ &= 30 \text{ cubic feet}\end{aligned}$$

12. $P = 4s = 4 \cdot 4 \text{ inches} = 16 \text{ inches}$

13. $$\begin{aligned}V &= lwh\\ &= 3 \text{ feet} \cdot 3 \text{ feet} \cdot 2 \text{ feet}\\ &= 18 \text{ cubic feet}\end{aligned}$$

14. $$\begin{aligned}P &= 2l + 2w\\ &= 2 \cdot 18 \text{ feet} + 2 \cdot 13 \text{ feet}\\ &= 36 \text{ feet} + 26 \text{ feet}\\ &= 62 \text{ feet}\end{aligned}$$

15. $$\begin{aligned}A &= lw\\ &= 123.8 \text{ feet} \cdot 80 \text{ feet}\\ &= 9904 \text{ square feet}\end{aligned}$$

$$\begin{aligned}\text{amount of insecticide} &= 0.02 \cdot 9904\\ &= 198.08\end{aligned}$$

198.08 ounces of insecticide are required.

16.
$$\begin{array}{r} 23 \\ 12\overline{)280} \\ -\underline{24} \\ 40 \\ -\underline{36} \\ 4 \end{array}$$
280 in. = 23 ft 4 in.

17. $2\frac{1}{2}\text{ gal} = \frac{5}{2}\text{ gal}\cdot\frac{4\text{ qt}}{1\text{ gal}} = 10\text{ qt}$

18. $30\text{ oz} = 30\text{ oz}\cdot\frac{1\text{ lb}}{16\text{ oz}} = 1.875\text{ lb}$

19. $2.8\text{ tons} = 2.8\text{ tons}\cdot\frac{2000\text{ lb}}{1\text{ ton}} = 5600\text{ lb}$

20. $38\text{ pt} = 38\text{ pt}\cdot\frac{1\text{ gal}}{8\text{ pt}} = \frac{19}{4}\text{ gal} = 4\frac{3}{4}\text{ gal}$

21. 40 mg = 0.04 g

22. 2.4 kg = 2400 g

23. 3.6 cm = 36 mm

24. 4.3 dg = 0.43 g

25. 0.83 L = 830 ml

26.
$$\begin{array}{r} 3\text{ qt }1\text{ pt} \\ \underline{+\ 2\text{ qt }1\text{ pt}} \\ 5\text{ qt }2\text{ pt} \end{array} = 1\text{ gal }1\text{ qt} + 1\text{ qt} = 1\text{ gal }2\text{ qt}$$

27.
$$\begin{array}{rcr} 8\text{ lb }6\text{ oz} & = & 7\text{ lb }22\text{ oz} \\ \underline{-\ 4\text{ lb }9\text{ oz}} & = & \underline{-\ 4\text{ lb }\ 9\text{ oz}} \\ & & 3\text{ lb }13\text{ oz} \end{array}$$

28.
$$\begin{array}{r} 2\text{ ft }9\text{ in.} \\ \underline{\times\quad 3} \\ 6\text{ ft }27\text{ in.} \end{array} = 6\text{ ft} + 2\text{ ft }3\text{ in.} = 8\text{ ft }3\text{ in.}$$

29.
$$\begin{array}{rl} & 2\text{ gal}\quad 3\text{ qt} \\ 2\big) & 5\text{ gal}\quad 2\text{ qt} \\ & \underline{-4\text{ gal}} \\ & 1\text{ gal} = \underline{4\text{ qt}} \\ & \qquad 6\text{ qt} \\ & \qquad \underline{-6\text{ qt}} \\ & \qquad 0 \end{array}$$

30. 8 cm = 80 mm or 14 mm = 1.4 cm
$$\begin{array}{r} 80\text{ mm} \\ \underline{-\ 14\text{ mm}} \\ 66\text{ mm} \end{array} \qquad \begin{array}{r} 8.0\text{ cm} \\ \underline{-\ 1.4\text{ cm}} \\ 6.6\text{ cm} \end{array}$$

31. 1.8 km = 1800 m or 456 m = 0.456 km
$$\begin{array}{r} 1800\text{ m} \\ \underline{+\,456\text{ m}} \\ 2256\text{ m} \end{array} \qquad \begin{array}{r} 1.800\text{ km} \\ \underline{+\,0.456\text{ km}} \\ 2.256\text{ km} \end{array}$$

32.
$$\begin{aligned} C &= \frac{5}{9}\cdot(84-32) \\ &= \frac{5}{9}\cdot 52 \\ &= \frac{260}{9} \\ &= 28.\overline{8} \\ &\approx 28.9 \end{aligned}$$
84°F ≈ 28.9°C

33.
$$\begin{aligned} F &= 1.8\cdot 12.6 + 32 \\ &= 22.68 + 32 \\ &= 54.68 \end{aligned}$$
12.6°C = 54.7°F

34. $8.4\text{ m}\cdot\frac{1}{3} = 2.8\text{ m}$
8.4 m − 2.8 m = 5.6 m
The maples will be 5.6 meters tall after the cutting.

35. 20 gal = 19 gal 4 qt
$$\begin{array}{r} 19\text{ gal }4\text{ qt} \\ \underline{-\ 15\text{ gal }1\text{ qt}} \\ 4\text{ gal }3\text{ qt} \end{array}$$
4 gal 3 qt of oil remains.

36. F = 1.8(41) + 32 = 73.8 + 32 = 105.8
Her fever is 105.8°F.

37. 340 cm = 3.4 m

$$\begin{array}{r} 88.0 \text{ m} \\ +\ 3.4 \text{ m} \\ \hline 91.4 \text{ m} \end{array}$$

The span is 91.4 m long.

38.

$$\begin{array}{r} 2 \text{ ft } 9 \text{ in.} \\ \times \quad 6 \\ \hline 12 \text{ ft } 54 \text{ in.} \end{array} = 12 \text{ ft} + 4 \text{ ft } 6 \text{ in.}$$

$$= 16 \text{ ft } 6 \text{ in. or } 16\frac{1}{2} \text{ ft}$$

16 ft 6 inches or $16\frac{1}{2}$ feet of material is needed for 6 scarves.

39.

$$\begin{array}{r} 1 \\ 16\overline{)\ 29} \\ -16 \\ \hline 13 \end{array}$$

29 ounces = 1 lb 13 oz

40. 320 grams = 0.32 kg

41.

$$4667 \text{ gal} = \frac{4667}{1} \text{ gal} \cdot \frac{4 \text{ qt}}{1 \text{ gal}} \cdot \frac{2 \text{ pt}}{1 \text{ qt}}$$

$$= 4667 \cdot 4 \cdot 2 \text{ pt}$$

$$= 37{,}336 \text{ pt}$$

Thus, 37,336 pints of ice cream were used.

Chapter 10

Pretest

1. $(7y^2 - 15y + 9) + (-8y + 17)$
$= 7y^2 - 15y - 8y + 9 + 17$
$= 7y^2 - 23y + 26$

2. $(9b - 5) - (-8b + 7) = (9b - 5) + (8b - 7)$
$= 9b + 8b - 5 - 7$
$= 17b - 12$

3. $(2z^4 - z^3 + 5z) - (-3z^4 + 6z^2 + 2z)$
$= (2z^4 - z^3 + 5z) + (3z^4 - 6z^2 - 2z)$
$= 2z^4 + 3z^4 - z^3 - 6z^2 + 5z - 2z$
$= 5z^4 - z^3 - 6z + 3z$

4. $6t^3 - 5t + 18$ when $t = -2$.
$6(-2)^3 - 5(-2) + 18 = 6(-8) + 10 + 18$
$= -48 + 10 + 18$
$= -38 + 18$
$= -20$

5. $9n^6 \cdot 4n^{12} = (9 \cdot 4)(n^6 \cdot n^{12})$
$= 36n^{6+12}$
$= 36n^{18}$

6. $4x \cdot 5x \cdot 6x = (4 \cdot 5 \cdot 6)(x \cdot x \cdot x) = 120x^3$

7. $(t^{18})^3 = t^{18 \cdot 3} = t^{54}$

8. $(3n^2)^4 = 3^4(n^2)^4 = 81n^{2 \cdot 4} = 81n^8$

9. $(3a^2bc^3)^5 \cdot (8ab^4c)^2$
$= 3^5(a^2)^5b^5(c^3)^5 \cdot 8^2a^2(b^4)^2c^2$
$= 243a^{10}b^5c^{15} \cdot 64a^2b^8c^2$
$= (243 \cdot 64)(a^{10} \cdot a^2)(b^5 \cdot b^8)(c^{15} \cdot c^2)$
$= 15,552a^{12}b^{13}c^{17}$

10. $2d(9d^4 - 5d^2 + 11)$
$= 2d \cdot 9d^4 - 2d \cdot 5d^2 + 2d \cdot 11$
$= 18d^5 - 10d^3 + 22d$

11. $(x + 6)(x + 3) = x(x + 3) + 6(x + 3)$
$= x \cdot x + x \cdot 3 + 6 \cdot x + 6 \cdot 3$
$= x^2 + 3x + 6x + 18$
$= x^2 + 9x + 18$

12. $(y - 4)(2y + 5) = y(2y + 5) - 4(2y + 5)$
$= y \cdot 2y + y \cdot 5 - 4 \cdot 2y - 4 \cdot 5$
$= 2y^2 + 5y - 8y - 20$
$= 2y^2 - 3y - 20$

13. $(3x - 2)^2 = (3x - 2)(3x - 2)$
$= 3x(3x - 2) - 2(3x - 2)$
$= 3x \cdot 3x - 3x \cdot 2 - 2 \cdot 3x - 2(-2)$
$= 9x^2 - 6x - 6x + 4$
$= 9x^2 - 12x + 4$

14. $(n + 10)(n - 10)$
$= n(n - 10) + 10(n - 10)$
$= n \cdot n - n \cdot 10 + 10 \cdot n - 10 \cdot 10$
$= n^2 - 10n + 10n - 100$
$= n^2 - 100$

15. $(4a+1)(2a^2-a+7) = 4a(2a^2-a+7)+1(2a^2-a+7)$
$= 4a \cdot 2a^2 - 4a \cdot a + 4a \cdot 7 + 1 \cdot 2a^2 - 1 \cdot a + 1 \cdot 7$
$= 8a^3 - 4a^2 + 28a + 2a^2 - a + 7$
$= 8a^3 - 2a^2 + 27a + 7$

16. $18 = 2 \cdot 3 \cdot 3$
$45 = 3 \cdot 3 \cdot 5$
$\text{GCF} = 3 \cdot 3 = 9$

17. $y^9 = y^3 \cdot y^6$
$y^3 = y^3 \cdot 1$
$y^8 = y^3 \cdot y^5$
$\text{GCF} = y^3$

18. $6 = 2 \cdot 3$
$14 = 2 \cdot 7$
$18 = 2 \cdot 3 \cdot 3$
$\text{GCF} = 2$
$m^5 = m \cdot m^4$
$m = m \cdot 1$
$m^4 = m \cdot m^3$
$\text{GCF} = m$
GCF of $6m^5$, $14m$, $18m^4 = 2 \cdot m = 2m$

19. $10y^2 = 2 \cdot 5y^2$
$6y = 2 \cdot 3y$
$14 = 2 \cdot 7$
$10y^2 + 6y - 14 = 2 \cdot 5y^2 + 2 \cdot 3y - 2 \cdot 7$
$= 2(5y^2 + 3y - 7)$

20. $8n^6 = 4n^3 \cdot 2n^3$
$12n^5 = 4n^3 \cdot 3n^2$
$20n^3 = 4n^3 \cdot 5$
$8n^6 - 12n^5 + 20n^3 = 4n^3 \cdot 2n^3 - 4n^3 \cdot 3n^2 + 4n^3 \cdot 5$
$= 4n^3(2n^3 - 3n^2 + 5)$

Section 10.1

Practice Problems

1. $(2y+7)+(9y-14) = (2y+9y)+(7-14)$
$= (11y)+(-7)$
$= 11y - 7$

2. $(5x^2 + 4x - 3) + (x^2 - 6x)$
$= 5x^2 + x^2 + 4x - 6x - 3$
$= 6x^2 - 2x - 3$

3. $(7z^2 - 4.2z + 11) + (-9z^2 - 1.9z + 4)$
$= 7z^2 - 9z^2 - 4.2z - 1.9z + 11 + 4$
$= -2z^2 - 6.1z + 15$

4.
$$\begin{array}{r} 7z^2 - 4.2z + 11 \\ \underline{-\ 9z^2 - 1.9z +\ 4} \\ -2z^2 - 6.1z + 15 \end{array}$$

5. $-(7y^2 + 4y - 6)$
$= -1(7y^2 + 4y - 6)$
$= -1(7y^2) + (-1)(4y) + (-1)(-6)$
$= -7y^2 - 4y + 6$

6. $(3b - 2) - (7b + 23) = (3b - 2) + (-7b - 23)$
$= 3b - 7b - 2 - 23$
$= -4b - 25$

7. $(-4x^2 + 20x + 17) - (3x^2 - 12x)$
$= (-4x^2 + 20x + 17) + (-3x^2 + 12x)$
$= -4x^2 - 3x^2 + 20x + 12x + 17$
$= -7x^2 + 32x + 17$

8.
$$\begin{array}{r} -4x^2 + 20x + 17 \\ \underline{-(3x^2 - 12x)\qquad} \end{array} \qquad \begin{array}{r} -4x^2 + 20x + 17 \\ \underline{-3x^2 + 12x\qquad} \\ -7x^2 + 32x + 17 \end{array}$$

9. $2y^3 + y^2 - 6 = 2(3)^3 + (3)^2 - 6$
$= 2(27) + (9) - 6$
$= 54 + 9 - 6$
$= 57$

10. When $t = 1$ second:
$-16t^2 + 530 = -16(1)^2 + 530$
$= -16 + 530$
$= 514$
The height of the object at 1 second is 514 feet.
When $t = 4$ seconds:
$-16t^2 + 530 = -16(4)^2 + 530$
$= -256 + 530$
$= 274$
The height of the object at 4 seconds is 274 feet.

Exercise Set 10.1

1. $(2x + 3) + (-7x - 27) = (2x - 7x) + (3 - 27)$
$= -5x + (-24)$
$= -5x - 24$

3.
$$\begin{array}{r} -4z^2 - 6z + 1 \\ \underline{+\ -5z^2 + 4z + 5} \\ -9z^2 - 2z + 6 \end{array}$$

5. $(12y - 20) + (9y^2 + 13y - 20)$
$= 9y^2 + (12y + 13y) + (-20 - 20)$
$= 9y^2 + 25y - 40$

7. $(4.3a^4 + 5) + (-8.6a^4 - 2a^2 + 4)$
$= (4.3a^4 - 8.6a^4) + (-2a^2) + (5 + 4)$
$= -4.3a^4 - 2a^2 + 9$

9. $(5a - 6) - (a + 2)$
$= (5a - 6) + (-a - 2)$
$= (5a - a) + (-6 - 2)$
$= 4a - 8$

11. $(3x^2 - 2x + 1) - (5x^2 - 6x)$
$= (3x^2 - 2x + 1) + (-5x^2 + 6x)$
$= (3x^2 - 5x^2) + (-2x + 6x) + 1$
$= -2x^2 + 4x + 1$

13. $(10y^2 - 7) - (20y^3 - 2y^2 - 3)$
$= (10y^2 - 7) + (-20y^3 + 2y^2 + 3)$
$= (-20y^3) + (10y^2 + 2y^2) + (-7 + 3)$
$= -20y^3 + 12y^2 - 4$

15. $\begin{array}{r} 2x+12 \\ \underline{-\ (3x-4)} \end{array} \qquad \begin{array}{r} 2x+12 \\ \underline{+\ -3x+\ 4} \\ -x+16 \end{array}$

17. $\begin{array}{r} 13y^2-6y-14 \\ \underline{-\ (5y^2+4y-\ 6)} \end{array} \qquad \begin{array}{r} 13y^2-\ 6y-14 \\ \underline{+\ -5y^2-\ 4y+\ 6} \\ 8y^2-10y-\ 8 \end{array}$

19. $(9x^2 - 6) + (-5x^2 + x - 10)$
$= (9x^2 - 5x^2) + x + (-6 - 10)$
$= 4x^2 + x - 16$

21. $(21y - 4.6) - (36y - 8.2)$
$= (21y - 4.6) + (-36y + 8.2)$
$= (21y - 36y) + (-4.6 + 8.2)$
$= -15y + 3.6$

23. $(b^3 - 2b^2 + 10b + 11) + (b^2 - 3b - 12)$
$= b^3 + (-2b^2 + b^2) + (10b - 3b) + (11 - 12)$
$= b^3 - b^2 + 7b - 1$

25. $\left(3z + \frac{6}{7}\right) - \left(3z - \frac{3}{7}\right) = \left(3z + \frac{6}{7}\right) + \left(-3z + \frac{3}{7}\right)$
$= (3z - 3z) + \left(\frac{6}{7} + \frac{3}{7}\right)$
$= \frac{9}{7}$

27. $-3x + 7$
Let $x = 2$.
$-3(2) + 7 = -6 + 7 = 1$

29. $x^2 - 6x + 3$
Let $x = 2$.
$(2)^2 - 6(2) + 3 = 4 - 12 + 3 = -5$

31. $\frac{3x^2}{2} - 14$
Let $x = 2$.
$\frac{3(2)^2}{2} - 14 = \frac{3 \cdot 4}{2} - 14$
$= \frac{12}{2} - 14$
$= 6 - 14$
$= -8$

33. $2x + 10$
Let $x = 5$.
$2(5) + 10 = 10 + 10 = 20$

35. x^2
Let $x = 5$.
$(5)^2 = 25$

37. $2x^2 + 4x - 20$
Let $x = 5$.
$2(5)^2 + 4(5) - 20 = 2(25) + 20 - 20 = 50$

39. $16t^2$
Let $t = 6$.
$16(6)^2 = 16(36) = 576$
The object falls 576 feet in 6 seconds.

41. $3000 + 20x$
Let $x = 10$.
$3000 + 20(10) = 3000 + 200 = 3200$
It costs \$3200 to manufacture 10 file cabinets.

43. When $t = 3$ seconds:
$1053 - 16t^2 = 1053 - 16(3)^2$
$= 1053 - 16 \cdot 9$
$= 1053 - 144$
$= 909$
The height of the object above the river after 3 seconds is 909 feet.

45. $3^4 = 3 \cdot 3 \cdot 3 \cdot 3 = 81$

47. $(-5)^2 = (-5) \cdot (-5) = 25$

49. $x \cdot x \cdot x = x^3$

51. $2 \cdot 2 \cdot a \cdot a \cdot a \cdot a = 2^2 a^4$

53. $(2x + 1) + (x + 11) + (5x - 10)$
$= (2x + x + 5x) + (1 + 11 - 10)$
$= 8x + 2$
The perimeter is $(8x + 2)$ inches.

55. $(7x - 10) - (3x + 5) = (7x - 10) + (-3x - 5)$
$= (7x - 3x) + (-10 - 5)$
$= 4x - 15$
The unknown length is $(4x - 15)$ units.

57.
$$\begin{array}{r} 3x^2 + \underline{\quad}x - \underline{\quad} \\ + \ \underline{\quad}x^2 - \ 6x + \ 2 \\ \hline 5x^2 + 14x - \ 4 \end{array}$$

$$\begin{bmatrix} (3 + \underline{\quad})x^2 = 5x^2 \\ (3 + \underline{2})x^2 = 5x^2 \end{bmatrix}$$

$$\begin{bmatrix} (\underline{\quad} - 6)x = 14x \\ (\underline{20} - 6)x = 14x \end{bmatrix}$$

$$\begin{bmatrix} (-\underline{\quad} + 2) = -4 \\ (-\underline{6} + 2) = -4 \end{bmatrix}$$

$$\begin{array}{r} 3x^2 + \underline{20}x - \underline{6} \\ + \ \underline{2}x^2 - \ 6x + 2 \\ \hline 5x^2 + 14x - 4 \end{array}$$

59. $7a^4 - 6a^2 + 2a - 1$
$= 7(1.2)^4 - 6(1.2)^2 + 2(1.2) - 1$
$= 7(2.0736) - 6(1.44) + 2.4 - 1$
$= 14.5152 - 8.64 + 2.4 - 1$
$= 7.2752$

61. When $t = 8$ seconds:
$1053 - 16t^2 = 1053 - 16(8)^2$
$= 1053 - 16 \cdot 64$
$= 1053 - 1024$
$= 29$
The height of the object above the river after 8 seconds is 29 feet.
When $t = 9$ seconds:
$1053 - 16t^2 = 1053 - 16(9)^2$
$= 1053 - 16 \cdot 81$
$= 1053 - 1296$
$= -243$
The height of the object above the river after 9 seconds is –243 feet. The object hits the water between 8 and 9 seconds.

Section 10.2

Practice Problems

1. $z^4 \cdot z^8 = z^{4+8} = z^{12}$

2. $7y^5 \cdot 3y^9 = (7 \cdot 3)(y^5 \cdot y^9)$
$= 21y^{5+9}$
$= 21y^{14}$

3. $(-7r^6s^2)(-3r^2s^5) = (-7 \cdot -3)(r^6 \cdot r^2)(s^2 \cdot s^5)$
$= 21r^{6+2}s^{2+5}$
$= 21r^8s^7$

4. $9y^4 \cdot 3y^2 \cdot y = (9 \cdot 3)(y^4 \cdot y^2 \cdot y^1)$
$= 27y^{4+2+1}$
$= 27y^7$

5. $(z^3)^{10} = z^{3 \cdot 10} = z^{30}$

6. $(z^4)^5 \cdot (z^3)^7 = z^{20} \cdot z^{21} = z^{20+21} = z^{41}$

7. $(3b)^4 = 3^4b^4 = 81b^4$

8. $(4x^2y^6)^3 = 4^3(x^2)^3(y^6)^3 = 64x^6y^{18}$

9. $(2x^2y^4)^4 \cdot (3x^6y^9)^2$
$= 2^4(x^2)^4(y^4)^4 \cdot 3^2(x^6)^2(y^9)^2$
$= 16x^8y^{16} \cdot 9x^{12}y^{18}$
$= (16 \cdot 9)(x^8 \cdot x^{12})(y^{16} \cdot y^{18})$
$= 144x^{20}y^{34}$

Exercise Set 10.2

1. $x^5 \cdot x^9 = x^{5+9} = x^{14}$

3. $a^6 \cdot a = a^{6+1} = a^7$

5. $3z^3 \cdot 5z^2 = (3 \cdot 5)(z^3 \cdot z^2) = 15z^5$

7. $-4x \cdot 10x = (-4 \cdot 10)(x \cdot x) = -40x^2$

9. $(-5x^2y^3)(-5x^4y) = (-5)(-5)(x^2 \cdot x^4)(y^3 \cdot y)$
$= 25x^6y^4$

11. $(7ab)(4a^4b^5) = (7 \cdot 4)(a \cdot a^4)(b \cdot b^5)$
$= 28a^5b^6$

13. $2x \cdot 3x \cdot 5x = (2 \cdot 3 \cdot 5)(x \cdot x \cdot x) = 30x^3$

15. $a \cdot 4a^{11} \cdot 3a^5 = (4 \cdot 3)(a \cdot a^{11} \cdot a^5) = 12a^{17}$

17. $(x^5)^3 = x^{5 \cdot 3} = x^{15}$

19. $(z^2)^{10} = z^{2 \cdot 10} = z^{20}$

21. $(b^7)^6 \cdot (b^2)^{10} = b^{7 \cdot 6} \cdot b^{2 \cdot 10}$
$= b^{42} \cdot b^{20}$
$= b^{42+20}$
$= b^{62}$

23. $(3a)^4 = 3^4 \cdot a^4 = 81a^4$

25. $(a^{11}b^8)^3 = a^{11 \cdot 3} \cdot b^{8 \cdot 3} = a^{33}b^{24}$

27. $(11x^3y^6)^2 = 11^2(x^3)^2(y^6)^2 = 121x^6y^{12}$

29. $(-3y)(2y^7)^3 = (-3y) \cdot 2^3(y^7)^3$
$= (-3y) \cdot 8y^{21}$
$= (-3)(8)(y^1 \cdot y^{21})$
$= -24y^{22}$

31. $(4xy)^3(2x^3y^5)^2 = 4^3x^3y^3 \cdot 2^2(x^3)^2(y^5)^2$
$= 64x^3y^3 \cdot 4x^6y^{10}$
$= (64 \cdot 4)(x^3 \cdot x^6)(y^3 \cdot y^{10})$
$= 256x^9y^{13}$

33. $7(x-3) = 7x - 21$

35. $-2(3a+2b) = -6a - 4b$

37. $9(x+2y-3) = 9x + 18y - 27$

39. area $= s^2 = (4x^6)^2 = 4^2(x^6)^2 = 16x^{12}$
The area is $16x^{12}$ square inches.

41. area $= \frac{1}{2} \cdot b \cdot h$
$= \frac{1}{2} \cdot (6a^3b^4) \cdot (4ab)$
$= \left(\frac{1}{2} \cdot 6 \cdot 4\right)(a^3 \cdot a)(b^4 \cdot b)$
$= 12a^4b^5$
The area is $12a^4b^5$ square meters.

43. $(14a^7b^6)^3(9a^6b^3)^4$
$= 14^3(a^7)^3(b^6)^3 \cdot 9^4(a^6)^4(b^3)^4$
$= 2744a^{21}b^{18} \cdot 6561a^{24}b^{12}$
$= (2744 \cdot 6561)(a^{21}a^{24})(b^{18}b^{12})$
$= 18,003,384a^{45}b^{30}$

45. $(8.1x^{10})^5 = 8.1^5(x^{10})^5 = 34,867.84401x^{50}$

47. Answers may vary.

Integrated Review

1. $(7x+1)+(-3x-2) = 7x - 3x + 1 - 2$
$= 4x - 1$

2. $(14y-6)+(19y-2) = 14y + 19y - 6 - 2$
$= 33y - 8$

3. $(7x+1)-(-3x-2) = (7x+1)+(3x+2)$
$= 7x + 3x + 1 + 2$
$= 10x + 3$

4. $(14y-6)-(19y-2) = (14y-6)+(-19y+2)$
$= 14y - 19y - 6 + 2$
$= -5y - 4$

5. $(a^3+1)+(2a^3+5a-9)$
$= a^3 + 2a^3 + 5a + 1 - 9$
$= 3a^3 + 5a - 8$

6. $(1.2y^2 - 3.6y) + (0.6y^2 + 1.2y - 5.6)$
$= 1.2y^2 + 0.6y^2 - 3.6y + 1.2y - 5.6$
$= 1.8y^2 - 2.4y - 5.6$

7. $(3.5x^2 - 0.5x) - (5.3x^2 - 2.9x + 1.7)$
$= (3.5x^2 - 0.5x) + (-5.3x^2 + 2.9x - 1.7)$
$= 3.5x^2 - 5.3x^2 - 0.5x + 2.9x - 1.7$
$= -1.8x^2 + 2.4x - 1.7$

8. $(2a^3 - 6a^2 + 11) - (6a^3 + 6a^2 + 11)$
$= (2a^3 - 6a^2 + 11) + (-6a^3 - 6a^2 - 11)$
$= 2a^3 - 6a^3 - 6a^2 - 6a^2 + 11 - 11$
$= -4a^3 - 12a^2$

9. $(8x + 1) - (2x - 6) = (8x + 1) + (-2x + 6)$
$= 8x - 2x + 1 + 6$
$= 6x + 7$

10. $(5x^2 + 2x - 10) - (3x^2 - x + 2)$
$= (5x^2 + 2x - 10) + (-3x^2 + x - 2)$
$= 5x^2 - 3x^2 + 2x + x - 10 - 2$
$= 2x^2 + 3x - 12$

11. $2x - 7 = 2(-3) - 7 = -6 - 7 = -13$

12. $x^2 + 5x + 2 = (-3)^2 + 5(-3) + 2$
$= 9 - 15 + 2$
$= -4$

13. $x^7 \cdot x^{11} = x^{7+11} = x^{18}$

14. $x^6 \cdot x^6 = x^{6+6} = x^{12}$

15. $y^3 \cdot y = y^{3+1} = y^4$

16. $a \cdot a^{10} = a^{1+10} = a^{11}$

17. $(x^7)^{11} = x^{7 \cdot 11} = x^{77}$

18. $(x^6)^6 = x^{6 \cdot 6} = x^{36}$

19. $(2x)^4 = 2^4 x^4 = 16x^4$

20. $(3y)^3 = 3^3 y^3 = 27y^3$

21. $(-6xy^2)(2xy^5) = (-6 \cdot 2)(x^1 \cdot x^1)(y^2 \cdot y^5)$
$= -12x^{1+1}y^{2+5}$
$= -12x^2y^7$

22. $(-4a^2b^3)(-3ab) = (-4 \cdot -3)(a^2 \cdot a^1)(b^3 \cdot b^1)$
$= 12a^{2+1}b^{3+1}$
$= 12a^3b^4$

23. $(x^9y^5)^4 = (x^9)^4(y^5)^4 = x^{36}y^{20}$

24. $(a^{10}b^{12})^2 = (a^{10})^2(b^{12})^2 = a^{20}b^{24}$

25. $(10x^2y)^2 \cdot (3y) = 10^2(x^2)^2(y)^2 \cdot (3y)$
$= 100x^4y^2 \cdot 3y$
$= 300x^4y^3$

26. $(8y^3z)^2 \cdot (2z^5) = 8^2(y^3)^2(z)^2 \cdot (2z^5)$
$= 64y^6z^2 \cdot 2z^5$
$= 128y^6z^7$

Section 10.3

Practice Problems

1. $3y(7y^2 + 5) = 3y \cdot 7y^2 + 3y \cdot 5 = 21y^3 + 15y$

2. $5r(8r^2 - r + 11) = 5r \cdot 8r^2 + 5r(-r) + 5r \cdot 11$
$= 40r^3 - 5r^2 + 55r$

3. $(b + 7)(b + 5) = b(b + 5) + 7(b + 5)$
$= b \cdot b + b \cdot 5 + 7 \cdot b + 7 \cdot 5$
$= b^2 + 5b + 7b + 35$
$= b^2 + 12b + 35$

4. $(5x-1)(5x+4) = 5x(5x+4) - 1(5x+4)$
$= 5x \cdot 5x + 5x \cdot 4 - 1 \cdot 5x - 1 \cdot 4$
$= 25x^2 + 20x - 5x - 4$
$= 25x^2 + 15x - 4$

5. $(6y-1)^2 = (6y-1)(6y-1)$
$= 6y(6y-1) - 1(6y-1)$
$= 6y \cdot 6y + 6y(-1) - 1(6y) - 1(-1)$
$= 36y^2 - 6y - 6y + 1$
$= 36y^2 - 12y + 1$

6. $(2x+5)(x^2+4x-1)$
$= 2x(x^2+4x-1) + 5(x^2+4x-1)$
$= 2x \cdot x^2 + 2x \cdot 4x + 2x(-1) + 5 \cdot x^2 + 5 \cdot 4x + 5(-1)$
$= 2x^3 + 8x^2 - 2x + 5x^2 + 20x - 5$
$= 2x^3 + 13x^2 + 18x - 5$

7.
$$\begin{array}{r} x^2 + 4x - 1 \\ \underline{\qquad\qquad 2x + 5} \\ 5x^2 + 20x - 5 \\ \underline{2x^3 + 8x^2 - 2x \qquad} \\ 2x^3 + 13x^2 + 18x - 5 \end{array}$$

Exercise Set 10.3

1. $3x(9x^2-3) = 3x \cdot 9x^2 + 3x \cdot (-3)$
$= (3 \cdot 9)(x \cdot x^2) + (3)(-3)(x)$
$= 27x^3 + (-9x)$
$= 27x^3 - 9x$

3. $-5a(4a^2 - 6a + 1)$
$= (-5a)(4a^2) + (-5a)(-6a) + (-5a)(1)$
$= (-5)(4)(a \cdot a^2) + (-5)(-6)(a \cdot a) + (-5a)$
$= -20a^3 + 30a^2 - 5a$

5. $7x^2(6x^2 - 5x + 7)$
$= (7x^2)(6x^2) + (7x^2)(-5x) + (7x^2)(7)$
$= (7 \cdot 6)(x^2 \cdot x^2) + (7)(-5)(x^2 \cdot x) + (7 \cdot 7)x^2$
$= 42x^4 - 35x^3 + 49x^2$

7. $(x+3)(x+10) = x(x+10)+3(x+10)$
$= x \cdot x + x \cdot 10 + 3 \cdot x + 3 \cdot 10$
$= x^2 + 10x + 3x + 30$
$= x^2 + 13x + 30$

9. $(2x-6)(x+4) = 2x(x+4) - 6(x+4)$
$= 2x \cdot x + 2x \cdot 4 - 6 \cdot x - 6 \cdot 4$
$= 2x^2 + 8x - 6x - 24$
$= 2x^2 + 2x - 24$

11. $(6a+4)^2 = (6a+4)(6a+4)$
$= 6a(6a+4) + 4(6a+4)$
$= 6a \cdot 6a + 6a \cdot 4 + 4 \cdot 6a + 4 \cdot 4$
$= 36a^2 + 24a + 24a + 16$
$= 36a^2 + 48a + 16$

13. $(a+6)(a^2 - 6a + 3) = a(a^2 - 6a + 3) + 6(a^2 - 6a + 3)$
$= a \cdot a^2 + a(-6a) + a \cdot 3 + 6 \cdot a^2 + 6(-6a) + 6 \cdot 3$
$= a^3 - 6a^2 + 3a + 6a^2 - 36a + 18$
$= a^3 - 33a + 18$

15. $(4x-5)(2x^2 + 3x - 10) = 4x(2x^2 + 3x - 10) - 5(2x^2 + 3x - 10)$
$= 4x \cdot 2x^2 + 4x \cdot 3x + 4x(-10) - 5 \cdot 2x^2 - 5 \cdot 3x - 5(-10)$
$= 8x^3 + 12x^2 - 40x - 10x^2 - 15x + 50$
$= 8x^3 + 2x^2 - 55x + 50$

17. $(x^3 + 2x + x^2)(3x + 1 + x^2) = x^3(3x + 1 + x^2) + 2x(3x + 1 + x^2) + x^2(3x + 1 + x^2)$
$= x^3 \cdot 3x + x^3 \cdot 1 + x^3 \cdot x^2 + 2x \cdot 3x + 2x \cdot 1 + 2x \cdot x^2 + x^2 \cdot 3x + x^2 \cdot 1 + x^2 \cdot x^2$
$= 3x^4 + x^3 + x^5 + 6x^2 + 2x + 2x^3 + 3x^3 + x^2 + x^4$
$= x^5 + 4x^4 + 6x^3 + 7x^2 + 2x$

19. $10r(-3r+2) = 10r(-3r) + 10r(2)$
$= -30r^2 + 20r$

21. $-2y^2(3y + y^2 - 6)$
$= -2y^2 \cdot 3y + (-2y^2) \cdot y^2 + (-2y^2)(-6)$
$= -6y^3 - 2y^4 + 12y^2$

23. $(x+2)(x+12) = x(x+12)+2(x+12)$
$= x \cdot x + x \cdot 12 + 2 \cdot x + 2 \cdot 12$
$= x^2 + 12x + 2x + 24$
$= x^2 + 14x + 24$

25. $(2a+3)(2a-3)$
$= 2a(2a-3)+3(2a-3)$
$= 2a \cdot 2a + 2a(-3) + 3 \cdot 2a + 3(-3)$
$= 4a^2 - 6a + 6a - 9$
$= 4a^2 - 9$

27. $(x+5)^2 = (x+5)(x+5)$
$= x(x+5)+5(x+5)$
$= x \cdot x + x \cdot 5 + 5 \cdot x + 5 \cdot 5$
$= x^2 + 5x + 5x + 25$
$= x^2 + 10x + 25$

29. $\left(b+\frac{3}{5}\right)\left(b+\frac{4}{5}\right) = b\left(b+\frac{4}{5}\right)+\frac{3}{5}\left(b+\frac{4}{5}\right)$
$= b^2 + \frac{4}{5}b + \frac{3}{5}b + \frac{3}{5} \cdot \frac{4}{5}$
$= b^2 + \frac{7}{5}b + \frac{12}{25}$

31. $(6x+1)(x^2+4x+1)$
$= 6x(x^2+4x+1)+1(x^2+4x+1)$
$= 6x^3 + 24x^2 + 6x + x^2 + 4x + 1$
$= 6x^3 + 25x^2 + 10x + 1$

33. $(7x+5)^2 = (7x+5)(7x+5)$
$= 7x(7x+5)+5(7x+5)$
$= 49x^2 + 35x + 35x + 25$
$= 49x^2 + 70x + 25$

35. $(2x-1)^2 = (2x-1)(2x-1)$
$= 2x(2x-1)-1(2x-1)$
$= 4x^2 - 2x - 2x + 1$
$= 4x^2 - 4x + 1$

37. $(2x^2-3)(4x^3+2x-3)$
$= 2x^2(4x^3+2x-3)-3(4x^3+2x-3)$
$= 8x^5 + 4x^3 - 6x^2 - 12x^3 - 6x + 9$
$= 8x^5 - 8x^3 - 6x^2 - 6x + 9$

39.
$$\begin{array}{r} x^3 + x^2 + x \\ \times \qquad x^2 + x + 1 \\ \hline x^3 + x^2 + x \\ x^4 + x^3 + x^2 \quad \\ x^5 + x^4 + x^3 \qquad \\ \hline x^5 + 2x^4 + 3x^3 + 2x^2 + x \end{array}$$

41.
$$\begin{array}{r} 2z^2 - z + 1 \\ \times \qquad 5z^2 + z - 2 \\ \hline -4z^2 + 2z - 2 \\ 2z^3 - z^2 + z \quad \\ 10z^4 - 5z^3 + 5z^2 \qquad \\ \hline 10z^4 - 3z^3 \qquad + 3z - 2 \end{array}$$

43. $50 = 2 \cdot 5 \cdot 5 = 2 \cdot 5^2$

45. $72 = 2 \cdot 2 \cdot 2 \cdot 3 \cdot 3 = 2^3 \cdot 3^2$

47. $200 = 2 \cdot 2 \cdot 2 \cdot 5 \cdot 5 = 2^3 \cdot 5^2$

49. $(y-6)(y^2+3y+2)$
$= y(y^2+3y+2)-6(y^2+3y+2)$
$= y^3 + 3y^2 + 2y - 6y^2 - 18y - 12$
$= y^3 - 3y^2 - 16y - 12$
The area is $(y^3 - 3y^2 - 16y - 12)$ square feet.

51. $(x^2-1)^2 - x^2 = (x^2-1)(x^2-1) - x^2$
$= x^2(x^2-1) - 1(x^2-1) - x^2$
$= x^4 - x^2 - x^2 + 1 - x^2$
$= x^4 - 3x^2 + 1$
The area of the shaded figure is $(x^4 - 3x^2 + 1)$ square meters.

53. Answers may vary.

Section 10.4

Practice Problems

1. $42 = 2 \cdot 3 \cdot 7$
   $28 = 2 \cdot 2 \cdot 7$
   The common prime factors are 2 and 7.
   The GCF is $2 \cdot 7 = 14$.

2. In the list z^7, z^8 and z^1, the smallest exponent to which z is raised is 1. Thus, the GCF is $z^1 = z$.

3. The GCF of 6, 3, and 15 is 3.
   The GCF of a^4, a^5, and a^2 is a^2.
   Thus, the GCF of $6a^4$, $3a^5$, and $15a^2$ is $3a^2$.

4. The GCF of $10y^7$ and $5y^9$ is $5y^7$.
   $10y^7 + 5y^9 = 5y^7 \cdot 2 + 5y^7 \cdot y^2$
   $= 5y^7(2 + y^2)$

5. The GCF of $4z^2 - 12z + 2$ is 2.
   $4z^2 - 12z + 2 = 2 \cdot 2z^2 - 2 \cdot 6z + 2 \cdot 1$
   $= 2(2z^2 - 6z + 1)$

6. $-3y^2 - 9y + 15x^2$
   $= (-3) \cdot y^2 + (-3)(3y) + (-3)(-5x^2)$
   $= -3(y^2 + 3y - 5x^2)$

Exercise Set 10.4

1. $48 = 2 \cdot 2 \cdot 2 \cdot 2 \cdot 3$
   $15 = 3 \cdot 5$
   GCF = 3

3. $60 = 2 \cdot 2 \cdot 3 \cdot 5$
   $72 = 2 \cdot 2 \cdot 2 \cdot 3 \cdot 3$
   GCF $= 2 \cdot 2 \cdot 3 = 12$

5. $12 = 2 \cdot 2 \cdot 3$
   $20 = 2 \cdot 2 \cdot 5$
   $36 = 2 \cdot 2 \cdot 3 \cdot 3$
   GCF $= 2 \cdot 2 = 4$

7. $8 = 2 \cdot 2 \cdot 2$
   $32 = 2 \cdot 2 \cdot 2 \cdot 2 \cdot 2$
   $100 = 2 \cdot 2 \cdot 5 \cdot 5$
   GCF $= 2 \cdot 2 = 4$

9. $y^7 = y^2 \cdot y^5$
   $y^2 = y^2$
   $y^{10} = y^2 \cdot y^8$
   GCF $= y^2$

11. $a^5 = a^5$
    $a^5 = a^5$
    $a^5 = a^5$
    GCF $= a^5$

13. $x^3y^2 = x \cdot x^2 \cdot y^2$
    $xy^2 = x \cdot y^2$
    $x^4y^2 = x \cdot x^3y^2$
    GCF $= x \cdot y^2 = xy^2$

15. $3 = 3$
    $5 = 5$
    $10 = 2 \cdot 5$
    GCF = 1

    $x^4 = x \cdot x^3$
    $x^7 = x \cdot x^6$
    $x = x$
    GCF $= x$

    GCF $= 1 \cdot x = x$

17. $2 = 2$
    $14 = 2 \cdot 7$
    $18 = 2 \cdot 3 \cdot 3$
    GCF = 2

    $z^3 = z^3$
    $z^5 = z^3 \cdot z^2$
    $z^3 = z^3$
    GCF $= z^3$

    GCF $= 2z^3$

19. $3y^2 = 3y \cdot y$
    $18y = 3y \cdot 6$
    GCF $= 3y$
    $3y^2 + 18y = 3y \cdot y + 3y \cdot 6 = 3y(y + 6)$
    Check:
    $3y(y + a) = 3y \cdot y + 3y \cdot 6 = 3y^2 + 18y$

21. $10a^6 = 5a^6 \cdot 2$
$5a^8 = 5a^6 \cdot a^2$
$\text{GCF} = 5a^6$
$10a^6 - 5a^8 = 5a^6 \cdot 2 - 5a^6 \cdot a^2$
$= 5a^6(2 - a^2)$
Check:
$5a^6(2 - a^2) = 5a^6 \cdot 2 - 5a^6 \cdot a^2$
$= 10a^6 - 5a^8$

23. $4x^3 = 4x \cdot x^2$
$12x^2 = 4x \cdot 3x$
$20x = 4x \cdot 5$
$\text{GCF} = 4x$
$4x^3 + 12x^2 + 20x = 4x \cdot x^2 + 4x \cdot 3x + 4x \cdot 5$
$= 4x(x^2 + 3x + 5)$
Check:
$4x(x^2 + 3x + 5) = 4x \cdot x^2 + 4x \cdot 3x + 4x \cdot 5$
$= 4x^3 + 12x^2 + 20x$

25. $z^7 = z^5 \cdot z^2$
$6z^5 = z^5 \cdot 6$
$\text{GCF} = z^5$
$z^7 - 6z^5 = z^5 \cdot z^2 - z^5 \cdot 6 = z^5(z^2 - 6)$
Check:
$z^5(z^2 - 6) = z^5 \cdot z^2 - z^5 \cdot 6 = z^7 - 6z^5$

27. $35 = 7 \cdot 5$
$14y = 7 \cdot 2y$
$7y^2 = 7 \cdot y^2$
$\text{GCF} = 7$
$-35 + 14y - 7y^2$
$= -7 \cdot 5 + 7 \cdot 2y - 7 \cdot y^2$
$= 7(-5 + 2y - y^2)$ or $-7(5 - 2y + y^2)$
Check:
$7(-5 + 2y - y^2) = 7 \cdot (-5) + 7 \cdot 2y - 7 \cdot y^2$
$= -35 + 14y - 7y^2$

29. $12a^5 = 12a^5$
$36a^6 = 12a^5 \cdot 3a$
$\text{GCF} = 12a^5$
$12a^5 - 36a^6 = 12a^5 \cdot 1 - 12a^5 \cdot 3a$
$= 12a^5(1 - 3a)$
Check:
$12a^5(1 - 3a) = 12a^5 \cdot 1 - 12a^5 \cdot 3a$
$= 12a^5 - 36a$

31. 30% of 120 is what number?
$30\% \cdot 120 = x$
$0.3 \cdot 120 = x$
$36 = x$

33. $80\% = \frac{80}{100} = \frac{4}{5}$

35. $\frac{3}{8} = \frac{x}{100}$
$3 \cdot 100 = 8 \cdot x$
$300 = 8x$
$x = \frac{300}{8} = 37.5$
$\frac{3}{8} = 37.5\%$

37. area on the left: $x \cdot x = x^2$
area on the right: $2 \cdot x = 2x$
total area: $x^2 + 2x$
Notice that $x(x + 2) = x^2 + 2x$.

39. Answers may vary.

Chapter 10 Review

1. $(2b + 7) + (8b - 10) = 2b + 7 + 8b - 10$
$= 2b + 8b + 7 - 10$
$= 10b - 3$

2. $(7s - 6) + (14s - 9) = 7s - 6 + 14s - 9$
$= 7s + 14s - 6 - 9$
$= 21s - 15$

3. $(3x+0.2)-(4x-2.6)$
$=(3x+0.2)+(-4x+2.6)$
$=3x-4x+0.2+2.6$
$=-x+2.8$

4. $(10y-6)-(11y+6)$
$=(10y-6)+(-11y-6)$
$=10y-11y-6-6$
$=-y-12$

5. $(4z^2+6z-1)+(5z-5)$
$=4z^2+6z+5z-1-5$
$=4z^2+11z-6$

6. $(17a^3+11a^2+a)+(14a^2-a)$
$=17a^3+11a^2+14a^2+a-a$
$=17a^3+25a^2$

7. $\left(9y^2-y+\frac{1}{2}\right)-\left(20y^2-\frac{1}{4}\right)$
$=\left(9y^2-y+\frac{1}{2}\right)+\left(-20y^2+\frac{1}{4}\right)$
$=9y^2-20y^2-y+\frac{1}{2}+\frac{1}{4}$
$=-11y^2-y+\frac{3}{4}$

8. $(x^2-6x+1)-(x-2)$
$=(x^2-6x+1)+(-x+2)$
$=x^2-6x-x+1+2$
$=x^2-7x+3$

9. $5x^2$ when $x=3$
$5(3)^2=5\cdot 9=45$

10. $2-7x$ when $x=3$
$2-7(3)=2-21=-19$

11. $p=2(10x-2)+2(3x+16)$
$=20x-4+6x+32$
$=(26x+28)$ feet

12. $x^{10}\cdot x^{14}=x^{10+14}=x^{24}$

13. $y\cdot y^6=y^{1+6}=y^7$

14. $4z^2\cdot 6z^5=(4\cdot 6)(z^2\cdot z^5)=24z^7$

15. $(-3x^2y)(5xy^4)=(-3\cdot 5)(x^2\cdot x)(y\cdot y^4)$
$=-15x^3y^5$

16. $(a^5)^7=a^{5\cdot 7}=a^{35}$

17. $(x^2)^4\cdot(x^{10})^2=x^{2\cdot 4}\cdot x^{10\cdot 2}$
$=x^8\cdot x^{20}$
$=x^{8+20}$
$=x^{28}$

18. $(9b)^2=9^2b^2=81b^2$

19. $(a^4b^2c)^5=(a^4)^5\cdot(b^2)^5\cdot(c)^5$
$=a^{20}b^{10}c^5$

20. $(7x)(2x^5)^3=7x\cdot 2^3\cdot(x^5)^3$
$=7x\cdot 8\cdot x^{15}$
$=(7\cdot 8)(x\cdot x^{15})$
$=56x^{16}$

21. $(3x^6y^5)^3(2x^6y^5)^2$
$=\left(3^3\cdot(x^6)^3\cdot(y^5)^3\right)\cdot\left(2^2\cdot(x^6)^2\cdot(y^5)^2\right)$
$=(27\cdot x^{18}\cdot y^{15})\cdot(4\cdot x^{12}\cdot y^{10})$
$=(27\cdot 4)(x^{18}\cdot x^{12})(y^{15}\cdot y^{10})$
$=108x^{30}y^{25}$

22. area $=s^2=(9a^7)^2=9^2\cdot(a^7)^2=81a^{14}$
The area of the square is $81a^{14}$ square miles.

23. $2a(5a^2-6)=2a\cdot 5a^2+2a\cdot(-6)$
$=10a^3-12a$

24. $-3y^2(y^2-2y+1)$
$=-3y^2\cdot y^2+(-3y^2)(-2y)+(-3y^2)(1)$
$=-3y^4+6y^3-3y^2$

25. $(x+2)(x+6) = x(x+6)+2(x+6)$
$= x^2+6x+2x+12$
$= x^2+8x+12$

26. $(3x-1)(5x-9) = 3x(5x-9)-1(5x-9)$
$= 15x^2-27x-5x+9$
$= 15x^2-32x+9$

27. $(y-5)^2 = (y-5)(y-5)$
$= y(y-5)-5(y-5)$
$= y^2-5y-5y+25$
$= y^2-10y+25$

28. $(7a+1)^2 = (7a+1)(7a+1)$
$= 7a(7a+1)+1(7a+1)$
$= 49a^2+7a+7a+1$
$= 49a^2+14a+1$

29. $(x+1)(x^2-2x+3)$
$= x(x^2-2x+3)+1(x^2-2x+3)$
$= x^3-2x^2+3x+x^2-2x+3$
$= x^3-x^2+x+3$

30. $(4y^2-3)(2y^2+y+1)$
$= 4y^2(2y^2+y+1)-3(2y^2+y+1)$
$= 8y^4+4y^3+4y^2-6y^2-3y-3$
$= 8y^4+4y^3-2y^2-3y-3$

31.
$$\begin{array}{r} 3z^2+2z+1 \\ \times \qquad z^2+z+1 \\ \hline 3z^2+2z+1 \\ 3z^3+2z^2+z \quad \\ 3z^4+2z^3+z^2 \qquad \\ \hline 3z^4+5z^3+6z^2+3z+1 \end{array}$$

32. $A = lw$
$= (a+6)(a^2-a+1)$
$= a(a^2-a+1)+6(a^2-a+1)$
$= a^3-a^2+a+6a^2-6a+6$
$= a^3+5a^2-5a+6$
The area is (a^3+5a^2-5a+6) square centimeters.

33. $20 = 2 \cdot 2 \cdot 5$
$35 = 5 \cdot 7$
$\text{GCF} = 5$

34. $12 = 2 \cdot 2 \cdot 3$
$32 = 2 \cdot 2 \cdot 2 \cdot 2 \cdot 2$
$\text{GCF} = 2 \cdot 2 = 4$

35. $24 = 2 \cdot 2 \cdot 2 \cdot 3$
$30 = 2 \cdot 3 \cdot 5$
$60 = 2 \cdot 2 \cdot 3 \cdot 5$
$\text{GCF} = 2 \cdot 3 = 6$

36. $10 = 2 \cdot 5$
$20 = 2 \cdot 2 \cdot 5$
$25 = 5 \cdot 5$
$\text{GCF} = 5$

37. $x^3 = x^2 \cdot x$
$x^2 = x^2$
$x^{10} = x^2 \cdot x^8$
$\text{GCF} = x^2$

38. $y^{10} = y^7 \cdot y^3$
$y^7 = y^7$
$y^7 = y^7$
$\text{GCF} = y^7$

39. $xy^2 = xy \cdot y$
$xy = xy$
$x^3y^3 = xy \cdot x^2y^2$
$\text{GCF} = xy$

40. $a^5b^4 = a^5b^2 \cdot b^2$
$a^6b^3 = a^5b^2 \cdot ab$
$a^7b^2 = a^5b^2 \cdot a^2$
$\text{GCF} = a^5b^2$

41. $5a^3 = 5a \cdot a^2$
$10a = 5a \cdot 2$
$20a^4 = 5a \cdot 4a^3$
$\text{GCF} = 5a$

42. $12y^2z = 4y^2z \cdot 3$
$20y^2z = 4y^2z \cdot 5$
$24y^5z = 4y^2z \cdot 6y^3$
$\text{GCF} = 4y^2z$

43. $2x^2 = 2x \cdot x$
$12x = 2x \cdot 6$
$\text{GCF} = 2x$
$2x^2 + 12x = 2x \cdot x + 2x \cdot 6 = 2x(x+6)$

44. $6a^2 = 6a \cdot a$
$12a = 6a \cdot 2$
$\text{GCF} = 6a$
$6a^2 - 12a = 6a \cdot a - 6a \cdot 2 = 6a(a-2)$

45. $6y^4 = y^4 \cdot 6$
$y^6 = y^4 \cdot y^2$
$\text{GCF} = y^4$
$6y^4 - y^6 = y^4 \cdot 6 - y^4 \cdot y^2 = y^4(6 - y^2)$

46. $7x^2 = 7 \cdot x^2$
$14x = 7 \cdot 2x$
$7 = 7 \cdot 1$
$\text{GCF} = 7$
$$7x^2 - 14x + 7 = 7 \cdot x^2 - 7 \cdot 2x + 7 \cdot 1 = 7(x^2 - 2x + 1)$$

47. $5a^7 = a^3 \cdot 5a^4$
$a^4 = a^3 \cdot a$
$a^3 = a^3 \cdot 1$
$\text{GCF} = a^3$
$$5a^7 - a^4 + a^3 = a^3 \cdot 5a^4 - a^3 \cdot a + a^3 \cdot 1 = a^3(5a^4 - a + 1)$$

48. $10y^6 = 10y \cdot y^5$
$10y = 10y \cdot 1$
$\text{GCF} = 10y$
$10y^6 - 10y = 10y \cdot y^5 - 10 \cdot 1 = 10y(y^5 - 1)$

Chapter 10 Test

1. $(11x-3)+(4x-1) = (11x+4x)+(-3-1) = 15x-4$

2. $(11x-3)-(4x-1) = (11x-3)+(-4x+1) = (11x-4x)+(-3+1) = 7x-2$

3. $(1.3y^2+5y)+(2.1y^2-3y-3)$
$= (1.3y^2+2.1y^2)+(5y-3y)-3$
$= 3.4y^2+2y-3$

4. $(6a^2+2a+1)-(8a^2+a)$
$= (6a^2+2a+1)+(-8a^2-a)$
$= (6a^2-8a^2)+(2a-a)+1$
$= -2a^2+a+1$

5. $x^2-6x+1 = (8)^2-6(8)+1 = 64-48+1 = 17$

6. $y^3 \cdot y^{11} = y^{3+11} = y^{14}$

7. $(y^3)^{11} = y^{3 \cdot 11} = y^{33}$

8. $(2x^2)^4 = 2^4(x^2)^4 = 16x^8$

9. $(6a^3)(-2a^7) = (6)(-2)(a^3 \cdot a^7) = -12a^{10}$

10. $(p^6)^7(p^2)^6 = p^{6\cdot 7} \cdot p^{2\cdot 6}$
$= p^{42} \cdot p^{12}$
$= p^{42+12}$
$= p^{54}$

11. $(3a^4b)^2(2ba^4)^3 = 3^2a^8b^2 \cdot 2^3b^3a^{12}$
$= (9 \cdot 8)(a^8 \cdot a^{12})(b^2 \cdot b^3)$
$= 72a^{20}b^5$

12. $5x(2x^2 + 1.3) = 5x \cdot 2x^2 + 5x \cdot 1.3$
$= 10x^3 + 6.5x$

13. $-2y(y^3 + 6y^2 - 4)$
$= -2y \cdot y^3 - 2y \cdot 6y^2 - 2y \cdot (-4)$
$= -2y^4 - 12y^3 + 8y$

14. $(x-3)(x+2) = x(x+2) - 3(x+2)$
$= x^2 + 2x - 3x - 6$
$= x^2 - x - 6$

15. $(5x+2)^2 = (5x+2)(5x+2)$
$= 5x(5x+2) + 2(5x+2)$
$= 25x^2 + 10x + 10x + 4$
$= 25x^2 + 20x + 4$

16. $(a+2)(a^2 - 2a + 4)$
$= a(a^2 - 2a + 4) + 2(a^2 - 2a + 4)$
$= a^3 - 2a^2 + 4a + 2a^2 - 4a + 8$
$= a^3 + 8$

17. area:
$(x+7)(5x-2) = x(5x-2) + 7(5x-2)$
$= 5x^2 - 2x + 35x - 14$
$= 5x^2 + 33x - 14$
The area is $(5x^2 + 33x - 14)$ square inches.
perimeter:
$2(2x) + 2(5x-2) = 4x + 10x - 4 = 14x - 4$
The perimeter is $(14x - 4)$ inches.

18. $45 = 3 \cdot 3 \cdot 5$
$60 = 2 \cdot 2 \cdot 3 \cdot 5$
$\text{GCF} = 3 \cdot 5 = 15$

19. $6y^3 = 3y^3 \cdot 2$
$9y^5 = 3y^3 \cdot 3y^2$
$18y^4 = 3y^3 \cdot 6y$
$\text{GCF} = 3y^3$

20. $3y^2 = 3y \cdot y$
$15y = 3y \cdot 5$
$3y^2 - 15y = 3y \cdot y - 3y \cdot 5 = 3y(y-5)$

21. $10a^2 = 2a \cdot 5a$
$12a = 2a \cdot 6$
$10a^2 + 12a = 2a \cdot 5a + 2a \cdot 6 = 2a(5a+6)$

22. $6x^2 = 6 \cdot x^2$
$12x = 6 \cdot 2x$
$30 = 6 \cdot 5$
$6x^2 - 12x - 30 = 6 \cdot x^2 - 6 \cdot 2x - 6 \cdot 5$
$= 6(x^2 - 2x - 5)$

23. $7x^6 = x^3 \cdot 7x^3$
$6x^4 = x^3 \cdot 6x$
$x^3 = x^3 \cdot 1$
$7x^6 - 6x^4 + x^3 = x^3 \cdot 7x^3 - x^3 \cdot 6x + x^3 \cdot 1$
$= x^3(7x^3 - 6x + 1)$